Fortschritte der Chemie organischer Naturstoffe

Progress in the Chemistry of Organic Natural Products

41

Founded by L. Zechmeister
Edited by W. Herz, H. Grisebach, G. W. Kirby

Authors:
J. W. Daly, D. Ferreira, St. J. Gould,
E. Haslam, D. J. Robins, D. G. Roux,
St. M. Weinreb

Springer-Verlag
Wien New York 1982

Dr. W. HERZ, Professor of Chemistry, Department of Chemistry,
The Florida State University, Tallahassee, Florida, U.S.A.

Prof. Dr. H. GRISEBACH, Biologisches Institut II, Lehrstuhl für Biochemie der Pflanzen,
Albert-Ludwigs-Universität, Freiburg i. Br., Federal Republic of Germany

G. W. KIRBY, Sc. D., Regius Professor of Chemistry, Chemistry Department,
The University, Glasgow, Scotland

With 37 Figures

© 1982 by Springer-Verlag/Wien
Softcover reprint of the hardcover 1st edition 1982

Library of Congress Catalog Card Number AC 39-1015

ISSN 0071-7886

ISBN-13: 978-3-7091-8658-9   e-ISBN-13: 978-3-7091-8656-5
DOI: 10.1007/978-3-7091-8656-5

Contents

List of Contributors

DALY, Dr. J. W., Laboratory of Bioorganic Chemistry, National Institute of Arthritis, Diabetes, and Digestive and Kidney Diseases, National Institutes of Health, Bethesda, MD 20205, U.S.A.

FERREIRA, Prof. Dr. D., Department of Chemistry, The University of the Orange Free State, P.O. Box 339, Bloemfontein 9300, Republic of South Africa.

GOULD, Dr. ST. J., School of Pharmacy. University of Connecticut, Storrs, CT 06268, U.S.A.

HASLAM, Prof. Dr. E., Department of Chemistry, The University of Sheffield, Sheffield S3 7HF, U.K.

ROBINS, Dr. D. J., Department of Chemistry, University of Glasgow, Glasgow G 12 8QQ, Scotland.

ROUX, Prof. Dr. D. G., Department of Chemistry, The University of the Orange Free State, P.O. Box 339, Bloemfontein 9300, Republic of South Africa.

WEINREB, Prof. Dr. ST. M., Department of Chemistry, The Pennsylvania State University, University Park, PA 16802, U.S.A.

The Metabolism of Gallic Acid and Hexahydroxydiphenic Acid in Higher Plants

By E. Haslam, Department of Chemistry, University of Sheffield, Sheffield, U. K.

With 10 Figures

Contents

This essay is dedicated to Professor T. S. Stevens, F. R. S., teacher, colleague and friend, on the occasion of his 80th birthday, 8th October 1980.

1. Polyphenol Biosynthesis in Higher Plants — An Overview

Three classes of phenolic metabolite overwhelmingly predominate in the leaves of vascular plants (2). These are respectively:

(i) proanthocyanidins — procyanidins (**1**, R = H) and prodelphinidins (**1**, R = OH) (25),

(ii) glycosylated flavonols (principally those of kaempferol, quercetin and myricetin) (*20*),

and (iii) esters, amides and glycosides of the hydroxycinnamic acids (principally those of *p*-coumaric, caffeic, ferulic and sinapic acids) (*31*).

Although the distribution and accumulation of phenolic constituents in different tissues of a particular plant may differ quite markedly in both a quantitative and qualitative sense the capacity for the synthesis of phenolic materials in the leaf of a plant is nevertheless frequently an accurate reflection of the synthetic capacity of the plant as a whole (e. g. stem, rhizomes, roots and fruit). It also provides the most convenient basis for the comparison of phenolic biosynthesis in different plants and plant families. BATE-SMITH (*2*) used the pattern of distribution of the three principal classes of phenolic metabolites (i, ii, iii) in the leaves of plants as a taxonomic guide for the classification of plants. Many correlations were made but in particular he noted that between the presence of leucoanthocyanidins (*syn*. proanthocyanidins) and a plant's "woody" habit of growth. BATE-SMITH (*2*) also drew attention to the ability of particular plant families to metabolise phenols containing the vicinal-trihydroxyaryl (pyrogallol) group. These two synthetic capabilities were accorded the characters *a* and *b* respectively and plants were classified in the groups ab, a_0b, ab_0 and a_0b_0.

(1, R = H - Procyanidins)
(1, R = OH - Prodelphinidins)

The biosynthetic origin of the three classes of phenolic metabolite (i, ii, iii) is generally assumed to be associated with the development of a vascular character in plants and with the development of the ability to synthesise the structural polymer lignin by the diversion of L-phenylalanine and L-tyrosine (in the Gramineae) from protein synthesis. Plants use one or more of the *p*-hydroxycinnamyl alcohols (**2, 3, 4**) — derived by reduction of the coenzyme A ester of the corresponding cinnamic acid (*100*) — as precursors to lignin (*17*). The intermediate coenzyme A esters then form the points of biosynthetic departure for the distinctive classes of phenolic metabolite (i, ii, iii) and in this context the role of some of these substances may be as storage products or as shunt metabolites of the *Co A* esters. A general outline of lignin biosynthesis is shown in Scheme 1 and incorporated into the scheme are reactions showing the derivation of phenolic metabolites (ii, iii) from the coenzyme A ester of *p*-coumaric acid.

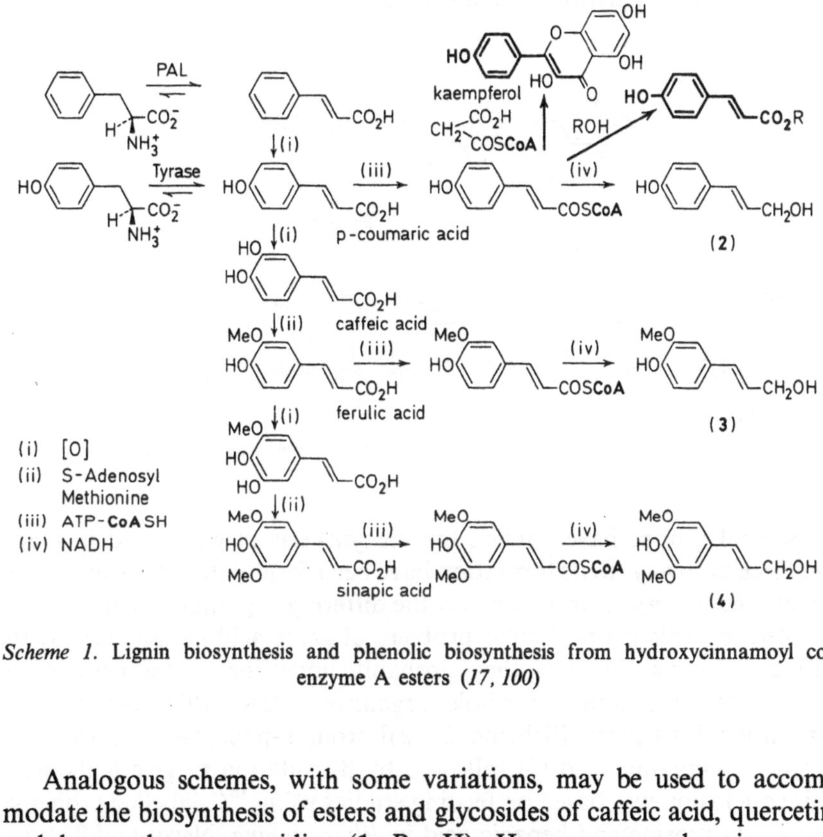

Scheme 1. Lignin biosynthesis and phenolic biosynthesis from hydroxycinnamoyl coenzyme A esters (*17, 100*)

Analogous schemes, with some variations, may be used to accommodate the biosynthesis of esters and glycosides of caffeic acid, quercetin and hence the procyanidins (**1**, R = H). However several serious uncertainties arise when the origin of flavonoids such as myricetin and the prodelphinidins (**1**, R = OH) is considered. 3,4,5-Trihydroxycinnamic acid (**5**) is **not** an intermediate in lignin synthesis and, except as its various o-methyl ethers, it has not been encountered in nature. Its role as a putative precursor of the vicinal trihydroxyaryl (pyrogallol) group in

1*

flavonoids such as myricetin and (1, R = OH) has therefore been questioned. In the context of this uncertainty BATE-SMITH (2) made an acute observation. He found, in his taxonomic work, that hexahydroxy-diphenic* acid [determined as its characteristic dilactone ellagic acid, (9), after acid hydrolysis] was very widely distributed in ester form (8) in plants and this led BATE-SMITH to suggest that it was the taxonomic equivalent of the "missing acid"-3,4,5-trihydroxycinnamic acid (5). Although direct experimental evidence is still awaited there is strong circumstantial evidence *(vide infra)* to suggest that the esters (8) are derived in plants by oxidative coupling of esters of gallic acid (7) and hence that gallic acid (6) is the real taxonomic equivalent of the "missing" 3,4,5-trihydroxycinnamic acid (5).

(5)   (6)
                              gallic acid

(7)   (8)   (9)
         hexahydroxydiphenic acid   ellagic acid

Several groups have studied the biosynthesis of gallic acid in plants and three pathways to its formation have been formulated. To a significant extent these divergent results reflect the differing experimental approaches adopted towards the particular problem of gallic acid metabolism by the different groups. As such they probably underline a weakness of the isotopic tracer method in whole organisms. ZENK (99) formulated a conventional pathway [Scheme 2, (a)] from L-phenylalanine to 3,4,5-trihydroxycinnamic acid (5) followed by β-oxidation to give gallic acid. This conclusion was based on feeding studies using ^{14}C-labelled L-phenyl-alanine, L-tyrosine and benzoic acid in *Rhus typhina*. NEISH and TOWERS (55) favoured a variation [Scheme 2, (b)] in which β-oxidation occurred at the caffeic acid stage to give protocatechuic acid which is then further

---

* This nomenclature is used throughout this review as a convenient trivial name for the 6,6-dicarbonyl-2,2′,3,3′,4,4′-hexahydroxybiphenyl radical.

hydroxylated to give gallic acid. Finally work with the mould *Phycomyces blakesleeanus* and later the plants *Rhus typhina* and various *Acer* sp. using ^{14}C labelled D-glucose and shikimic acid demonstrated (*10, 12, 36*) that a third route [Scheme 2, (c)] — the direct dehydrogenation of 3-dehydroshikimic acid — existed to gallic acid. CONN and SWAIN (*9*) supported this latter conclusion in work with *Geranium* sp. and in passing it may be noted that route (c) parallels the *in vitro* conversion of 3-dehydroshikimic acid to gallic acid by a variety of mild oxidising agents such as Cu^{2+}.

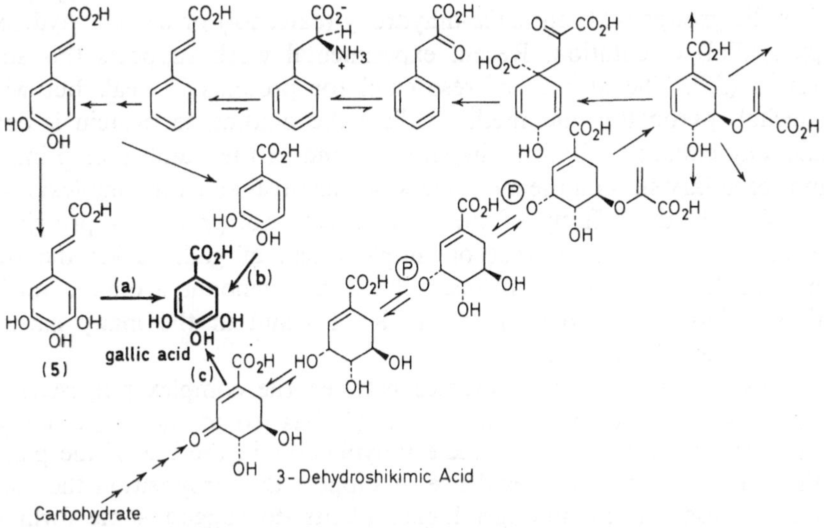

Scheme 2. Biosynthetic pathways to gallic acid in plants (*9, 10, 12, 36, 55, 99*)

The biosynthetic origins of gallic acid (**6**) and its taxonomic implications are thus two problems in the biochemistry of plant phenols which remain to be fully resolved. The metabolism of gallic acid in higher plants however poses other questions of equal significance. In contrast to other natural phenolic esters, such as those of *p*-coumaric acid and caffeic acid which are invariably mono- or bis-esters of polyols and esters of *o*- and *p*-hydroxybenzoic acid and protocatechuic acid which are relatively rarely encountered in plants, esters of gallic acid occur widely ranging from simple mono-esters such as β-D-glucogallin and theogallin to the complex polyesters with D-glucose whose molecular weights extend to at least 2,000. These esters are unique to the plant kingdom and with the proanthocyanidins (**1**) they together constitute the vegetable tannins of the earlier botanical and chemical literature (*109*).

Such molecules (which incidentally are preferably referred to as complex polyphenols), possess the distinctive property of precipitating natural macromolecules (particularly proteins) from solution. This phenomenon has long been recognised as responsible for the astringency of unripe fruit and various beverages, for the impaired nutritional quality of some cereals and herbiage crops and for the inactivation of enzymes and viruses. It has likewise been widely applied (e.g. in medicine, in the conversion of hide to leather and in the purification of enzymes). Overwhelming circumstantial evidence points unmistakably to the fact that these properties derive from the accumulation, within a molecule of moderate molecular size, of a substantial number of unconjugated phenolic groups with an *ortho*-dihydroxy (catechol) or *ortho*-trihydroxy (pyrogallol) orientation. Recent experimental work supports this suggestion (*53*). The affinity of resorcinol for proteins is weak but with catechol, pyrogallol and methyl gallate the binding to protein is considerably enhanced, both in its strength and in the number of primary and secondary sites on the molecule which are available for complexation. Significantly the affinity of the isolated catechol group for protein is broadly comparable to that of simple esters of gallic acid: the two functionalities are found respectively as the pendant aromatic nuclei of the ubiquitous procyanidins (**1**, R = H) and in the many natural galloyl esters found in plants.

This concordance in properties between the complex polyesters of gallic acid and the proanthocyanidins (**1**) has led to the suggestion of a similar functional role for these polyphenols in the life of the plant. Whilst there is little firm evidence to support this proposition the most general observations amongst higher plants do suggest some form of metabolic link between these two forms of polyphenol synthesis. Where there is found a high level of proanthocyanidin synthesis in plants that of galloyl esters is small or absent and similarly when the metabolism of galloyl esters is maximal then that of proanthocyanidins is small or absent. In this context it is interesting to note the cases of the very few plants which metabolise galloyl esters of flavan-3-ols (catechins and gallocatechins) — e.g. leaves of *Camellia sinensis* (tea) and fruit pods of *Ceratonia siliqua* (carob). The synthesis of the flavan-3-ols has been intimately linked to that of the proanthocyanidins (*25*) but in these plant tissues whilst there are substantial quantities of the flavan-3-ols formed there are miniscule quantities of proanthocyanidins to be found.

Because of their importance in earlier days to commerce the plants which metabolise substantial quantities of the polyesters of gallic acid and hexahydroxydiphenic acid have been the subject of the most intensive chemical study and analysis (*109*). A watershed in this work was the investigation of EMIL FISCHER (*14*) on the constituents of Chinese and Aleppo galls and studies of this nature continue to the present day. Particular attention and scrutiny has been devoted to the plants listed in Table 1, and this work, most notably that of SCHMIDT and MAYER in Heidelberg, has provided a wealth of valuable information upon particular forms of association of gallic acid and hexahydroxydiphenic acid, usually with D-glucose (*78, 80, 109*).

Table 1. *Sources of Complex Esters of Gallic Acid and Hexahydroxydiphenic Acid of Commercial Importance*

Anacardiaceae	
Rhus typhina; R. coriaria — leaves	(Sumach)
Rhus semialata — galls	(Chinese)
Leguminosae	
Caesalpinia coriaria — fruit pods	(Divi-divi)
C. spinosa — fruit, fruit pods	(Tara)
C. brevifolia — fruit, fruit pods	(Algarobilla)
Fagaceae	
Castanea sp. — wood, bark, leaves	(Chestnut)
Quercus aegilops — acorn cups	(Valonea)
Quercus sp. — wood, bark	(Oak)
Quercus infectoria — galls	(Turkish, Aleppo)
Combretaceae	
Terminalia chebula — fruit	(Myrabolans)

BATE-SMITH's (*2*) original observations on the phenolic constituents of higher plants and their taxonomic significance were based on the examination of plants from nearly 200 plant families and he drew attention to the wide occurrence of gallic acid and hexahydroxydiphenic acid as phenolic metabolites in plants. The attention which chemists have given to the individual plant sources in Table 1 has necessarily given a restricted and specialised view of gallic acid metabolism and has circumscribed our knowledge of its wider role in the plant kingdom. In the present account the author has drawn on unpublished material to discuss the metabolism of gallic acid in the plant kingdom as a whole. Wherever possible the subject is presented in an order based on biogenetic considerations which, to the author, appear to provide the best framework for such a discussion. These studies are of course mere prologue; ultimately the search must be for an understanding of the role of complex polyphenols such as (**1**) and the polyesters of gallic acid in the economy of the plant cell.

2. Metabolites of Gallic Acid

2.1 Simple Esters — Occurrence and Structural Analysis

Gallic acid is almost invariably found in plant tissues in ester form (*109*). Various simple esters of gallic acid have been described from plant sources (Table 1 and Table 5) and these metabolites are in many ways analogous to the hydroxycinnamoyl esters occurring in plants (*31*). These esters are thus formed by association with sugars, polyols, glycosides and other phenols. In contrast to the hydroxycinnamic acids however (*100*), fully authenticated examples of the occurrence of N-acyl derivatives (with amines, amino-acids and alkaloids) of gallic acid have not yet been described in the literature.

Table 2. *Naturally Occurring Galloyl Esters of Phenols and Phenolic Glycosides*

Ester	Plant	Plant Family	Ref.
1. (−)-Epicatechin-3-gallate	*Camellia sinensis*	Theaceae	(*6*)
2. (−)-Epigallocatechin-3-gallate	*Camellia sinensis*	Theaceae	(*6*)
3. (+)-Catechin-3-gallate	*Bergenia crassifolia*	Saxifragaceae	(*23*)
	Bergenia cordifolia	Saxifragaceae	(*23*)
4. (−)-Epicatechin-3,5-digallate	*Camellia sinensis*	Theaceae	(*11*)
5. (−)-Epigallocatechin-3,5-digallate	*Camellia sinensis*	Theaceae	(*11*)
6. 2-o-Galloylquinol-β-D-glucoside	*Bergenia crassifolia*	Saxifragaceae	
7. 6-o-Galloylquinol-β-D-glucoside	*Bergenia cordifolia*	Saxifragaceae	(*7*)
8. *p*-Galloyloxyphenyl-β-D-glucoside	*Arctostaphylos uva-ursi*	Ericaceae	
9. 2′,6′-Dihydroxy-4′-galloyloxymethyl-phenyl-β-D-glucoside (Cretanin)	*Castanea crenata*	Fagaceae	(*63*)
10. 6-Acetyl-2′,6′-dihydroxy-4′-galloyl-oxymethylphenyl-β-D-glucoside	*Castanea crenata*	Fagaceae	(*66*)
11. 4-(4′-Hydroxyphenyl)-2-butanon-4′-o-β-D-(6′′-o-galloyl)-glucoside (lindleyin)	*Aconiun lindleyi*	Crassulaceae	(*16*)
12. Quercetin-3-o-(6′′-o-galloyl)-β-D-glucoside	*Tellima grandiflora*	Saxifragaceae	(*4*)
13. Quercetin-3-o-(6′′-o-galloyl)-β-D-galactoside	*Euphorbia* sp.	Euphorbiaceae	(*101*)
14. Quercetin-3-o-(2′′-o-galloyl)-β-D-galactoside	*Euphorbia* sp.	Euphorbiaceae	(*101*)
15. Myricetin-3-o-(6′′-o-galloyl)-β-D-glucoside	*Tolmiea menziesii*	Saxifragaceae	(*5*)
16. Naringenin-7-o-(6′′-o-galloyl)-β-D-glucoside	*Acacia farnesiana*	Leguminosae	(*96*)

Miscellaneous Esters

1. 5-o-Galloylquinic acid (theogallin)	*Camellia sinensis*	Theaceae	(*69*)
2. 2-o-Cinnamoyl-1-o-galloyl-β-D-glucoside	*Rhizoma rhei*	Polygonaceae	(*47*)

The structure of cynarin from the artichoke has been shown to be that of 1,3-bis-o-caffeoylquinic acid and isochlorogenic acid is similarly a complex of at least three bis-caffeoylquinic acids (*31*). Bis-caffeoyl esters as phenolic metabolites are otherwise rare and tris-caffeoyl esters have not (to the author's knowledge), been recorded. These same strictures apply to the other naturally occurring hydroxycinnamic acids. In contrast gallic acid forms, for example with D-glucose with which it is most frequently associated, polyesters whose molecular composition ranges from one molecule of the sugar to one molecule of gallic acid to those in which one molecule of sugar is combined with seven or eight molecules of gallic acid. This characteristic also contrasts sharply with the other naturally occurring hydroxybenzoic acids, such as *p*-hydroxybenzoic and salicylic acids, whose occurrence is restricted and whose esters closely resemble in their molecular composition those of the hydroxycinnamic acids (*31*). Gallic acid is also noteworthy in at least one other respect, *i.e.* its occurrence esterified to itself in depside form (**10**).

(10)

The examination of plant extracts to determine the presence of esters of gallic acid is considerably facilitated by hplc and by two dimensional paper chromatography which gives a "fingerprint" of the plant's phenolic metabolism (*1, 29*). Galloyl esters are detected by their violet (mono, bis) — dark blue (tris, etc.) absorption in U.V. light which is enhanced by fuming with ammonia, and by various chromogenic sprays (*1*). Specific for galloyl esters is KIO_3 soln. which gives a rose-pink colour and reacts with gallic acid to form the characteristic orange of purpurogallin carboxylic acid (*3, 106*). R_F Values in the butanol-acetic acid-water system are broadly comparable for a whole range of galloyl glucose esters (0.30 − 0.50) but R_F values in aqueous acetic acid are very sensitive to the degree of esterification and decrease with increasing galloylation (Fig. 1).

Although derivatisation (methyl ether, acetate) is usually a prerequisite to obtain micro-analytical data, structural information is generally most easily obtained with the free phenolic forms of galloyl esters. Hydrolysis with the adaptive enzyme tannase [obtained by growth of *Aspergillus niger* on tannic acid (*21*)] is a specific means of cleavage of natural esters to yield gallic acid and the polyol (usually D-glucose).

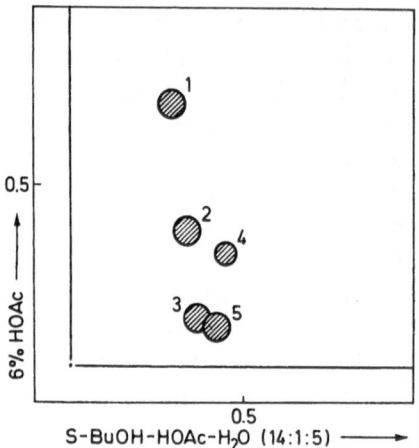

Fig. 1. Paper chromatogram of galloyl-D-glucose esters

R_F values

	BAW	6% HOAc
1. β-D-Glucogallin	0.30	0.75
2. 3,6-Di-o-galloyl-D-Glucose	0.35	0.31
3. β-1,3,6-Tri-o-galloyl-D-Glucose	0.40	0.13
4. 2,3,4,6-Tetra-o-galloyl-D-Glucose	0.45	0.29
5. β-Penta-o-galloyl-D-Glucose	0.42	0.12

With appropriate controls the enzyme may be used to quantitatively determine the ratio of gallic acid to the polyol. ^{1}H Nmr spectroscopy (220 or 360 MHz) with spin-decoupling usually provides a complete structural analysis for natural galloyl-D-glucopyranose derivatives. When the anomeric centre is acylated the galloyl group almost invariably adopts the β-configuration and the sugar usually adopts the C-1 ($^4C_1$) conformation (34, 105). Immediate recognition of these two features is given by the characteristic low field doublet for H-1 (δ 6.39 − 5.7 ppm TMS, J = 9.5 Hz) in the ^{1}H nmr spectrum. The position of this signal in the ^{1}H nmr spectrum is remarkably sensitive to the presence of other galloyl ester groups in the molecule and is a measure of the index of galloylation in the D-glucose series, Table 3.

Galloyl groups are distinguished by a series of 2-proton singlets (d$_6$ acetone, δ 6.95 − 7.25 ppm TMS) and similar information derives from ^{13}C nmr measurements. The position of the ^{13}C resonances due to the galloyl ester carbonyl groups is solvent sensitive and in prototropic media these resonances occur further downfield from TMS relative to the corresponding signals of spectra measured in d$_6$ acetone (Table 4). The

Table 3. *Chemical Shift of the Anomeric Proton (H-1) in β-1-O-Galloyl-D-Glucopyranose Derivatives (d_6 Acetone, 220 MHz). [Taken from (26)]*

Derivative	(ppm TMS)
β-penta-o-galloyl-D-glucose	6.39
β-1,2,3,6-tetra-o-galloyl-D-glucose	6.18
β-1,2,4,6-tetra-o-galloyl-D-glucose	6.10
β-1,2,6-tri-o-galloyl-D-glucose	5.90
β-1,3,6-tri-o-galloyl-D-glucose	5.89
β-1,6-di-o-galloyl-D-glucose	5.77
β-1-o-galloyl-D-glucose	5.70

Table 4. *Chemical Shift of the Galloyl Ester Carbonyl Groups (CO) in β-Penta-O-Galloyl-D-Glucose, (11). [Taken from (26)]*

Solvent	^{13}C δ values (ppm TMS)
d_4 methanol	167.8, 167.1, 166.9 (2), 166.1
d_6 acetone	166.4, 165.7 (3), 165.0

relative intensities of these signals and the multiplicity provides a good measure of the extent of esterification of the glucose nucleus by gallic acid. The ^{13}C nmr signals of the individual carbon atoms in the D-glucose residue accord generally with the relative positions of the ^{13}C signals for the carbon atoms in β-D-glucose itself. In all cases the anomeric carbon occurs at lowest field (δ 93.4–95.5 ppm TMS) and C-6 at highest field (δ 62.9–63.8 ppm TMS) in acetone and methanol (Fig. 2). ^{13}C Resonances of the carbon atoms of the aryl rings of the galloyl ester groups are also readily recognised and identified. Assignments are made relative to those of methyl gallate (Fig. 3). In polygalloyl esters [e.g. β-penta-

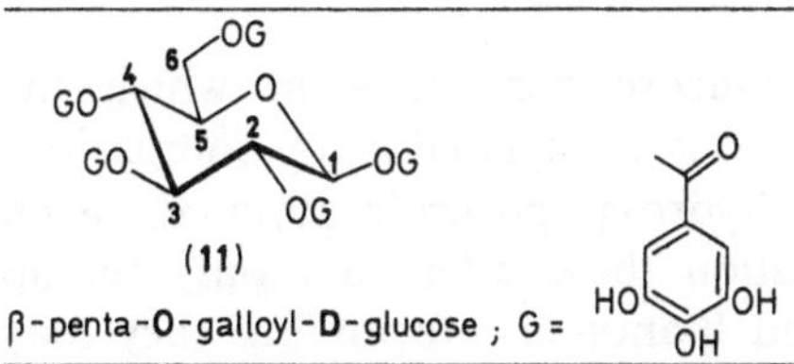

	Glucose Carbon	δ (ppm TMS)	
		d_6 acetone	d_4 methanol
	C-1	93.4	93.7
	C-2	71.9	72.1
	C-3	73.5	74.2
	C-4	69.5	69.7
	C-5	74.1	74.2
	C-6	62.9	63.0

(11)
β-penta-O-galloyl-D-glucose ; G =

Fig. 2. Chemical shift of the carbon atoms of D-glucose in β-penta-o-galloyl-D-glucose (11). [Taken from (26)]

	Aryl C atom	δ ppm	
		d₆ acetone	Calculated*
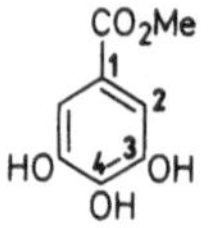 (structure 12)	C-1	121.4	125.7
	C-2	110.0	109.5
	C-3	146.4	144.3
	C-4	139.7	133.9

* STOTHERS (94)

Fig. 3. Chemical shift of the carbon atoms of the aromatic ring in methyl gallate (12).
[Taken from (26)]

O-galloyl-D-glucose (11)] the ^{13}C resonances due to C-2 and C-3 in the individual galloyl groups are not resolved at 25.15 MHz and almost invariably occur as singlets at 109.8 – 110.6 and 145.9 – 146.3 (δ ppm TMS) respectively. The ^{13}C resonances due to C-1 and C-4 of the various galloyl groups are however generally resolved at 25.15 MHz and, in an analogous manner to the multiplicity of signals due to the ester carbonyl group carbon atoms, they provide a convenient index of the number of galloyl ester groups attached to the D-glucose core.

A further means for ready identification of a galloyl ester is by mass spectrometry of the methylated (diazomethane) ester. The mass spectral fragmentation pattern usually displays the ion (13) as base peak, m/e 195 (22, 63).

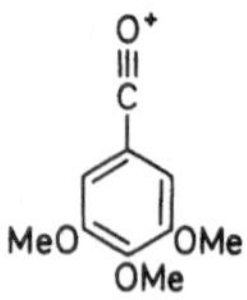

(13, m/e 195)

Structural analysis of galloyl-D-glucose derivatives in which the anomeric hydroxyl group (C-1 position) is unacylated is, by comparison with those derivatives in which this hydroxyl group is acylated, much more difficult to carry out. In solution these esters not only set up an equilibrium mixture of the α- and β-anomeric forms but they may assume alternative conformations of the glucopyranose ring to the normally preferred C-1 ($^4C_1$) conformation (26, 32, 34). Derivatisation

(e. g. acetylation) similarly usually produces a mixture of α and β derivatives and, unless these are separable, spectroscopic identification (particularly ^{1}H nmr) is complicated by the doubling of signals in the spectrum.

The structure of a particular galloyl-D-glucose derivative may be finally confirmed by synthesis. FISCHER and BERGMANN, in their earlier synthetic work in this area, introduced the use of tri-o-acetyl galloyl chloride as a means of synthesising galloyl esters (13). SCHMIDT and his collaborators (80) and later HAWORTH (29) have exploited the alternative acylating agent tri-o-benzylgalloyl chloride (15) to achieve the same purpose. In this way various galloyl-D-glucose esters have been prepared [α-glucogallin (85); 2-o-galloyl-D-glucose (84); 3-o-galloyl-D-glucose (70); 6-o-galloyl-D-glucose (70); and 3,6-di-o-galloyl-D-glucose (70)], and in particular β-D-glucogallin (85); β-1,3,6-tri-o-galloyl-D-glucose (83); and β-penta-o-galloyl-D-glucose (1); — all naturally occurring esters of gallic acid — have been synthesised. The synthesis of β-1,3,6-trigalloyl-D-glucose (14) by SCHMIDT and KLINGER (83) is shown in Scheme 3.

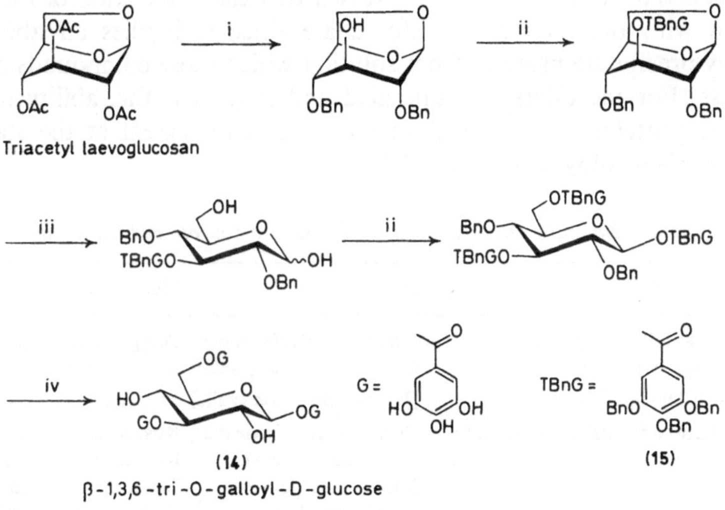

Reagents: (i) BnCl-KOH; (ii) TBnGCl (15)-pyridine; (iii) (CF$_3$CO)$_2$O H$^+$; (iv) H$_2$-PdCl$_2$

Scheme 3. Synthesis of β-1,3,6-Tri-o-galloyl-D-glucose (83)

β-D-Glucogallin (β-1-o-galloyl-D-glucose) was first isolated as its crystalline hydrate from the roots of Chinese rhubarb (Rheum officinale) in 1903 by GILSON (15) and although in the interim period several galloyl-D-glucose derivatives were obtained as degradation products of

natural products it was half a century later before further simple
galloyl esters of D-glucose were isolated and described as natural products
in their own right. SCHMIDT and his collaborators then described (90)
the isolation of β-penta-o-galloyl-D-glucose (11) and β-1,3,6-tri-o-galloyl-
D-glucose (14) from Myrabolans (*Terminalia chebula,* fruit) and later
(91) β-1,3,4,6-tetra-o-galloyl-D-glucose from Algarobilla (*Caesalpinia
brevifolia,* fruit pods). In more recent work β-penta-o-galloyl-D-glucose
(11) has been isolated and identified in a range of plants [Table 5 (26)]
where it appears to act as a key intermediate in the further metabolism
of gallic acid *(vide infra).* The various simple unmodified naturally
occurring esters of gallic acid with D-glucose (and other sugars) which
have now been recognised in higher plants are shown in Table 5. It is
interesting to note amongst the partially esterified derivatives the pre-
disposition for the ester groups to be located at positions 1, 2 and 6 on
the D-glucose core and this feature may also be noted amongst esters
listed earlier in Table 2. Also included in Table 5 are Aceritannin
[structure revised to 2,6-di-o-galloyl-1,5-anhydro-D-glucitol (19, 56)] and
Hamameli tannin which have been recognised for many years as natural
products. However it should be stressed that the association of the word
tannin with these esters is unfortunate since it implies an ability to
readily precipitate proteins from solution which these compounds do not
possess. For the esters of gallic acid and D-glucose this ability to precipitate proteins only appears to any significant extent at the stage of
β-1,3,6-tri-o-galloyl-D-glucose (24).

Table 5. *Naturally Occurring Galloyl Esters of D-Glucose and Other Sugars*

Galloyl Ester	Plant	Plant Family	Ref.
1. β-1-o-Galloyl-D-glucose (β-D-glucogallin)	*Rheum officinale* (root)	Polygonaceae	(15)
	Ceratonia siliqua (pods)	Leguminosae	(57)
2. β-1,6-Di-o-galloyl-D-glucose	*Quercus infectoria* (galls)	Fagaceae	(18)
3. β-1,2,6-Tri-o-galloyl-D-glucose	*Rubus fructicosus* (leaves)	Rosaceae	(26)
	Rubus idaeus (leaves)	Rosaceae	(26)
	Rosa canina (leaves)	Rosaceae	(26)
	Fuchsia sp. (leaves)	Onagraceae	(26)
	Epilobium angustifolium (leaves)	Onagraceae	(26)
4. β-1,3,6-Tri-o-galloyl-D-glucose	*Terminalia chebula* (fruit)	Combretaceae	(90)
5. β-1,2,3,6-Tetra-o-galloyl-D-glucose	*Fuchsia* sp. (leaves)	Onagraceae	(26)
	Epilobium angustifolium (leaves)	Onagraceae	(26)
	Quercus infectoria (galls)	Fagaceae	(18)
	Ceratonia siliqua (pods)	Leguminosae	(37)

Table 5 *(continued)*

Galloyl Ester	Plant	Plant Family	Ref.
6. β-1,2,4,6-Tetra-o-galloyl-D-glucose	*Bergenia crassifolica* (roots and rhizomes)	Saxifragaceae	*(26)*
	Bergenia cordifolia (roots and rhizomes)	Saxifragaceae	*(26)*
7. β-1,3,4,6-Tetra-o-galloyl-D-glucose	*Caesalpinia brevifolia* (pods)	Leguminosae	*(91)*
8. β-Penta-o-galloyl-D-glucose	*Terminalia chebula* (fruit)	Combretaceae	*(90)*
	Acer platanoides (leaves)	Aceraceae	*(26)*
	Acer campestre (leaves)	Aceraceae	*(26)*
	Quercus infectoria (galls)	Fagaceae	*(18)*
	Quercus borealis (leaves)	Fagaceae	*(26)*
	Rhus semialata (galls)	Anacardiaceae	*(26)*
	Rhus coriaria (leaves)	Anacardiaceae	*(26)*
	Rhus typhina (leaves)	Anacardiaceae	*(26)*
	Cotinus coggyria (leaves)	Anacardiaceae	*(26)*
	Fuchsia sp. (leaves)	Onagraceae	*(26)*
	Epilobium angustifolium (leaves)	Onagraceae	*(26)*
	Rubus idaeus (leaves)	Rosaceae	*(26)*
	Rubus fructicosus (leaves)	Rosaceae	*(26)*
	Rosa canina (leaves)	Rosaceae	*(18)*
	Geranium robertanium (leaves)	Geraniaceae	*(26)*
9. 2,6-Di-o-galloyl-1,5-anhydro-D-glucitol (Aceritannin)	*Acer ginnale* (leaves)	Aceraceae ⎫	
	Acer saccharinum (leaves)	Aceraceae ⎬	*(19, 56, 67)*
	Acer tartaricum (leaves)	Aceraceae ⎭	
10. 2',5-Di-o-galloyl-D-hamamelose (Hamameli tannin)	*Hamamelis virginiana* (bark)	Hamamelidaceae ⎫	
		⎬	*(40)*
	Quercus rubra (bark)	Fagaceae ⎪	
	Castanea sativa (bark)	Fagaceae ⎭	

Using phenolic biosynthesis in the leaves of a plant as a point of reference the metabolism of gallic acid for many plants begins and ends with their ability to synthesise esters of gallic acid with D-glucose (and occasionally other sugars), phenols and phenolic glycosides. For a great many other plants, however, this is not the limit of the range of metabolites associated with gallic acid and in this particular context those plants may be noted which are able to biosynthesise β-penta-o-galloyl-D-glucose (Table 5). These are typical of a very wide range of plants which are apparently able to subsequently metabolise this intermediate further.

The patterns of further metabolism of β-penta-o-galloyl-D-glucose (11) which have been discerned in the leaves of plants (26) are broadly divisible into at least three groups (A, B, C; Scheme 4). Attention is directed in this review to these three major patterns but it should be noted that this scheme probably does not yet represent an all embracing classification of gallic acid metabolism in higher plants. Previously (Table 1) those plant species were noted which provided in earlier times extracts of commercial importance: whilst several points of contact between the phenolic metabolites of these particular plants and Scheme 4 are self evident *(vide infra)* others — notably *Terminalia chebula, Caesalpinia coriaria* and *Caesalpinia brevifolia* — quite clearly produce a range of phenolic metabolites which, on the basis of the available evidence, are extensions of categories B and C. Additionally it is also clear that specialised organs of a particular plant (e.g. fruit, fruit pod, bark) may well metabolise other phenols which are derived from one (or more) of the "primary" phenolic metabolites formed by routes A, B or C.

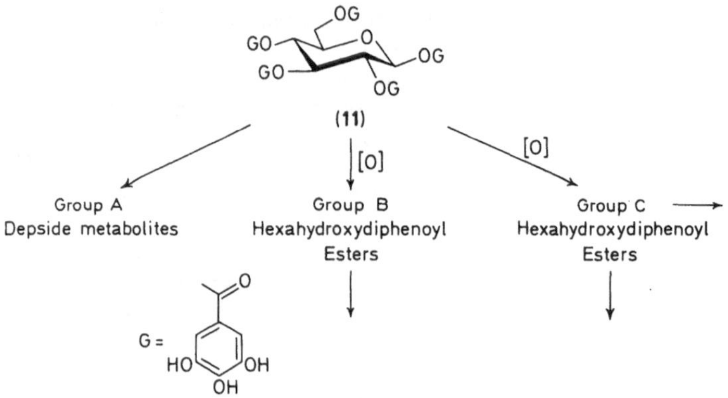

Scheme 4. The further metabolism of β-penta-o-galloyl-D-glucose in the leaves of higher plants (26)

2.2 Depside Metabolites — Group A

The ability to produce metabolites in which additional gallic acid molecules are esterified to a pre-existing galloyl ester in the form of depsides is limited to a comparatively few plant families. Many of the products of this form of gallic acid metabolism were grouped together in the earlier literature under the generic term "gallotannin" (14, 29). The depsidically linked galloyl ester residues are specifically cleaved under very mild conditions (methanol, r.t., pH 6.0 or refluxing methanol) leaving other galloyl esters unaffected. This methanolysis reaction is undoubtedly facilitated by participation of the phenolic group adjacent to the depside link (29) (Scheme 5). The reaction provided HAWORTH and his collaborators with a ready means of degradation and structural analysis for this type of molecule (1, 8, 22, 35). Combined with hplc and paper chromatography it undoubtedly provides the most ready means of identification of this type of metabolite in plant extracts (26).

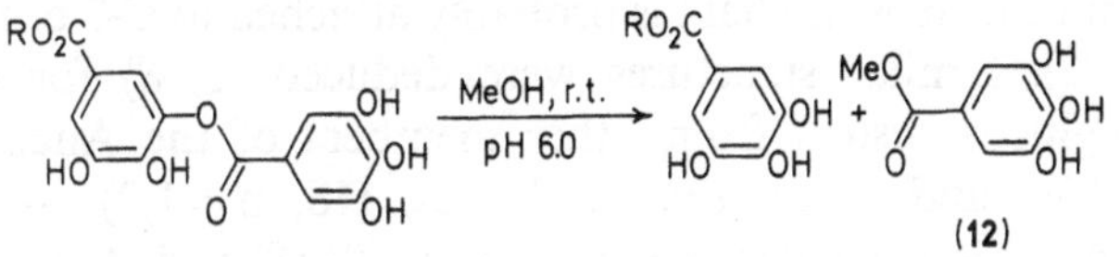

Scheme 5. The methanolysis of depside galloyl esters (1, 29)

In the period 1910–1930 several distinguished chemists — EMIL FISCHER, KARL FREUDENBERG and PAUL KARRER — made substantial contributions to early ideas on the "gallotannins". But progress was severely hampered by the slow realisation that all the plant extracts consisted of mixtures of closely related metabolites and other phenols whose presence made separation and purification of the desired materials extremely difficult (97). Many of these problems have been relieved in recent years by application of various forms of chromatography and by concomitant use of spectroscopic methods for identification.

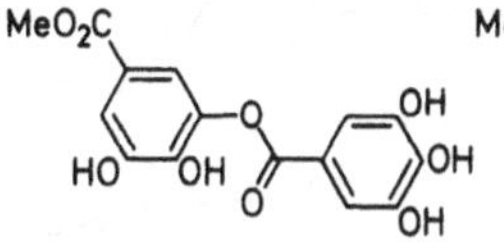

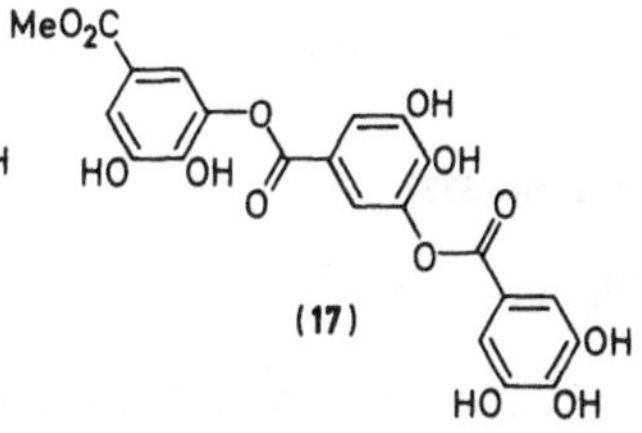

The structure of the complex polyphenol derived from the twig galls of *Rhus semialata* ("Chinese gallotannin" or "tannic acid"), as a core of β-penta-o-galloyl-D-glucose (11) to which other (approximately 5) galloyl groups are linked in depside fashion, was first proposed by FISCHER and FREUDENBERG (14). Methanol reacts with this complex galloyl ester to preferentially strip away the depsidically linked galloyl ester groups. This gives as major products β-penta-o-galloyl-D-glucose (11) and methyl gallate (12), (29). At an intermediate stage during the methanolysis methyl-*m*-digallate (16) and methyl-*m*-trigallate (17) may be detected and identified (8, 29). Molecular weight determinations (vapour pressure osmometry) gave a number average molecular weight value of 1250 ± 60. Of this molecular weight the β-penta-o-galloyl-D-glucose core accounts for 940 and the residue of relative molecular mass ~ 310 corresponds to *approximately* two additional depsidically linked galloyl ester groups. In order to account for the formation of (16) and (17) during the methanolysis it was suggested (8, 29) that the depsidically linked galloyl ester groups were attached to the core (11) to give (on average) a tri-tetra galloyl depside chain (18, n = 1, 2). Additional refinement was given to this structure by the application of ^{1}H nmr to suggest that the depside chain is probably attached to C-2 of the D-glucose residue (22). Similar structures were deduced (1, 8) for the complex galloyl-D-glucose esters from other members of the Anacardiaceae — *Rhus typhina* and *Rhus coriaria*, leaves, (18, n = 1,2) — and for the esters from *Arctostaphylos uva-ursi* (18, n = 0) (8) and *Anogeissus latifolia* (leaves) (68).

Spectroscopic measurements combined with the use of the methanolysis reaction have provided evidence for the occurrence of the complex ester in other plants. The chemical shift of the aromatic protons of depsidically linked galloyl ester groups occurs downfield from those of simple aliphatic galloyl esters (d_6 acetone or d_6 DMSO, δ values from TMS, depside: 7.25 − 7.70, simple ester: 6.9 − 7.20). It is possible using this measurement to estimate the average value of the number n for the ester isolated from a particular plant. Methanolysis followed by a quantitative hplc analysis of the products (11) and (12) confirms the nature of this analysis.

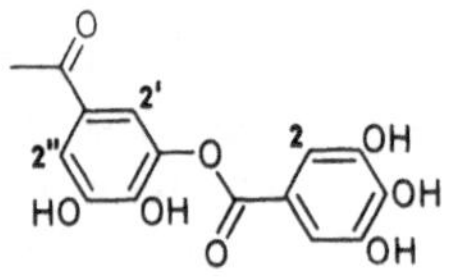

	(δ) Chemical Shift (ppm TMS) d_6 acetone	
C-2		110.8
C-2′		117.7
C-2″		114.8

Fig. 4. ^{13}C Chemical shift of aryl carbon atoms of depside galloyl esters

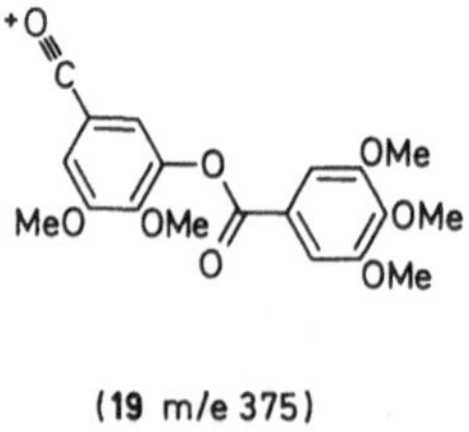

(19, m/e = 375)

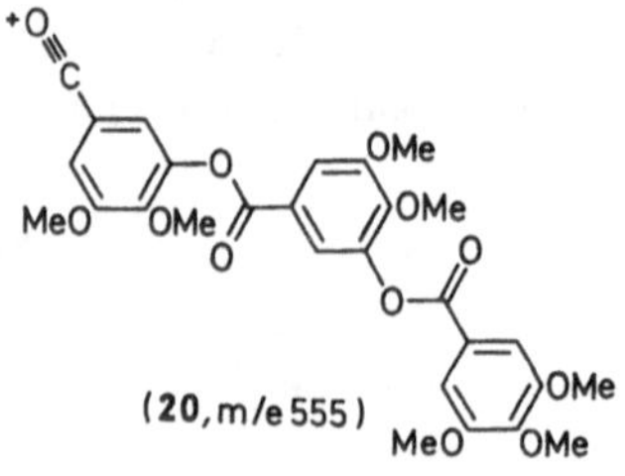

(20, m/e = 555)

Fig. 5. Fragment ions from methylated (18); (22)

The presence of depsidically linked galloyl ester groups in (18) is also indicated by ^{13}C nmr and by mass spectrometry. In the ^{13}C nmr spectrum the change in chemical shift of C-2 in the aromatic ring is characteristic (Fig. 4) and the mass spectral fragmentation pattern of the fully methylated (18) shows ions at m/e 375 (19) where n = 0 and m/e 375 (19) and m/e 555 (20) where n = 1, Fig. 5 (22).

Although the methanolysis reaction was much used in earlier work (29) to identify the β-penta-o-galloyl-D-glucose core of (18) nmr measurements also provide an immediate means of recognition of this group. The presence of the depsides causes little perturbation in the ^{1}H nmr spectrum to the protons of the glucose nucleus compared to the pattern displayed in β-penta-o-galloyl-D-glucose (11) and the ^{13}C chemical shift of the glucose carbon atoms again show little change from those of (11) (Fig. 6). These measurements show that the complex ester (18) adopts the C-1 (^{4}C$_1$) D-glucose conformation with all ester groups equatorially disposed. Molecular models show that this molecule has a flexible disc-like shape with the phenolic groups disposed towards the periphery of the disc.

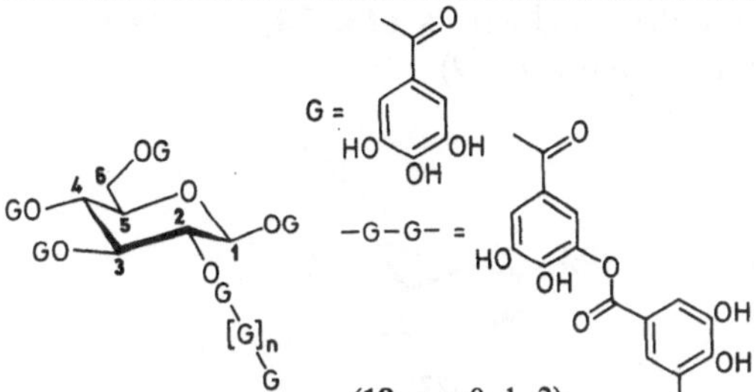

	δ Chemical shift (d$_6$ acetone) ppm from TMS	
C-1		93.4
C-2		71.8
C-3		74.0
C-4		69.4
C-5		73.9
C-6		62.9

Fig. 6. ^{13}C Chemical shift of the glucose carbon atoms in the ester (18), (26)

To date the ester (18) is the most widely encountered in the plant kingdom and it is worth noting that in plants which metabolise this phenol it overwhelmingly dominates the phenolic extract. The ester (18) has very similar R$_F$ values to β-penta-o-galloyl-D-glucose (11) (Fig. 1), and is distinguished by preparative chromatography on Sephadex LH-20. Table 6 shows plants in which this complex phenol has been identified.

Table 6. *Complex Galloyl Esters* (**18**) *Based Upon* β-*Penta-o-Galloyl-D-Glucose* (**11**)

Ester	Plant	Plant Family	Ref.
(**18**, n = 1 − 2)	*Rhus semialata* (galls, leaf)	Anacardiaceae	(*1, 22*)
(**18**, n = 1 − 2)	*Rhus coriaria* (leaf)	Anacardiaceae	(*1, 22*)
(**18**, n = 1 − 2)	*Rhus typhina* (leaf)	Anacardiaceae	(*1, 22*)
(**18**, n = 1 − 2)	*Cotinus coggyria* (leaf, stem)	Anacardiaceae	(*26*)
(**18**)	*Anogeissus latifolia* (leaf)	Combretaceae	(*68*)
(**18**, n = 0)	*Acer platanoides* (leaf)	Aceraceae	(*26, 106*)
(**18**, n = 0)	*Acer campestre* (leaf)	Aceraceae	(*26, 106*)
(**18**, n = 0)	*Acer rubrum* (leaf)	Aceraceae	(*26, 106*)
(**18**, n = 0)	*Arctostaphylos uva-ursi* (leaf)	Ericaceae	(*8*)
(**18**, n = 1, 2)	*Pelargonium* sp. (leaf, stem)	Geraniaceae	(*26*)
(**18**, n = 0 − 1)	*Paeonia officinalis* (leaf)	Paeoniaceae	(*26*)
(**18**, n = 1 − 2)	*Hamamelis mollis* (leaf)	Hamameliadaceae	(*26*)
	Parrottia persica (leaf)	Hamameliadaceae	(*26*)

Although the principal form in which gallic acid is found in depside linkage is the ester (**18**) other esters of this type have been described. The galls of various oak species *(Quercus infectoria, Quercus lusitanica)* thus yield an ester (**21**) in which the depsidically linked galloyl groups are at position 6 on the β-1,3,4,6-tetra-o-galloyl-D-glucose core (*1*), and the fruit pods of *Caesalpinia spinosa* yield the ester (**22**) which is based on a 3,4,5-tri-o-galloyl quinic acid core (*35*). More recently a mixture of the two tri-o-galloyl-1,5-anhydro-D-glucitols (**23, 24**) has been obtained from the leaves of *Acer saccharinum* (*19*).

(21)   (22)   (23)   (24)   G =

2.3 Metabolites Formed by Oxidative Coupling of Galloyl Esters — Groups B and C

2.3.1 Hexahydroxydiphenoyl Esters

In his taxonomic surveys of the 1950's BATE-SMITH (2) drew particular attention to the occurrence of ellagic acid (9) in the acid hydrolysates of dicotyledonous plant extracts. From this he inferred that its presumed precursor — hexahydrodiphenic acid, bound in ester form (8) — was widely distributed in the plant kingdom. Subsequent work has confirmed the correctness of this assumption. Notable work in this particular area of plant chemistry has been carried out by SCHMIDT (80, 46) and MAYER (109) in their classical work on the ellagitannins (plant sources in Table 1, Algarobilla, Divi-divi, Myrabolans, Valonea, Chestnut). The following discussion will emphasize the fundamental knowledge which has emerged from these observations but will also seek to place the contributions of the Heidelberg School in the context of more recent observations over the plant kingdom as a whole and within the emerging biogenetic framework.

Hexahydroxydiphenic acid in a bound form (8) is readily detected in a plant extract by an old but very distinctive colour test — the Procter-Paessler reaction. Esters of type (8) react with nitrous acid to give a carmine or rose red, changing to brown-green then purple and finally indigo blue (78). The chemical basis of this reaction has not been discussed but it has been employed by BATE-SMITH (3) for the quantitative determination of (8) in plant extracts, and it forms the basis of a very useful spray reagent for the detection of (8) by paper chromatography (26).

Although there is no firm experimental proof it is generally assumed, following the hypothesis of SCHMIDT and MAYER (81), that esters of type (8) are formed by oxidative coupling of suitably placed galloyl ester groups in a suitable polygalloyl ester precursor. One specific chirality is imposed on the biphenyl system of the hexahydroxydiphenoyl ester group as it is formed and this chirality is determined by the need of the hexahydroxydiphenoyl group to bridge particular positions in the alcohol portion of the ester. In all cases so far discussed by SCHMIDT and MAYER and by other workers the hexahydroxydiphenoyl group bridges positions within the D-glucose molecule. Methylation of hexahydroxydiphenoyl esters (8) followed by hydrolysis gives hexamethoxydiphenic acid (25). SCHMIDT (72, 76, 80) has shown that rotation around the biphenyl linkage of this acid is sufficiently slow at ambient temperatures to allow the isolation of the acid (25) in the chiral form in which it occurs. Thus it has been demonstrated that the hexahydroxydiphenic acid which bridges the 3,6 positions of D-glucopyranose is the dextrorotary form, whilst that which bridges the 2,3 or 4,6 positions of D-glucopyranose is the laevorotatory isomer (*77, 102*). SCHMIDT (*72, 76*) has prepared the optically active forms of both hexamethoxydiphenic acid (**25**) and hexabenzyloxydiphenic acid (**26**) (Scheme 6). The racemic forms were resolved via their quinidine and cinchonine salts. The half-life for racemisation of the methoxy acid (**25**) was determined as $14^3/_4$ hr in refluxing aqueous alkali and for the hydroxy acid (**27**) — produced by hydrogenolysis of (**26**) — $4^3/_4$ hr at 20°.

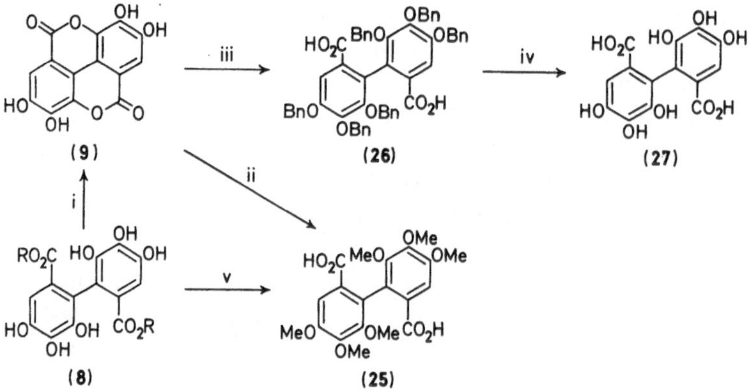

Reagents:. (i) H^+; (ii) $Me_2SO_4-OH^-$; (iii) $BnCl-K_2CO_3$, $BnCl-OH^-$
(iv) H_2-Pd; (v) CH_2N_2, OH^-

Scheme 6. The preparation of hexamethoxydiphenic acid and hexahydroxydiphenic acid, (*72, 76*)

MISLOW and DJERASSI (*54*), by applying the amide rule of the glyconic acids to the biphenyl system of the (+)-hexahydroxydiphenoyl group in corilagin (**28**), (β-1-O-galloyl-3,6-hexahydroxydiphenoyl-D-glucose) (*71*) proposed that the (+)-hexahydroxydiphenoyl group had the s-configuration. However OKUDA (*62*) has very recently correlated (+)-hexamethoxy-diphenic acid (**25**) from corilagin (**28**) with the same acid derived from the dibenzocyclo-octane lignan schizandrin (**29**) whose chirality had been shown to be R by comparison with the X-ray derived molecular structure of the lignan gomisin D (*33*) (Scheme 7).

It is interesting to note that the same conclusions, regarding the chirality of hexahydroxydiphenoyl groups bound to D-glucopyranose, follow from theoretical considerations. Two assumptions are necessary:

(i) that the ester carbonyl groups of the hexahydroxydiphenoyl residue have, like other ester groups (*22, 107, 108*), a preference for the

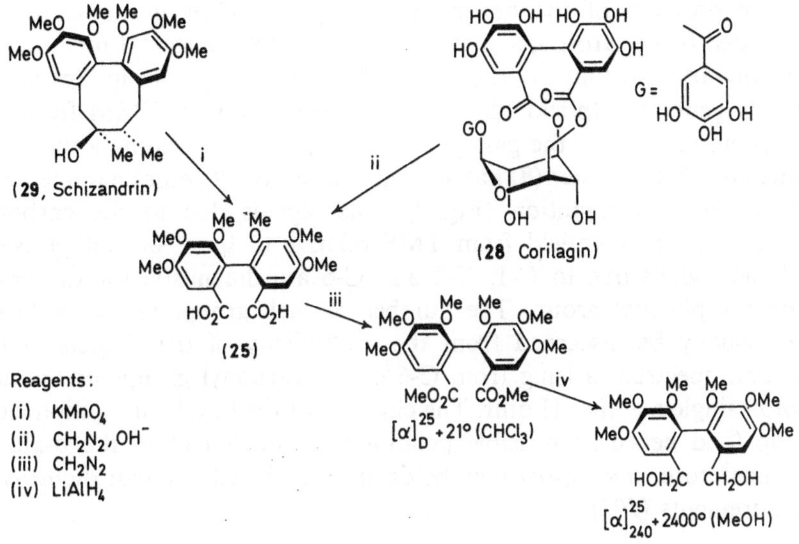

(29, Schizandrin)

(28 Corilagin)

(25)

$[\alpha]_D^{25}$ +21° (CHCl$_3$)

$[\alpha]_{240}^{25}$ +2400° (MeOH)

Reagents:
(i)   KMnO$_4$
(ii) CH$_2$N$_2$, OH$^-$
(iii) CH$_2$N$_2$
(iv) LiAlH$_4$

Scheme 7. Determination of the absolute configuration of (+)-hexamethoxydiphenic acid, (33, 62)

eclipsed conformation of the ester carbonyl oxygen and the adjacent hydrogen atom on the glucose ring (30),

and (ii) that in the hexahydroxydiphenoyl group there is a preference for the *anti* or opposed arrangement of the two carbonyl groups with the C-O dipoles aligned antiparallel (31).

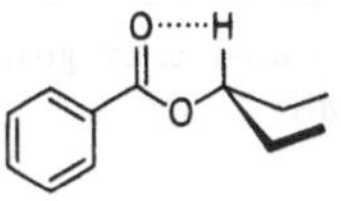

(30) 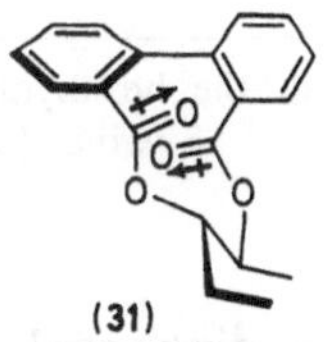   (31)

If these two features dominate the oxidative coupling reactions of galloyl esters of D-glucose then the geometrical restrictions imposed by the sugar ring lead directly to the conclusion that the chirality of the derived hexahydroxydiphenoyl group is determined by the sugar. Molecular models show that the 3,6-bridges [e.g. corilagin (28)] have the R-configuration and 4,6 or 2,3 hexahydroxydiphenoyl linkages (e.g. pedunculagin) have the alternative s configuration. Although hexahydroxydiphenoyl esters linked 2,4 have not yet been isolated from natural sources, this type of argument suggests that they should have the R-configuration.

The presence of the hexahydroxydiphenoyl ester group is readily delineated by spectroscopic means (*26, 32, 34, 98*). In the ^{1}H nmr spectra the two aryl protons of the hexahydroxydiphenoyl group appear as singlets generally upfield (d_6 acetone, δ 6.3 – 6.8 ppm TMS) from the two proton singlet of the galloyl ester group. Correspondingly the hexahydroxydiphenoyl ester (**8**) has several distinctive ^{13}C nmr characteristics which aid its recognition (Fig. 7). The signals due to the carbonyl groups appear downfield from TMS relative to those of galloyl esters and the signals due to C-1, C-2 and C-6 are diagnostic for the hexahydroxydiphenoyl group. The number of such groups in the molecule may usually be discerned from the multiplicity of the singlets in the ^{13}C nmr spectrum arising from C-6 or the carbonyl groups or the aryl proton singlets in the ^{1}H nmr. The chirality of the hexahydroxydiphenoyl group (and hence its possible positions of attachment to the D-glucopyranose core, *vide supra*) may be determined by cd (circular dichroism) measurements (*103*).

^{1}H chemical shift	^{13}C chemical shifts		
d_6 acetone		d_6 acetone	d_4-methanol
H-2 6.3 – 6.8	C-1	125.5 – 127.0	
	C-2	107.0 – 109.0	
	C-6	114.0 – 116.5	
	C-7	167.8 – 169.5	168.9 – 171.2

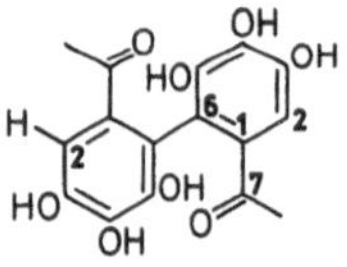

Fig. 7. Nmr characteristics of the hexahydroxydiphenoyl ester group (*26, 32, 34, 98*). δ values (ppm TMS)

The formation of a diphenoyl ester bridge in a galloyl-D-glucose derivatives gives the molecule much greater rigidity and this may result in constraints on the D-glucose residue to adopt an unfavourable shape or conformation. SCHMIDT and JOCHIMS (*34*), have demonstrated this feature in ^{1}H nmr studies. Thus for example corilagin [Scheme 6 (**28**)] may adopt a flexible boat conformation or the chair form (**28**) in which the bulky substituents are all axially disposed. The conformational restraints which result from the presence of hexahydroxydiphenoyl ester groups in a molecule usually results in changes in the ^{13}C chemical shift of the D-glucose carbon atom signals relative to the related galloyl-D-glucose derivative (Fig. 8).

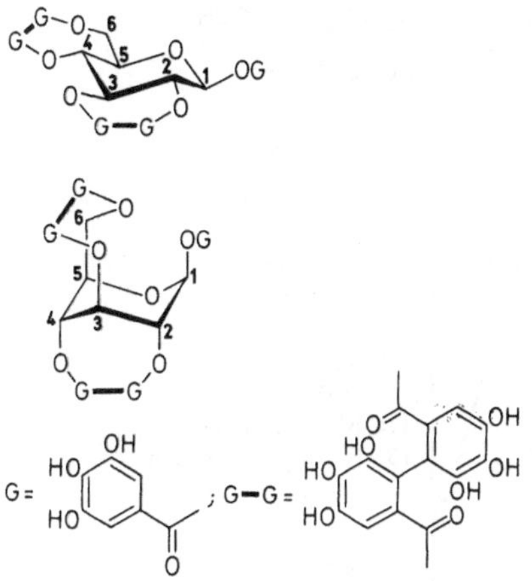

	Chemical shift (d$_6$ acetone, δ ppm TMS)	Change in Chemical Shift relative to β-penta-o-galloyl D-glucose (Δδ)
C-1	92.1	−1.2
C-2	73.4	+1.5
C-3	77.2	+3.7
C-4	69.2	−0.3
C-5	77.2	+3.1
C-6	63.1	+0.2
C-1	91.9	−1.5
C-2	67.8	−4.1
C-3	76.1	+2.6
C-4	67.8	−1.7
C-5	77.0	+2.9
C-6	65.3	+2.4

G = galloyl; G-G-hexahydroxydiphenoyl

Fig. 8. ^{13}C Chemical shift of the D-glucose carbon atoms in some hexahydroxydiphenoyl esters of D-glucose relative to β-penta-o-galloyl-D-glucose, (26, 19)

2.3.2 Dehydrohexahydroxydiphenoyl and Isohexahydroxydiphenoyl Esters

In terms of the biogenetic relationships to be outlined two important, and in chemical terms perhaps unexpected, structural variations of the hexahydroxydiphenoyl ester group were identified by SCHMIDT. These are the dehydrohexahydroxydiphenoyl group (32) [found in brevilagin 1 and 2 from the fruit pods of *Caesalpinia brevifolia* (86, 87)] and the *iso*-hexahydroxydiphenoyl ester group (33) [present in terchebin from the fruits of *Terminalia chebula* (88)]. Although terchebin remains the sole example of the occurrence of (33) in plants the ester (32) has been identified as a structural fragment of the yellow crystalline substance geraniin and the amorphous mallotusinic acid by OKUDA (58, 59, 60, 61, 62) in plants of the Geraniaceae and Euphorbiaceae. Geraniin has also been isolated from other plants (26) (e.g. members of the Aceraceae and various *Fuchsia* sp.).

Both (32) and (33) react with *o*-phenylenediamine to give, after hydrolysis, the crystalline phenazine (38) (Scheme 8). The latter was characterised as its dimethyl ether, diacetate and dibenzoate and hydrolysis and methylation gave the phenazine dimethyl ester (39) which was identified by synthesis (86, 87, 88). The hydrolytic reactions of both (32) and (33) are also significant. Acid hydrolysis of (33) gives ellagic

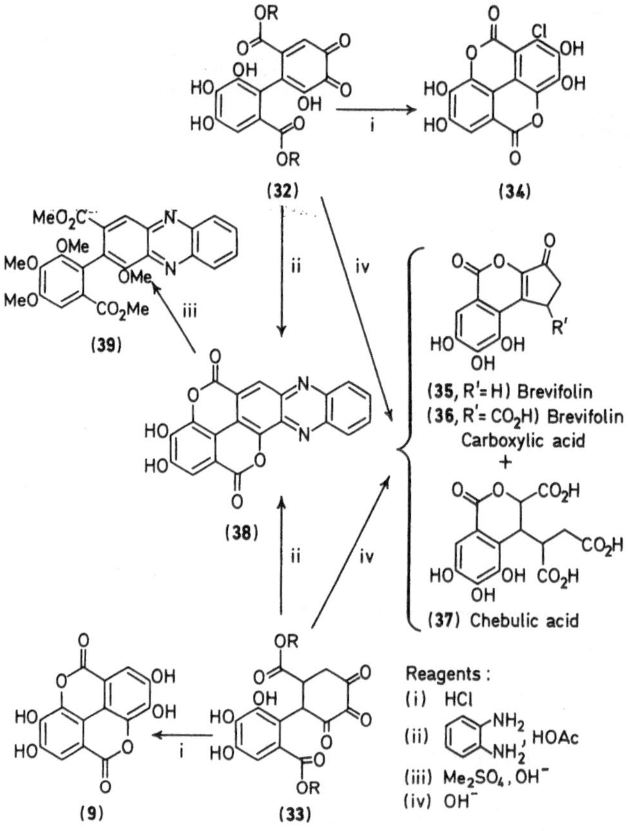

Scheme 8. Reactions of isohexahydroxydiphenoyl and dehydrohexahydroxydiphenoyl esters
(19, 58, 86, 87, 88, 89)

acid (9) but the action of concentrated hydrochloric acid on (32) gives chlorellagic acid (34). Both (32) and (33) undergo base-catalysed hydrolysis (89) to yield brevifolin (35), brevifolin carboxylic acid (36) and chebulic acid (37) — all of which are products which result from the ring fission or ring contraction of the non-aromatic ring of the ester groups (32) and (33).

Spectroscopically (^{1}H nmr and ^{13}C nmr) the presence of both the dehydrohexahydroxydiphenoyl ester group (32) and the isohexahydroxydiphenoyl ester (33) in a molecule leads to a complexity of signals which at the same time is both characteristic and distinctive but difficult to interpret satisfactorily in structural terms. In both cases isomerisation occurs in solution and leads to an equilibrium mixture of isomers and duplication of signals. The attainment of the position of equilibrium is accelerated in hydroxylic media. Although in the ^{1}H nmr spectra the most obvious changes occur in the protons of the ester groups (32, 33) transformations in the protons of the associated D-glucose are also observed. Bearing in mind the different solvents in which the experiments were conducted the data for (32) of OKUDA (58, 59) and for (33) of SCHMIDT (88, 34) show a disconcerting similarity, Fig. 9.

d$_6$ acetone	δ values (ppm TMS)		
before equilibration	6.43 (s)	5.16 (s)	7.13 (s)
after equilibration	6.53 (s)	5.16 (s)	7.13 (s)
	6.26 (d, J = 1.6 Hz)	4.72 (d, J = 1.5 Hz)	7.16 (s)

OKUDA (58, 59)

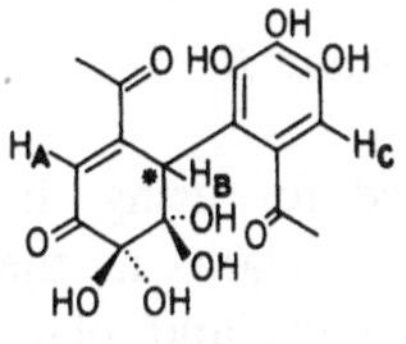

(32a)

d$_6$ DMSO	δ values (ppm TMS)		
after equilibration	6.24 (d, 1.3 Hz)	4.66 (d, 1.3 Hz)	7.22 (s)
	6.48 (s)	4.96 (s)	7.15 (s)

SCHMIDT (34, 88)

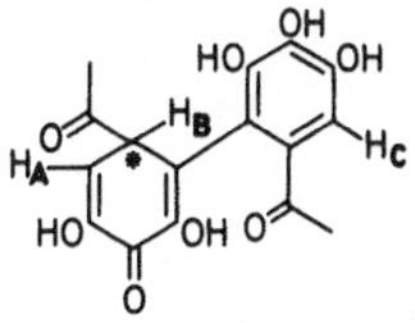

(33)

Fig. 9. Isomerisation of dehydrohexahydroxydiphenoyl esters (32) and isohexahydroxydiphenoyl esters (33) in solution. [1]H nmr chemical shift data (34, 58, 59, 88)

SCHMIDT (86, 87, 88) postulated that these changes were due to the isomerisation of the protons H$_B$ in (32a) and (33) and the creation of diastereoisomers differing in configuration at the allylic centres (*) and OKUDA (59) initially also favoured epimerisation at the methine C (*) to explain this phenomenon in (32). However in the presence of deuterium oxide **no** incorporation of deuterium is observed in geraniin [contains (32)] and the molecule is recovered unchanged after equilibration. Signals at δ 92.3, 92.5 (2) and 96.3 (ppm from TMS) in the ^{13}C nmr spectrum of geraniin [contains the bis-ester (32)] have been attributed by OKUDA (59, 62) to hydration of two carbonyl groups in the dehydrohexahydroxydiphenoyl group (32) to give structure (32a) (Fig. 9) in geraniin. In a very recent communication OKUDA (62) has put forward two hemiacetal structures for the dehydrohexahydroxydiphenoyl ester (32b) and (32c) to explain the isomerisation. Other possibilities which should perhaps be considered to explain the isomerisation process in these esters are changes in the position of hydration in the non-aromatic ring and conformational changes in the 11-membered ring of the ester (**32**) as it bridges the 2,4-positions in the D-glucopyranose ring.

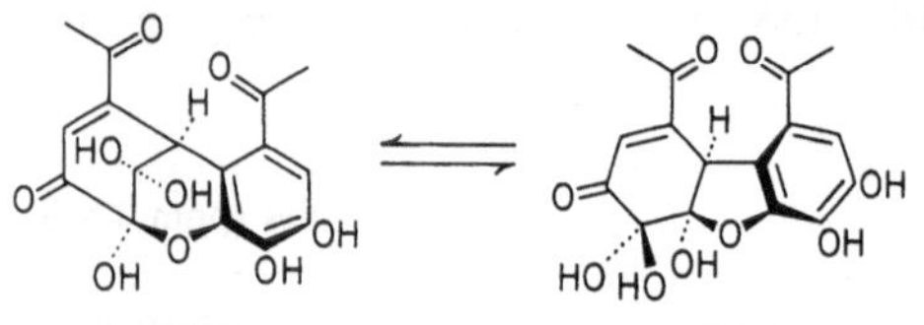

HO
OH
O
OH
OH
OH
H
O
O
HO
HO
OH
OH
OH

(32 b) (32c)

2.3.3 *Hexahydroxydiphenoyl Metabolites — Group B*

Earlier in this review good reasons were advanced for using the patterns of phenolic metabolism in the leaves of higher plants as the basis for a comparison of this activity in different plants and plant

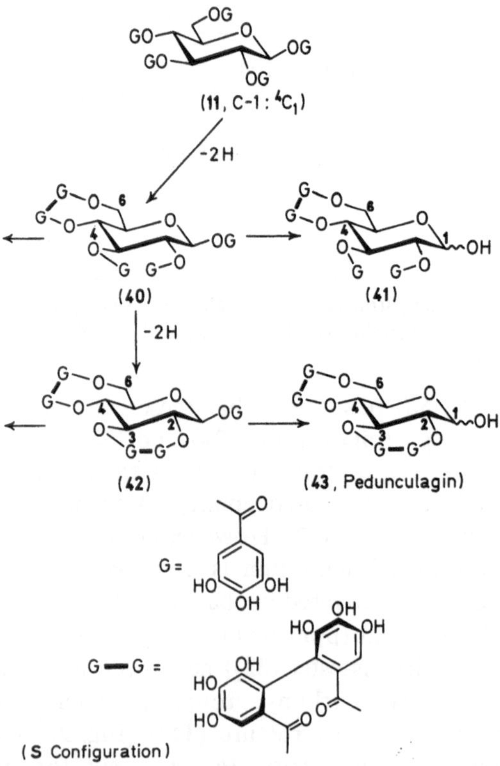

(11, C-1: $^4C_1$)
-2H
(40)
(41)
-2H
(42)
(43, Pedunculagin)
G =
G—G =
(S Configuration)

Scheme 9. The metabolism of hexahydroxydiphenoyl esters in higher plants — class B metabolites

References, pp. 42—46

families. Two dimensional paper chromatographic analysis of plant phenolic extracts has proved to be an invaluable aid in the delineation of the various patterns of metabolism of proanthocyanidins (*25*) and now of hexahydroxydiphenoyl esters and their derivatives in the plant kingdom (*26*). In over 100 plants which have been examined a very widely distributed metabolic fingerprint is that in which β-penta-o-galloyl-D-glucose **(11)** — it is assumed — is further transformed by oxidative coupling of pairs of adjacent galloyl ester groups (2, 3 and 4,6) on the β-D-glucopyranose core (Scheme 9). Some details of this pattern of metabolism were hinted at in earlier work by HILLIS and SIEKEL (*32*), by JURD (*104*) and by WILKINS and BOHM (*98*), and SCHMIDT earlier isolated and identified (*102*) pedunculagin **(43)** (2,3 : 4,6-bishexahydroxydiphenoyl-D-glucose) from oak galls. This ester **(43)** is a key metabolite in the phenolic fingerprint (Fig. 10).

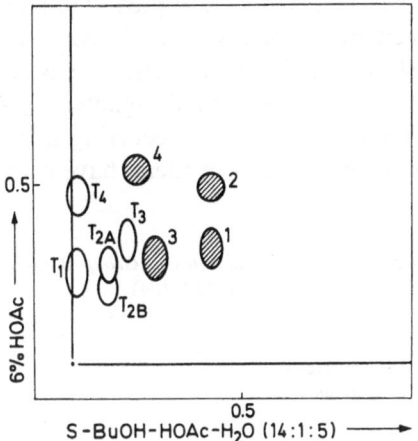

1. β-1,2,3-Tri-o-galloyl-4,6-[s]-hexahydroxydiphenoyl-D-glucose, **(40)**
2. 2,3-Di-o-galloyl-4,6-[s]-hexahydroxydiphenoyl-D-glucose, **(41)**
3. β-1-o-Galloyl-2,3 : 4,6-bis-[s]-hexahydroxydiphenoyl-D-glucose, **(42)**
4. 2,3 : 4,6-Bis-[s]-hexahydroxydiphenoyl-D-glucose (Pedunculagin), **(43)**

Fig. 10. The metabolism of hexahydroxydiphenoyl esters in higher plants — class B metabolites. Paper chromatographic fingerprint

The metabolites produced in this manner are referred to here as class B metabolites. Plants whose phenolic metabolism places them within this category furnish, along with β-penta-o-galloyl-D-glucose **(11)**, one or more of the hexahydroxydiphenoyl esters **(40, 41, 42)** and **(43)** as key phenolic metabolites (*26*). The ratio of these four metabolites varies quite markedly

from plant to plant, even within the same plant family, and with this
variation there is a parallel variation in the occurrence of the substances
denoted as $T_1 - T_4$ (Fig. 10). Thus for example in the plant family
Rosaceae the leaves of *Rosa canina* and *Filipendula ulmaria* metabolise
substantial quantities of the monohexahydroxydiphenoyl esters (**40, 41**)
and these are accompanied by the esters T_{2A}, T_{2B}. In the same plant
family *Rubus idaeus, Rubus fructicosus, Geum rivale* and *Potentilla* sp.
biosynthesise predominantly the bishexahydroxydiphenoyl esters (**42,
43**) and these occur alongside the ester T_1. These patterns of occurrence are common ones and suggest a relationship of the hexahydroxydiphenoyl esters (**40, 41, 42**) and (**43**) with the esters T_1 to T_4. Substance
T_1 has been examined in greatest detail (*18, 26*) and its occurrence parallels
the dominance of the ester (**42**) in the phenolic profile of the plant.
Refluxing in water or methanol gives both 2,3- and very small amounts
of 4,6-hexahydroxydiphenoyl-D-glucose but after short periods of hydrolysis the ester (**42**) may be isolated. Coupled with spectroscopic evidence
the ester T_1 has been provisionally formulated as a dimer of (**42**). An
analogous relationship is similarly thought to exist between the various
metabolites (**40, 41**) and the substances designated $T_2 - T_4$. These patterns
which have been sketched are widely occurring in plants and the plant
families in which they, or elements of them, have been detected are listed
in Table 7.

Table 7. *Hexahydroxydiphenoyl Ester Metabolites — Group B. Occurrence in Plant Families*
(*26, 98, 102, 104*)

Fagales	
	Betulaceae
	Fagaceae
Rosales	
	Rosaceae
	Saxifragaceae
Myrtales	
	Myrtaceae
	Onagraceae
	Combretaceae
Cornales	
	Cornaceae
Proteales	
	Elaeagnaceae
Juglandales	
	Juglandaceae

The structures of the hexahydroxydiphenoyl esters (**40, 41, 42, 43**)
were established by spectroscopic methods in accordance with the procedures outlined *(vide supra)* and by partial hydrolysis. Thus for example

pedunculagin (**43**), 2,3-hexahydroxydiphenoyl-D-glucose (**45**) and β-1-O-galloyl-2,3-hexahydroxydiphenoyl-D-glucose (**44**) were isolated and identified as partial hydrolytic products of the bishexahydroxydiphenoyl ester (**42**) (Scheme 10). Interestingly the ester (**44**) has also been identified as a phenolic metabolite of *Hippophae rhamnoides*.

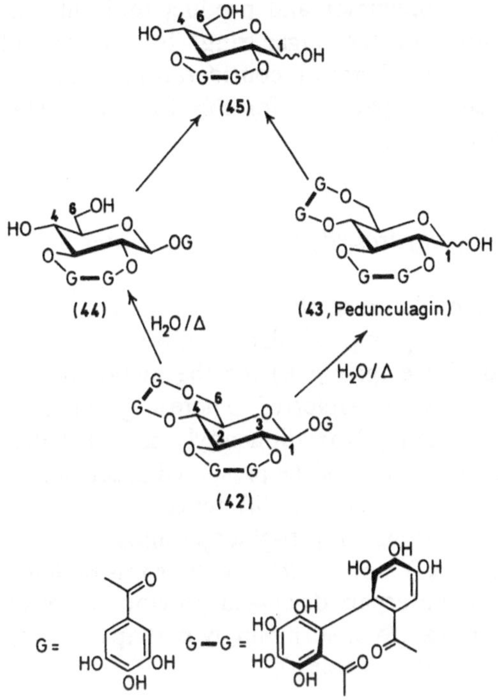

Scheme 10. Partial hydrolysis of β-1-O-galloyl-2,3 : 4,6-bishexahydroxydiphenoyl-D-glucose
(**42**), (*26*)

SCHMIDT (*102*) had earlier demonstrated that the chiralities of the hexahydroxydiphenoyl ester groups linked 2,3 and 4,6 to β-D-glucose in pedunculagin (**43**) were the same, corresponded to the laevorotary form of hexahydroxydiphenic acid and hence possess the s configuration *(vide supra)*. These observations are also borne out by cd studies of the various esters (**40, 41, 42, 43**) which are characterised by a distinctive couplet centred at ~200−210 nm with a positive maximum at 228 to 238 nm (*103*). The $\Delta\varepsilon_{max}$ values are approximately incremental (+30) for the number of biphenyl (hexahydroxydiphenoyl ester) linkages in the molecule (*103*). The $\Delta\varepsilon_{max}$ value for the ester T_1 is strongly suggestive of about 4 hexahydroxydiphenoyl ester groups, (*103*) in the molecule and hence of a dimeric structure of the type proposed.

Although the evidence is wholly circumstantial, the co-occurrence of the various hexahydroxydiphenoyl esters (**40, 41, 42, 43**) ($T_1 - T_4$, Fig. 10) alongside β-penta-o-galloyl-D-glucose (**11**) suggests that the latter galloyl ester is the biogenetic precursor of the hexahydroxydiphenoyl esters. This presumes that oxidative coupling takes place preferentially between galloyl ester groups 2 and 3, and 4 and 6 on the D-glucopyranose ring with both the precursor and product molecules adopting the C-1 ($^4C_1$) conformation of the sugar ring. Its widespread occurrence in plants, coupled with chemical considerations, suggest that it is the energetically preferred pathway for this form of metabolism of hexahydroxydiphenoyl esters.

2.3.4 Hexahydroxydiphenoyl Metabolites — Group C

Although the evidence at this stage is far from complete it is clear that a much smaller group of plants utilize a method, quite different from that noted above (group B) for the metabolism of hexahydroxydiphenoyl esters. This class (referred to here as group C) exhibits oxidative coupling to form hexahydroxydiphenoyl esters via the 1-C ($^1C_4$) conformation, or its equivalent, of the presumed precursor β-penta-o-galloyl-D-glucose (**11**). The free energy difference between the 1-C ($^1C_4$) and C-1 ($^4C_1$) conformations of β-D-glucopyranose has been estimated as -5.95 kcal (-24.9 kJ) mol^{-1} (*105*) and thus this pathway of metabolism, compared to its counterpart discussed previously above (group B), is a higher energy pathway. Not surprisingly perhaps its occurrence in plants is much more sporadic (Table 8).

Table 8. *Hexahydroxydiphenoyl Metabolites — Group C. Occurrence in Plant Families* (*26, 61*)

Hamamelidales	
	Cercidiphyllaceae
Ericales	
	Ericaceae
Myrtales	
	Onagraceae
Euphorbiales	
	Euphorbiaceae
Sapindales	
	Aceraceae
	Simaroubaceae
Geraniales	
	Geraniaceae
Cornales	
	Nyssaceae

In this mode of metabolism oxidative coupling occurs primarily between the 2,4 and the 3,6 positions and preliminary work (*19*) suggests that geraniin (**46**) is one of the key metabolites of this pathway and dominates the phenolic "fingerprint". Geraniin (**46**) was first described by OKUDA (*58, 59, 60, 61, 62*) in *Geranium* species. The dehydrohexahydroxydiphenoyl ester group (**32**) in geraniin was characterised by formation (Scheme 11) of the phenazine (**38**). The residual structural fragment from this reaction was the known hexahydroxydiphenoyl ester corilagin (**28**) in which the hexahydroxydiphenoyl group has the R configuration. OKUDA (*59*) located the dehydrohexahydroxydiphenoyl ester group in the manner shown on the basis of spectroscopic arguments. Biogenetic support for this formulation is given by the clearly implied relationship of structure (**46**) to those of chebulinic acid and chebulagic acid from Myrabolans [*vide infra (34, 95*)]. The 2,4:3,6-bishexahydroxydiphenoyl ester (**47**) has not yet been isolated from plants but it may be formed by catalytic reduction of geraniin (*19*). In water the 2,4-hexahydroxydiphenoyl ester group is, somewhat suprisingly, very rapidly cleaved from (**47**) to give corilagin (**28**).

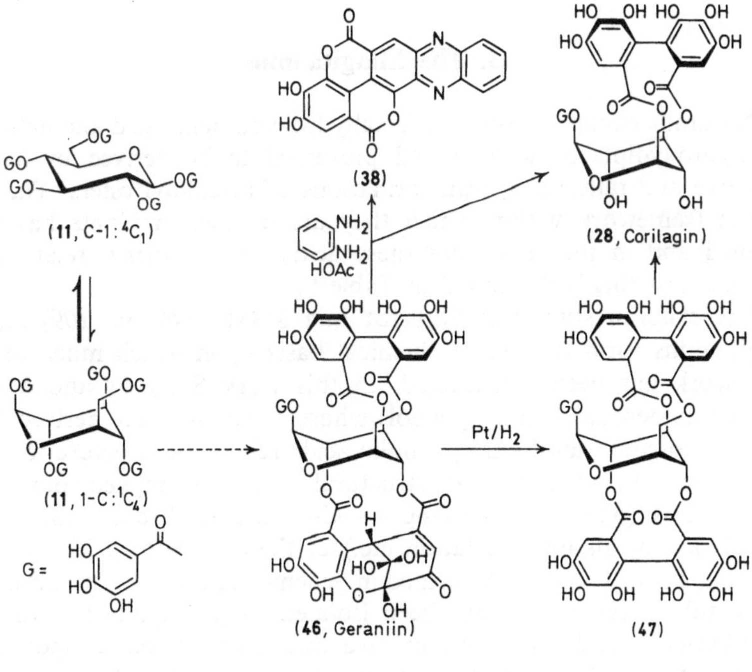

Scheme 11. Biosynthesis and some chemical reactions of geraniin (*19, 58 — 62*)

In contrast to the hexahydroxydiphenoyl ester groups which bridge the 2,3 : 4,6 and 3,6 positions of D-glucopyranose and in which there is a relatively small dihedral angle between the two aromatic rings, in the 2,4-linked hexahydroxydiphenoyl ester group the two aromatic rings of the biphenyl system are constrained to a virtually orthogonal relationship. This juxtaposition of the two aromatic rings facilitates direct or indirect (neighbouring group) participation of the phenolic groups *ortho* to the biphenyl linkage in hydrolytic reactions and this is probably the cause of the facile cleavage of the 2,4-hexahydroxydiphenoyl ester from the substrate (**47**). It is interesting to speculate, in the light of this observation, on the possible reasons for the metabolism of the dehydrohexahydroxydiphenoyl ester group in (**46**). Models show that formation of the ester (**32**) in the 2,4-position and consequently generation of a quaternary centre at one of the termini of the biphenyl link, relieves some of the constraint and renders participation of the neighbouring phenolic groups in hydrolysis less likely. This makes (**32**) more stable towards hydrolytic cleavage than (**8**) when bridging the 2,4-positions of D-glucopyranose.

Geraniin has been isolated and identified from the leaves of various *Acer* and *Fuchsia* species (**26**). Its occurrence in *Fuchsia* is interesting since it remains the sole example of a plant species to produce both group B and C hexahydroxydiphenoyl metabolites.

3. The Ellagitannins

Naturally occurring esters of hexahydroxydiphenic acid and dehydrohexahydroxydiphenic acid are all presumed to be derived by *in vivo* oxidative and tautomeric transformations of bisgalloyl esters. The biogenetic framework within which they are created in plants has been outlined and in the sense that these acids bear a formal relationship to gallic acid this is illustrated in Table 9.

The classical studies of SCHMIDT and MAYER (*46, 80, 109*) on the ellagitannins form the secure chemical basis upon which much of this framework has been constructed. In this work SCHMIDT and MAYER isolated numerous complex phenols whose structures were derived from a whole range of additional phenolic acids related in structure to gallic acid (Table 9). The formal structural and the implied biogenetic relationships (oxidative coupling — C-C and C-O bond formation, hydrolytic ring fission, oxidation etc.) of these acids to gallic acid are shown in Table 9 and are illustrated in Schemes 12 and 13. The structures of the substances which have been isolated — principally by SCHMIDT and MAYER — and which contain structural elements based upon these various additional carboxylic acids are shown in Table 10.

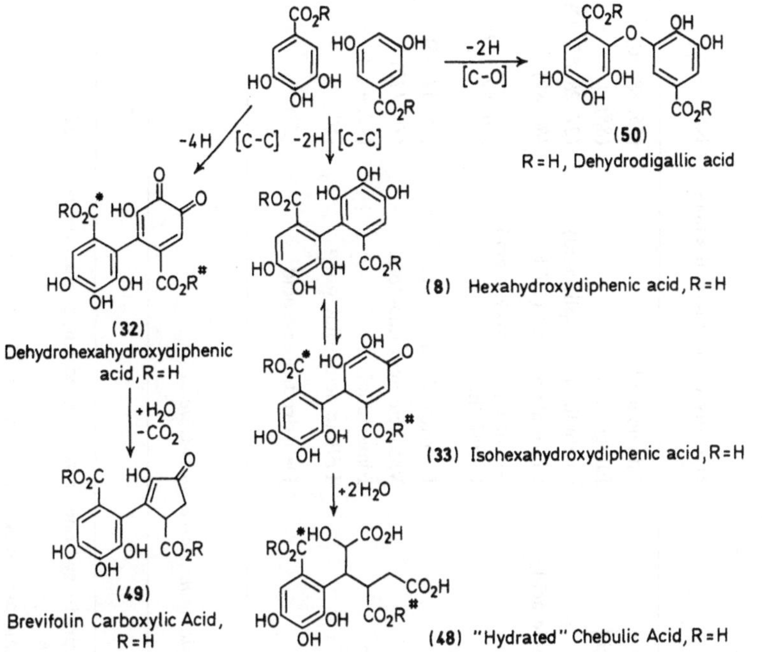

Scheme 12. Phenolic esters formed by oxidative coupling and other transformations of bis-galloyl esters

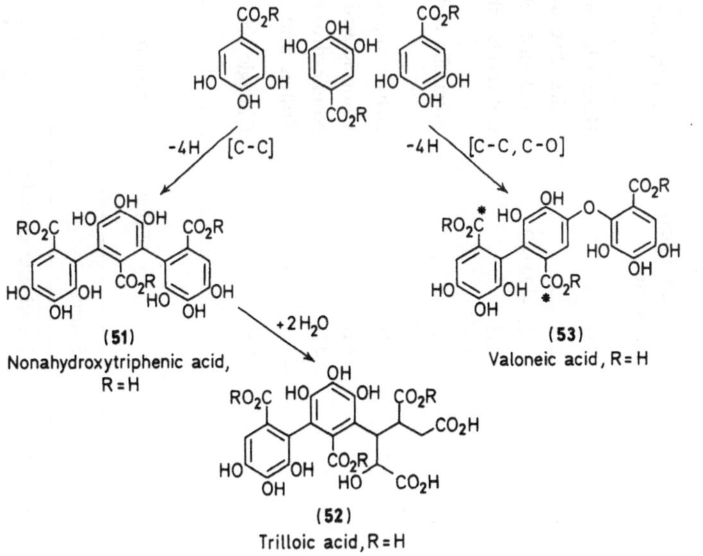

Scheme 13. Phenolic esters formed by oxidative coupling and other transformations of tris-galloyl esters

3*

Table 9. *Formal Relationships Between Gallic Acid and some Naturally Occurring Phenolcarboxylic Acids*

Number of gallic acid molecules	Formal chemical relationship	Phenolcarboxylic acid	Hydrolysis products of esters	Ref.
2	$-2H$	Hexahydroxydiphenic acid	Ellagic acid	(75, 78, 80)
2	$-2H$	Isohexahydroxydiphenic acid	Ellagic acid	(88)
2	$-2H, +2H_2O$	"Hydrated" chebulic acid	Chebulic acid (Split acid)	(27, 28, 82, 92)
2	$-4H$ $+H_2O$	Dehydrohexahydroxydiphenic acid	Chlorellagic acid (HCl)	(86, 87)
	$-4H$ $-CO_2$	Brevifolin carboxylic acid	Brevifolin carboxylic acid Brevifolin	(80, 89, 111)
2	$-2H$	Dehydrodigallic acid	Dehydrodigallic acid	(38, 64, 65)
3	$-4H$	Valonic acid	Valoneic acid dilactone	(79, 110)
3	$-4H$	Nonahydroxytriphenic acid	Flavogallol	(42, 43, 44)
3	$-4H, +2H_2O$	Trilloic acid	Trilloic acid trilactone	(50)
4	$-6H$ $-2H_2O$	Gallic acid	Dehydrodiellagic acid	(52)

Table 10. *The Structure of Some Ellagitannins Based on D-Glucose*

Compound	Source	Structure	Ref.
1. Corilagin (28)	*Terminalia chebula*	β-1-o-galloyl; R-(+)-hexahydroxydiphenic acid linked 3,6; D-glucopyranose conformation — 1B	(32, 71, 73)
2. Pedunculagin (43)	*Quercus pedunculata Q. sessiflora* (galls)	s-(−)-hexahydroxydiphenic acid linked 2,3 and 4,6; D-glucopyranose conformation — C-1	(32, 102)
3. Terchebin (55)	*Terminalia chebula* (Myrabolans)	β-1,3,6-tri-o-galloyl; Residue (+) form of (33) linked 2*, 4‡; D-gluco-pyranose conformation — B-3 or $1C \rightleftharpoons 1B$	(34, 88)

No.	Compound	Plant source	Description	References
4.	Chebulinic acid (56)	Terminalia chebula (Myrabolans)	β-1,3,6-tri-o-galloyl; Residue (48) linked 2*, 4‡; D-glucopyranose conformation — B-3 or 1c⇌1B	(27, 28, 82, 95)
5.	Chebulagic acid (57)	Terminalia chebula (Myrabolans)	β-1-o-galloyl; R-(+)-hexahydroxydiphenic acid linked 3,6; Residue (48) linked 2*, 4‡; D-glucopyranose conformation — 1c	(73, 78, 80, 82)
6.	Brevilagin 1	Caesalpinia brevifolia (Algarobilla)	Residue (−) form of (32) linked β-1, 3 and 4,6; D-glucopyranose conformation — 2B	(34, 86)
7.	Brevilagin 2	Caesalpinia brevifolia (Algarobilla)	s-(−)-hexahydroxydiphenic acid linked 4,6; Residue (−) form of (32) linked β-1,3; D-glucopyranose conformation — 2B⇌B-3	(34, 87)
8.	Vescalin	Quercus sessiflora	Based on nonahydroxytriphenic acid (51) linked 2,3,5 with C-glycoside link to c-1; Open chain form of D-glucose	(43)
9.	Castalin	Castanea sativa	Isomeric with Vescalin at anomeric centre C-1	(41)
10.	Vescalagin	Quercus sessiflora	s-(−)-hexahydroxydiphenic acid linked 4,6 to vescalin	(44, 45)
11.	Castalagin	Castanea sativa	Isomeric with vescalagin at anomeric centre C-1	(42, 45)
12.	Valolaginic acid	Quercus valonea	Based on trilloic acid (52) linked 2,3,5 with C-glycose link to C-1; s-(−)-hexahydroxydiphenic acid linked 4,6 to open chain form of D-glucose	(49)
13.	Isovalolaginic acid	Quercus valonea	Isomeric with valolaginic acid at anomeric centre C-1	(51)
14.	Castavalonenic acid	Quercus valonea	Based on nonahydroxytriphenic acid (51) linked 2,3,5 with C-glycosidic link to C-1; valoneic acid (53) linked 4,6; open chain form of D-glucose	(48)
15.	Punicalin	Punica granatum	Gallagic acid linked 2,6; D-glucopyranose	(52)
16.	Punicalagin	Punica granatum	s-(−)-hexahydroxydiphenic acid linked 3,4 to Punicalin	(52)
17.	Geraniin (46)	Geranium thunbergii Acer sp., Fuchsia sp.	β-1-o-galloyl; R-(+)-hexahydroxydiphenic acid linked 3,6; Dehydrohexahydroxydiphenic acid linked 2*, 4‡; D-glucopyranose conformation — 1c	(26, 58, 59, 62)
18.	Mallotinic acid	Mallotus japonicus	β-1-o-galloyl; valoneic acid (53) linked 3,6; D-glucopyranose	(60)
19.	Mallotusinic acid	Mallotus japonicus	Dehydrohexahydroxydiphenic acid (32) linked 2*, 4‡ to mallotinic acid; D-glucopyranose	(60)

In the great majority of cases the sources of plant materials employed in the isolation of these polyphenolic compounds have been the traditional sources of economically important ellagitannins (Table 1, for example — Myrabolans, Algarobilla, Valonea, Oak and Chestnut bark and Oak galls). It is clear, from paper chromatographic analysis (26), that these particular plant materials are specialised sources and in many cases a direct link cannot yet be discerned with the patterns of gallic acid metabolism found in the leaves of plants which have been discussed above. In this context the unique structures assigned to vescalin and vescalagin (43, 44), castalin and castalagin (41, 42), valolaginic acid and isovalolaginic acid (49, 51), which are based upon the open-chain conformation of the D-glucose molecule, are worthy of comment and particular attention. It is also interesting to note that all these latter polyphenols contain a distinctive structural feature — namely that of a c-glycoside — analogous to that of the c-glucoside bergenin (30). Where a relationship can be discerned (e. g. Chebulinic acid and Chebulagic acid in Myrabolans) this usually involves an extension of one of the basic patterns of gallic acid metabolism (A, B or C) outlined above.

When the structure of (+)-chebulic acid [(54), "split" acid] was finally elucidated by SCHMIDT and MAYER (78, 80) and by HAWORTH (27, 28) in the early 1950's it pointed the way to a putative biogenetic relationship with gallic acid and hexahydroxydiphenic acid. The relationship was fully elaborated by SCHMIDT and MAYER (81) and, with the further refinements later made to the structure of chebulinic acid (34, 56, 95), is shown in Schemes 12 and 14. The hypothesis is based on a suggestion that one of the aromatic residues of a bound hexahydroxydiphenoyl group can undergo isomerisation and hydrolytic fission to give the modified residue (48) which is part of the structure of chebulinic acid (56) and chebulagic acid (47). Chebulic acid as it is bound to D-glucose in these acids is in the form (48) but hydrolysis of either compound gives, after lactonisation, free chebulic acid (54). This form of metabolism occurs in the fruit of *Terminalia chebula*. The laevorotary form of chebulic acid has also been obtained (92) from Algarobilla extract (Table 1) but these are the only reported sources of products of this extended form of hexahydroxydiphenic acid metabolism. In the light of the earlier discussion *(vide supra)* the biosynthesis of the chebulinic acid (56) and chebulagic acid (57) structures is a modification of the pattern labelled Group C.

MAYER first isolated dehydrodigallic acid (50, R = H) from Sweet Chestnut leaves [*Castanea sativa* (38)] and more recently OZAWA and TAKINO (64, 65) have described two new phenolic glycosides chestanin (58) and chesnatin from *Castanea crenata* galls. Both chestanin and chesnatin contain the dehydrogallic acid structural fragment (50) esterified to a phenolic glycoside. Dehydrodigalloyl esters (50) can be visualised as being derived (Scheme 11) by C-O as opposed to C-C oxidative coupling of two galloyl ester groups; thus in the chestnut galls *(Castanea crenata)* chestanin (58) is probably derived by inter-molecular oxidative coupling of the galloyl ester groups in two molecules of the galloyl glycoside precursor cretanin (59) (Scheme 15).

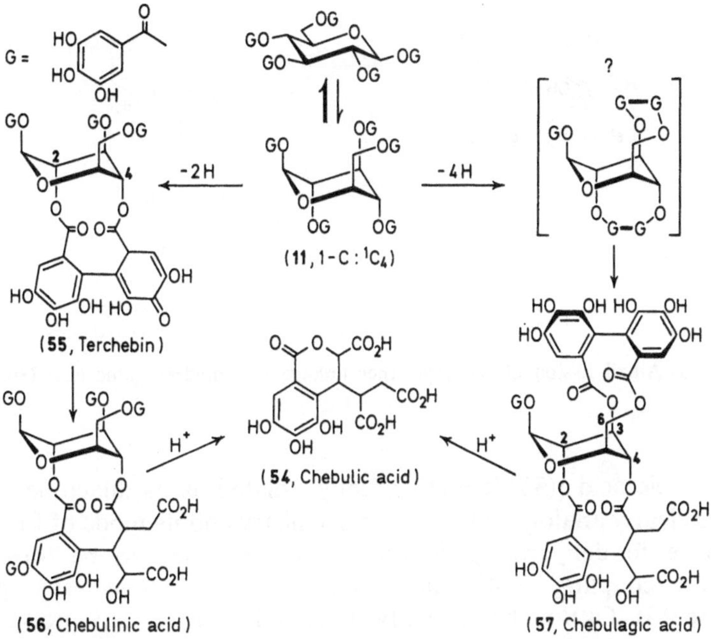

Scheme 14. Proposed pathways of biosynthesis of chebulinic acid (56) and chebulagic acid (57) from β-penta-O-galloyl-D-glucose (11), (81). Hydrolysis of (56 and 57) to give chebulic acid (54)

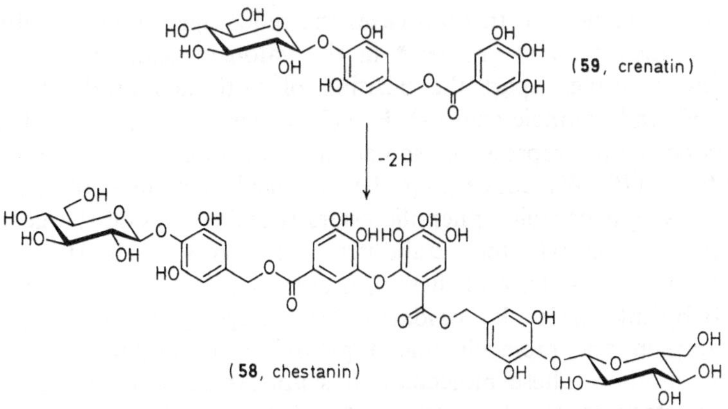

Scheme 15. Proposed pathway of biosynthesis of chestanin (58)

Perhaps the most interesting aspect of the chemistry of dehydro-digallic acid is the unusual fission of the diaryl ether linkage which takes place in aqueous alkali (Scheme 16) (*39*). The reaction may be formulated as a β-elimination process.

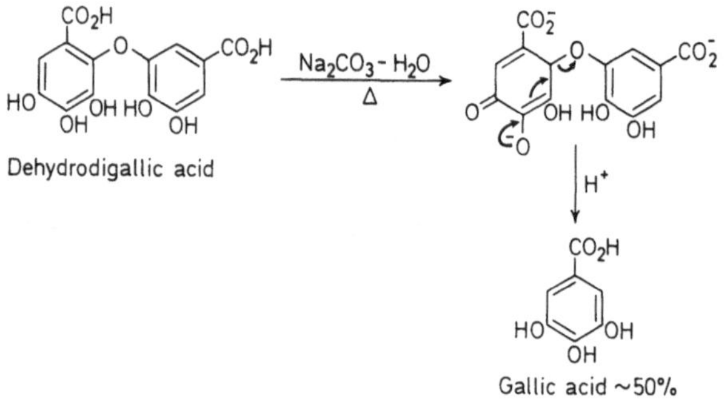

Scheme 16. Alkali fission of the aryl ether linkage of dehydrodigallic acid (**50**, R = H)

Valoneic acid [(**53**, R = H), usually isolated as its dilactone (*48, 61, 109*)] is clearly analogous both in its chemistry and its mode of formation in nature to dehydrodigallic acid. It is presumed to be formed by oxidative coupling (C-C and C-O) of three galloyl ester groups (Scheme 13). OKUDA has recently described (*61*) two new polyphenols from *Mallotus japonica*, mallotusinic acid and mallotinic acid (Table 9) which contain the valoneic acid residue bound at the 3,6 positions to D-glucopyranose and MAYER has similarly described (**48**) castavaloneic acid from *Quercus valonea* which has valoneic acid bound to the 4,6 positions of D-glucose (Table 12). In both cases the bidentage linkage to glucose occurs via the carboxyl groups * in the valoneic acid structure (**53**). These few examples apart the position of both dehydrodigallic acid (**50**, R = H) and valoneic acid (**53**, R = H) in the overall picture of gallic acid metabolism has represented something of an enigma, However recent observations (*19, 26*) concerning the chemical structures of the high molecular weight complex phenolic esters found alongside the group B hexahydroxydiphenoyl ester metabolites — the esters $T_1 - T_4$ (Fig. 10) — suggests that some at least of these molecules are derived from the esters (**40**—**43**) by intermolecular oxidative C-O coupling between a galloyl ester group in one molecule and a galloyl or hexahydroxydiphenoyl group in another. These molecules thus appear to be elaborated at a secondary stage in the biosynthesis by intermolecular C-O oxidative coupling of lower molecular weight species in much the same way as has been suggested for the biosynthesis of chestanin from crenatin (Scheme 14). Chemical, and biogenetic considerations such as these lead to the tentative proposal of the structure (60) for the complex ester T_1 characteristic of the group B hexahydroxydiphenoyl metabolites.

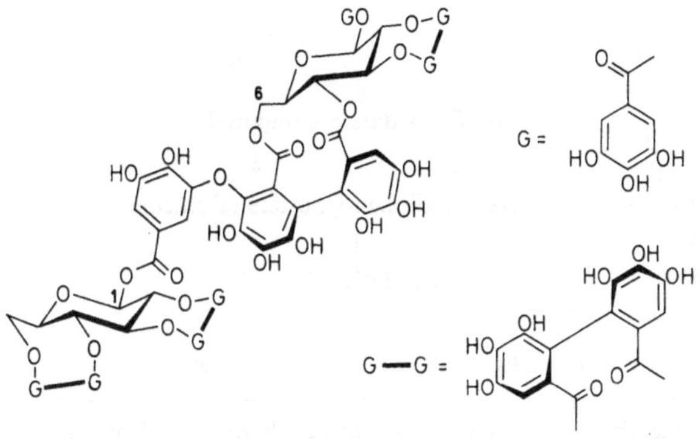

(60)

4. Postscript

"At last the secret is out, as it always must come in the end,
The delicious story is ripe to tell the intimate friend;"

Auden — Twelve Songs

The Swedish chemist SCHEELE first isolated gallic acid from gall nuts in 1786 at the dawn of the era of organic chemistry. As befits a molecule of such apparent simplicity it was almost 200 years before the first synthesis of gallic acid from wholly aliphatic precursors was announced (93). Correspondingly, but for rather less time, the role of gallic acid in the phenolic metabolism of higher plants has been something of an enigma. Although our knowledge and understanding of this area of plant biochemistry still retain a crepuscular quality some of nature's secrets have been revealed. Most commonly gallic acid is found in plants associated with the sugar D-glucose and the overall patterns of metabolism which have emerged are tentatively collated and summarised in Scheme 17.

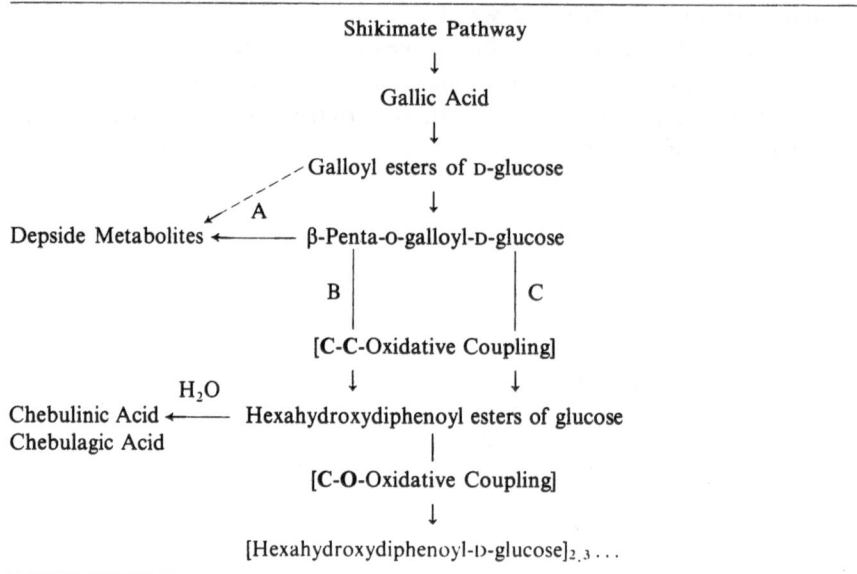

Shikimate Pathway

↓

Gallic Acid

↓

Galloyl esters of D-glucose

↓

Depside Metabolites ←———— β-Penta-o-galloyl-D-glucose

[C-C-Oxidative Coupling]

H₂O

Chebulinic Acid ←——— Hexahydroxydiphenoyl esters of glucose
Chebulagic Acid

[C-O-Oxidative Coupling]

↓

[Hexahydroxydiphenoyl-D-glucose]₂,₃ ...

Scheme 17. The metabolism of gallic acid and hexahydroxydiphenic acid in higher plants

References

1. ARMITAGE, R., G. S. BAYLISS, J. G. GRAMSHAW, E. HASLAM, R. D. HAWORTH, K. JONES, H. J. ROGERS, and T. SEARLE: Gallotannins Part 3. The Constitution of Chinese, Turkish, Sumach and Tara Tannins. J. Chem. Soc. **1961**, 1842.
2. BATE-SMITH, E. C.: The phenolic constituents of plants and their taxonomic significance. 1. Dicotyledons. J. Linnaen Soc. (Bot.) **58**, 95 (1962).
3. — Astringent tannins of *Acer* species. Phytochemistry **16**, 1421 (1977).
4. BOHM, B. A., F. W. COLLINS, and C. K. WILKINS: Flavonol Glycoside gallates from *Tellima grandiflora*. Phytochemistry **14**, 1099 (1975).
5. BOHM, B. A.: Flavonoids of *Tolmiea menziesii*. Phytochemistry **18**, 1079 (1979).
6. BRADFIELD, A. E., and M. PENNEY: The Catechins of Green Tea. Part 2. J. Chem. Soc. **1948**, 2239.
7. BRITTON, G., and E. HASLAM: Gallotannins Part 12. Phenolic constituents of *Arctostaphylos uva-ursi* L. Spreng. J. Chem. Soc. **1965**, 7312.
8. BRITTON, G., P. W. CRABTREE, J. E. STANGROOM, and E. HASLAM: Gallotannins Part 13. The Structure of Chinese Gallotannin: Evidence for a Polygalloyl Chain. J. Chem. Soc. (C) **1966**, 783.
9. CONN, E. E., and T. SWAIN: Biosynthesis of Gallic acid in Higher Plants. Chem. and Ind. **1961**, 592.
10. CORNTHWAITE, D. C., and E. HASLAM: Gallotannins Part 9. The Biosynthesis of Gallic acid in *Rhus typhina*. J. Chem. Soc. **1965**, 3008.
11. COXON, D. T., A. HOLMES, W. D. OLLIS, V. C. VORA, M. S. GRANT, and J. L. TEE: Flavonol Digallates in Green Tea Leaf. Tetrahedron **28**, 2819 (1972).
12. DEWICK, P. M., and E. HASLAM: Phenol Biosynthesis in Higher Plants — Gallic Acid. Biochem. J. **113**, 537 (1969).

13. FISCHER, E., and M. BERGMANN: Über das Tannin und die Synthese ähnlicher Stoffe V. Ber. Deutsche Chem. Ges. **51**, 1760 (1918).
14. FISCHER, E.: Untersuchungen über Depside und Gerbstoffe. Berlin: J. Springer. 1919.
15. GILSON, E. R.: Sur deux nouveaux glucotannoides. Compt. Rend. Acad. Sci. **136**, 385 (1903).
16. GONZALEZ, A. G., C. G. FRANCISCO, R. FRIERE, R. HERNANDEZ, J. SALAZAR, and E. SUAREZ: Lindleyin, a new Phenolic gallyl glucoside from *Aconium lindleyi*. Phytochemistry **15**, 344 (1976).
17. GROSS, S. R.: Recent Advances in the Chemistry and Biochemistry of Lignin in Recent Advances in Phytochemistry, Vol. 12 — Biochemistry of Plant Phenolics — Swain, Harborne and Van Sumere, p. 177. London and New York: Plenum Press. 1978.
18. GUPTA, R. K., S. AL-SHAFI, and E. HASLAM: Unpublished Observations, **1980**.
19. HADDOCK, E. A., and E. HASLAM: Unpublished Observations, **1980**.
20. HARBORNE, J. B.: Comparative Biochemistry of the Flavonoids. London: Academic Press. 1967.
21. HASLAM, E., and J. E. STANGROOM: The Esterase and Depsidase Activities of Tannase. Biochem. J. **99**, 28 (1966).
22. HASLAM, E.: Gallotannins Part 14. Structure of the Gallotannins. J. Chem. Soc. (C) **1967**, 1734.
23. — (+)-Catechin-3-o-gallate and a Polymeric Procyanidin from *Bergenia* sp. J. Chem. Soc. (C) **1969**, 1825.
24. — Polyphenol-Protein Interactions. Biochem. J. **139**, 285 (1974).
25. — Symmetry and Promiscuity in Procyanidin Biochemistry, Phytochemistry **16**, 1625 (1977).
26. HASLAM, E., and R. K. GUPTA: Unpublished Observations, **1980**.
27. HAWORTH, R. D., and L. B. DA SILVA: Chebulinic Acid. J. Chem. Soc. **1951**, 3511.
28. — — Chebulinic Acid Part 2. J. Chem. Soc. **1954**, 2611.
29. HAWORTH, R. D.: Some Problems in the Chemistry of the Gallotannins, Pedler Lecture. Proc. Chem. Soc. **1961**, 401.
30. HAY, J. E., and L. J. HAYNES: Bergenin, a-C-glycopyranosyl derivate of 4-o-Methylgallic Acid. J. Chem. Soc. **1958**, 2231.
31. HERRMAN, K.: Hydroxyzimtsäuren und Hydroxybenzoesäuren enthaltende Naturstoffe in Pflanzen. Fortschr. Chem. Org. Naturst. **35**, 73 (1978).
32. HILLIS, W. E., and M. SIEKEL: Hydrolysable tannins of *Eucalyptus delegatensis* wood. Phytochemistry **9**, 1115 (1970).
33. IKEYA, Y., H. TAGUCHI, I. YOSIOKA, and H. KOBAYASHI: The constituents of *Schizandra chinensis* Baile. Isolation and Structure Determination of five new lignans, Gomisin A, B, C, F and G and the Absolute Structure of Schizandrin. Chem. Pharm. Bull. **27**, 1383 (1979).
34. JOCHIMS, J. C., G. TAIGEL, and O. TH. SCHMIDT: Protonenresonanz — Spektrum und Konformationsbestimmung einiger natürlicher Gerbstoffe. Liebigs Annalen **717**, 169 (1968).
35. KEEN, P. C., R. D. HAWORTH, and E. HASLAM: Gallotannins Part 7. Tara Gallotannin. J. Chem. Soc. **1962**, 3814.
36. KNOWLES, P. F., R. D. HAWORTH, and E. HASLAM: Gallotannins Part 4. The Biosynthesis of Gallic Acid. J. Chem. Soc. **1961**, 1854.
37. MAGNOLATO, D.: Private Communication.
38. MAYER, W.: Dehydro-digallusäure. Liebigs Annalen **578**, 34 (1954).
39. MAYER, W., R. FIKENTSCHER, J. SCHMIDT, and O. TH. SCHMIDT: Über eine ungewöhnliche Spaltung von Diaryl-äthern. Chem. Ber. **93**, 2761 (1960).

40. MAYER, W., N. KUNZ, and F. LOEBICH: Die Struktur Hamamelitannins. Liebigs Annalen **688**, 232 (1965).
41. MAYER, W., A. EINWILLER, and J. C. JOCHIMS: Die Struktur des Castalins. Liebigs Annalen **707**, 182 (1967).
42. MAYER, W., H. SWITZ, and J. C. JOCHIMS: Die Struktur des Castalagins. Liebigs Annalen **721**, 186 (1969).
43. MAYER, W., F. KULLMAN, and G. SCHILLING: Die Struktur des Vescalins. Liebigs Annalen **747**, 51 (1971).
44. MAYER, W., H. SEITZ, J. C. JOCHIMS, K. SCHAUERTE, and G. SCHILLING: Struktur des Vescalagins. Liebigs Annalen **751**, 60 (1971).
45. MAYER, W., B. BILZER, and K. SCHAUERTE: Isolierung von Castalagin und Vescalagin aus Valoneagerbstoffen. Liebigs Annalen **754**, 149 (1971).
46. MAYER, W.: Otto Theodor Schmidt. Liebigs Annalen **1973**, 1759.
47. MAYER, W., G. SCHULTZ, S. WREDE, and G. SCHILLING: 2-o-Cinnamoyl-1-o-galloyl-β-D-glucopyranose aus *Rhizoma rhei*. Liebigs Annalen **1975**, 946.
48. MAYER, W., W. BILZER, and G. SCHILLING: Castavaloninsäure, Isolierung und Strukturmittelung. Liebigs Annalen **1976**, 876.
49. MAYER, W., A. GÜNTHER, H. BUSATH, W. BILZER, and G. SCHILLING: Valolaginsäure. Liebigs Annalen **1976**, 987.
50. MAYER, W., H. SCHICK, and G. SCHILLING: Trillosäure, eine neue Phenolcarbonsäure aus Valoneagerbstoffen. Liebigs Annalen **1976**, 2178.
51. MAYER, W., H. BUSATH, and H. SCHICK: Isovalolaginsäure. Liebigs Annalen **1976**, 2169.
52. MAYER, W., A. GORNER, and K. ANDRA: Punicalagin und Punicalin, zwei Gerbstoffe aus den Schalen der Granatäpfel. Liebigs Annalen **1977**, 1976.
53. MCMANUS, J., T. H. LILLEY, and E. HASLAM: The Association of Phenols with Proteins. J. Chem. Soc. Communications **1981**, 309
54. MISLOW, К., M. A. W. GLASS, R. E. O'BRIEN, P. RUTKIN, D. H. STEINBERG, J. WEISS, and C. DJERASSI: Configuration, Conformation and Rotary Dispersion of Optically-Active Biphenyls. J. Amer. Chem. Soc. **84**, 1455 (1962).
55. NEISH, A. C., G. H. N. TOWERS, D. CHEN, S. Z. EL-BASYOUNI, and R. K. IBRAHIM: The Biosynthesis of Hydroxybenzoic Acids in Higher Plants. Phytochemistry **3**, 485 (1964).
56. NIELSEN, B. J., N. F. LACOUR, S. R. JENSEN, and K. BOCK: The Structure of Acer Tannin. Phytochemistry **19**, 2033 (1980).
57. NISHIRA, H., and M. A. JOSLYN: The Galloyl Glucose Compounds in Green Carob Pods *(Ceratonia Siliqua)*. Phytochemistry **7**, 2147 (1968).
58. OKUDA, T., H. NAYESHIRO, and T. YOSHIDA: Constituents of *Geranium thunbergii* Zieb et Zucc. 4. Ellagitannins (2) Structure of Geraniin. Chem. Pharm. Bull. **25**, 1862 (1977).
59. OKUDA, R., H. NAYESHIRO, and K. SENO: Structure of Geraniin in the Equilibrium State. Tetrahedron Letters **1977**, 4421.
60. OKUDA, T., and K. SENO: Mallotusinic Acid and Mallotinic Acid, new hydrolysable tannins from *Mallotus japonicus*. Tetrahedron Letters **1978**, 139.
61. OKUDA, T., K. MORI, and T. HATANO: The Distribution of Geraniin and Mallotusinic Acid in the order Geraniales. Phytochemistry **19**, 547 (1980).
62. OKUDA, T., Y. YOSHIDA, and T. HATANO: Equilibrated Stereostructure of hydrated Geraniin and Mallotusinic Acid. Tetrahedron Letters **1980**, 2561.
63. OZAWA, T., D. KOBAYASHI, and Y. TAKINO: Structure of the New Phenolic glycosides MP-2 and MP-10 from Chestnut galls. Agr. Biol. Chem. **41**, 1257 (1977).
64. OZAWA, T., K. HAGA, N. ARAI, and Y. TAKINO: Structure of a New Phenolic glycoside from Chestnut galls. Agr. Biol. Chem. **42**, 1511 (1978).

65. OZAWA, T., N. ARAI, and Y. TAKINO: Structure of a New Phenolic Glycoside Chesnatin from Chestnut galls. Agr. Biol. Chem. **42**, 1907 (1978).
66. OZAWA, T., Y. ODAIRA, H. IMAGAWA, and Y. TAKINO: A new Phenolic glycoside Acetylcretanin and flavonoids from Chestnut galls. Agr. Biol. Chem. **44**, 581 (1980).
67. PERKIN, A. G., and Y. UYEDA: Occurrence of a Crystalline tannin in the leaves of the *Acer ginnala*. J. Chem. Soc. **1922**, 66.
68. REDDY, K. K., S. RAJADURAI, S. K. N. SASTRY, and Y. NAYUDAMMA: Studies of the Dhava tannins 1. The Isolation and Constitution of a Gallotannin from Dhava *(Anogeissus latifolia)*. Aust. J. Chem. **17**, 238 (1964).
69. ROBERTS, E. A. H., and M. J. MYERS: Theogallin, a polyphenol occurring in Tea. II. Identification as a Galloyl quinic Acid. J. Sci. Food Agric. **9**, 701 (1958).
70. SCHMIDT, O. TH., and A. SCHACH: Synthese der 3-Galloyl-glucose, 6-Galloyl-glucose und 3,6-Digalloyl-glucose. Liebigs Annalen **571**, 29 (1951).
71. SCHMIDT, O. TH., and R. LADEMANN: Corilagin, ein weiterer kristallisierter Gerbstoff aus Divi-Divi. Liebigs Annalen **571**, 232 (1951).
72. SCHMIDT, O. TH., and K. DEMMLER: Optisch aktive 2,3,4,2',3',4'-Hexamethoxydiphenyl-carbonsäure-6,6'. Liebigs Annalen **576**, 85 (1952).
73. SCHMIDT, O. TH., F. BLINN, and R. LADEMAN: Über der Bindung der Ellagsäure in Corilagin und Chebulagsäure. Liebigs Annalen **576**, 75 (1952).
74. SCHMIDT, O. TH., and D. M. SCHMIDT: Über das Vorkommen von Corilagin in Myrabolanen. Liebigs Annalen **578**, 31 (1953).
75. SCHMIDT, O. TH.: Ellagengerbstoffe. Leder **5**, 129 (1954).
76. SCHMIDT, O. TH., and K. DEMMLER: Racemische und optisch aktive 2,3,4,2',3',4'-Hexaoxydiphenyl-carbonsäure-6,6'. Liebigs Annalen **585**, 179 (1954).
77. SCHMIDT, O. TH., D. M. SCHMIDT, and J. HEROK: Die Konstitution und Konfiguration des Corilagins. Liebigs Annalen **587**, 67 (1954).
78. SCHMIDT, O. TH.: Natürliche Gerbstoffe, in: Moderne Methoden der Pflanzenanalyse, Vol. 3: Paech and Tracey, p. 517. Berlin-Göttingen-Heidelberg: Springer. 1955.
79. SCHMIDT, O. TH., and E. KOMAREK: Valoneasäure. Liebigs Annalen **591**, 156 (1955).
80. SCHMIDT, O. TH.: Gallotannine und Ellagen-Gerbstoff. Fortschritte der Chemie organischer Naturstoffe, Bd. XIII, p. 570. Wien: Springer. 1956.
81. SCHMIDT, O. TH., and W. MAYER: Natürliche Gerbstoffe. Angew. Chem. **68**, 103 (1956).
82. SCHMIDT, O. TH.: Über Chebulagsäure und Chebulinsäure. Leder **8**, 106 (1957).
83. SCHMIDT, O. TH., and G. KLINGER: Synthese der 1,3,6-Trigalloyl-glucose. Liebigs Annalen **609**, 199 (1957).
84. SCHMIDT, O. TH., and H. REUSS: 2-[-[*p*-Hydroxy-benzoyl]-glucose und 2-Galloyl-glucose. Liebigs Annalen **649**, 137 (1961).
85. SCHMIDT, O. TH., and H. SCHMADEL: Synthesen des α-Glucogallins und neue Synthesen des β-Glucogallins. Liebigs Annalen **649**, 149 (1961).
86. SCHMIDT, O. TH., R. SCHANZ, R. ECKERT, and R. WURMB: Brevilagin 1. Liebigs Annalen **706**, 131 (1967).
87. SCHMIDT, O. TH., R. SCHANZ, R. WURMB, and W. GROEBKE: Brevilagin 2. Liebigs Annalen **706**, 154 (1967).
88. SCHMIDT, O. TH., J. SCHULZ, and R. WURMB: Terchebin. Liebigs Annalen **706**, 169 (1967).
89. SCHMIDT, O. TH., R. WURMB, and J. SCHULZ: Hexahydroxydiphensäure (Ellagsäure), Brevifolin-carbonsäure und Chebulsäure aus Umwandlungsprodukten der Brevilagine und des Terchebins. Liebigs Annalen **76**, 180 (1967).
90. SCHMIDT, O. TH., J. SCHULZ, and H. FEISSER: Die Gerbstoffe der Myrabolanen. Liebigs Annalen **706**, 187 (1967).

91. SCHMIDT, O. TH., W. EBERT, and M. KOPP: 1,3,4,6-Tetragalloyl-β-D-glucose aus Algarobilla. Liebigs Annalen **729**, 251 (1969).
92. SCHMIDT, O. TH., and H. KOTTENHAHN: (−)-Chebulsäure aus Algarobilla. Liebigs Annalen **729**, 249 (1969).
93. SHIPCHANDLER, M. T., C. A. PETERS, and C. D. HURD: Syntheses of Gallic Acid and Pyrogallol. J. Chem. Soc. (Perkin 1) **1975**, 1400.
94. STOTHERS, J. B.: Carbon-13 NMR Spectroscopy. London and New York: Academic Press. **1972**, 197.
95. UDDIN, M., and E. HASLAM: Gallotannins Part 15. Some Observations on the Structure of Chebulinic Acid and its Derivatives. J. Chem. Soc. (C) **1967**, 2381.
96. WAGNER, H., M. A. IYENGAR, O. SELIGMAN, H. I. EL-SISSI, N. A. M. SALEH, and S. I. EL-NEGOUMY: Prunin-o-6″-gallate aus *Acacia farnesiana.* Phytochemistry **13**, 2843 (1974).
97. WHITE, T.: Tannins their Occurrence and Significance. J. Sci. Food Agric. **8**, 377 (1957).
98. WILKINS, C. K., and B. A. BOHM: Ellagitannins from *Tellima grandiflora.* Phytochemistry **15**, 211 (1976).
99. ZENK, M. H.: Zur Frage der Biosynthese von Gallusäure. Z. Naturforsch. **19 B**, 83 (1964).
100. — Recent Work on Cinnamyl COA derivatives, Biochemistry of Plant Phenolics, Recent Advances in Phytochemistry, Vol. 12: Swain, Harborne and Van Sumere, p. 139. New York and London: Plenum Press. 1978.
101. NAHRSTEDT, A., K. DUMKOW, B. JANISTYN, and R. POHL: Quercetin-Galactosid-Gallate in Euphorbiaceen. Tetrahedron Letters **1974**, 559.
102. SCHMIDT, O. TH., L. WURTELE, and A. HARREUS: Pedunculagin, eine 2,3:4,6-Di-[(−)-hexahydroxydiphenoyl]-glucose aus Knoppern. Liebigs Annalen **690**, 150 (1965).
103. SCOPES, M., and E. HASLAM: Unpublished Observations, **1980.**
104. JURD, L.: Plant Polyphenols III. The Isolation of a New Ellagitannin from the Pellicle of the Walnut. J. Amer. Chem. Soc. **80**, 2249 (1958).
105. FERRIER, R. J., and P. M. COLLINS: Monosaccharide Chemistry, p. 40. London: Penguin. 1972.
106. HASLAM, E.: Galloyl esters in the Aceraceae. Phytochemistry **4**, 495 (1965).
107. MATHIESON, A. MCL.: The Preferred Conformation of the Ester Group in Relation to Saturated Ring Systems. Tetrahedron Letters **1965**, 4137.
108. CULVENOR, C. J.: The Conformation of Esters and the "Acylation Shift" NMR Evidence from Pyrrolizidine Alkaloids. Tetrahedron Letters **1966**, 1091.
109. HASLAM, E.: Chemistry of Vegetable Tannins. London and New York: Academic Press. 1966.
110. SCHMIDT, O. TH., E. KOMAREK, and H. RENTEL: Synthese der Octamethylvalonea-säure. Liebigs Annalen **602**, 50 (1957).
111. SCHMIDT, O. TH., and K. BERNAUER: Brevifolin und Bervifolincarbonsäure. Liebigs Annalen **588**, 211 (1954).

(Received November 24, 1980)

The Direct Biomimetic Synthesis, Structure and Absolute Configuration of Angular and Linear Condensed Tannins

By D. G. Roux and D. Ferreira, Department of Chemistry, University of the Orange Free State, Bloemfontein, South Africa

Contents

Acknowledgements: The authors acknowledge major contributions, particularly by co-workers H. K. L. Hundt, J. J. Botha, J. H. van der Westhuizen, P. M. Viviers and D. A. Young, to the synthetic studies on condensed tannins. Collaboration by Dr. W. E. Hull, Bruker Analytische Messtechnik GmbH, Silberstreifen, D-7512 Rheinstetten-Forchheim in recording ^{1}H n.m.r. spectra and performing extensive spin decouplings at high magnetic field strengths (360, 400 and 500 MHz) and temperatures ($373 - 473°$ K) has made these studies on higher oligomers possible.

I. Introduction

The chemistry of condensed tannins has hitherto represented a relatively unattractive and therefore neglected area of study; one in which the weight of research effort involved is invariably disproportionate to the results achieved, in which the participating schools generally confine their approach to specific molecular species, and in which as yet no consensus has been reached regarding likely precursors.

The problems which beset those engaged in this field represent a combined function of the abnormal complexity of the gradational range of oligomers of increasing mass and affinity for substrates which typify most extracts rich in tannins, and the consequent problem of their isolation and purification, the high chirality of tannin oligomers, the need to contend with the phenomenon of dynamic rotational isomerism about interflavanoid bonds in the ^{1}H n.m.r. spectral interpretation of their derivatives, the lack of precise knowledge regarding the points of bonding at nucleophilic centres, and the obvious limitations of a hitherto predominantly analytical approach. The last of these reflects the need for a general method of synthesis which permits unambiguous proof of both structure and absolute configuration also at higher oligomeric levels.

With these objectives in mind we initiated a purely synthetic approach based on the premise that flavan-3,4-diols as source of electrophilic flavanyl-4-carbocations, and flavan-3-ols as nucleophiles (cf. 1, 2) represent the prime initiators of a process of repetitive condensation in which the immediate products also represent the sequent nucleophilic substrates.

II. Bonding Positions at Nucleophilic Centres

Our starting-point was the required definition of the bonding positions at either C-6 or C-8 of the presumed "terminal" (+)-catechin [(2R:3S)-2,3-trans-flavan-3,3',4',5,7-pentaol] unit which serves as initial nucleophile in most tannins under consideration. The solvent shift method for methoxyl function developed by Pelter et al. (3), although applicable to the methyl ethers of biflavonoids (i. e. compounds with 3- or 4-carbonyl function), was less reliable when applied to the methyl ether acetates of biflavanoids*, often leading to ambiguous or erroneous conclusions**.

---

* In order to achieve the desired purity tannin oligomers are usually separated successively as phenols, as methyl ethers and as methyl ether acetates by chromatography.

** Due presumably to solvation effects associated with 3-acetoxy groups (A. Pelter, personal communication).

References, pp. 74—76

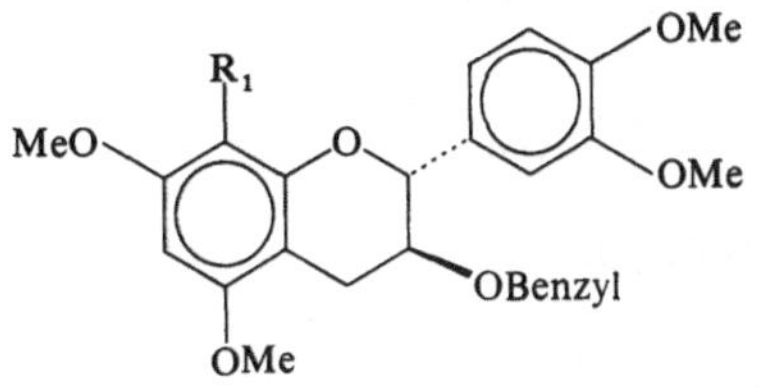

(1)

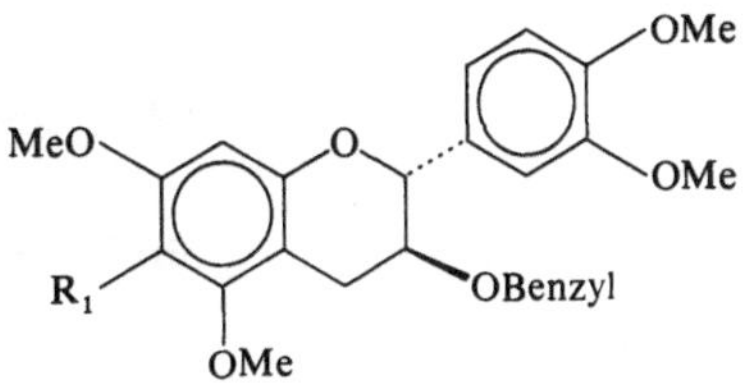

a) $R_1 = Br$
b) $R_1 = OH$
c) $R_1 = OAc$
d) $R_1 = COOMe$

(2)

Accordingly our approach was that of defining the absolute or relative shifts of protons attached to C-6 and C-8 in (+)-catechin. This required the synthesis of a range of 8- and 6-substituted derivatives (1 and 2 resp.) in order to determine the relative shifts of the residual 6- and 8-protons respectively on the A-ring. 8-Bromination of a (+)-catechin derivative occurs with relative ease and the substitution position (1 a) was confirmed by X-ray crystallographic analysis (4), but 6-bromination (2a) was achieved for the first time by 6,8-dibromination of a (+)-catechin tetramethyl ether derivative followed by selective debromination at C-8 (5) (Scheme 1). From these 6- and 8-monobrominated compounds an unique range of derivatives of varying 6- and 8-functionality [e. g. (1 and 2a — d)] were available via the corresponding lithio derivatives (Scheme 2). The absolute (5, 6), as well as the relative (6) differences in the chemical shifts of the residual A-ring protons were found to be diagnostic in determining the substitution positions in those biflavanoids in which the alternatives of 4,6- and 4,8-links are possible.

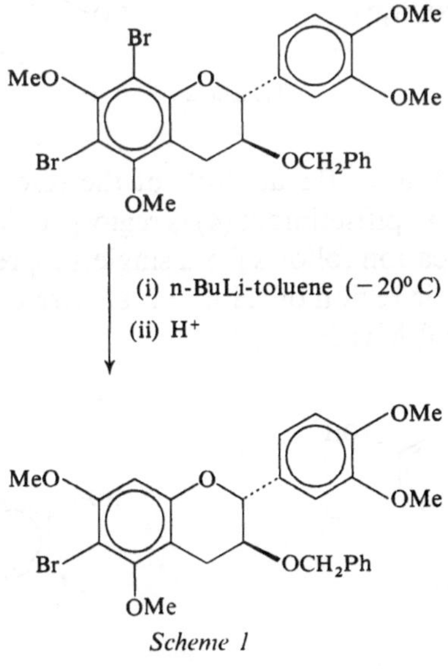

(i) n-BuLi-toluene ($-20°$ C)

(ii) H^+

Scheme 1

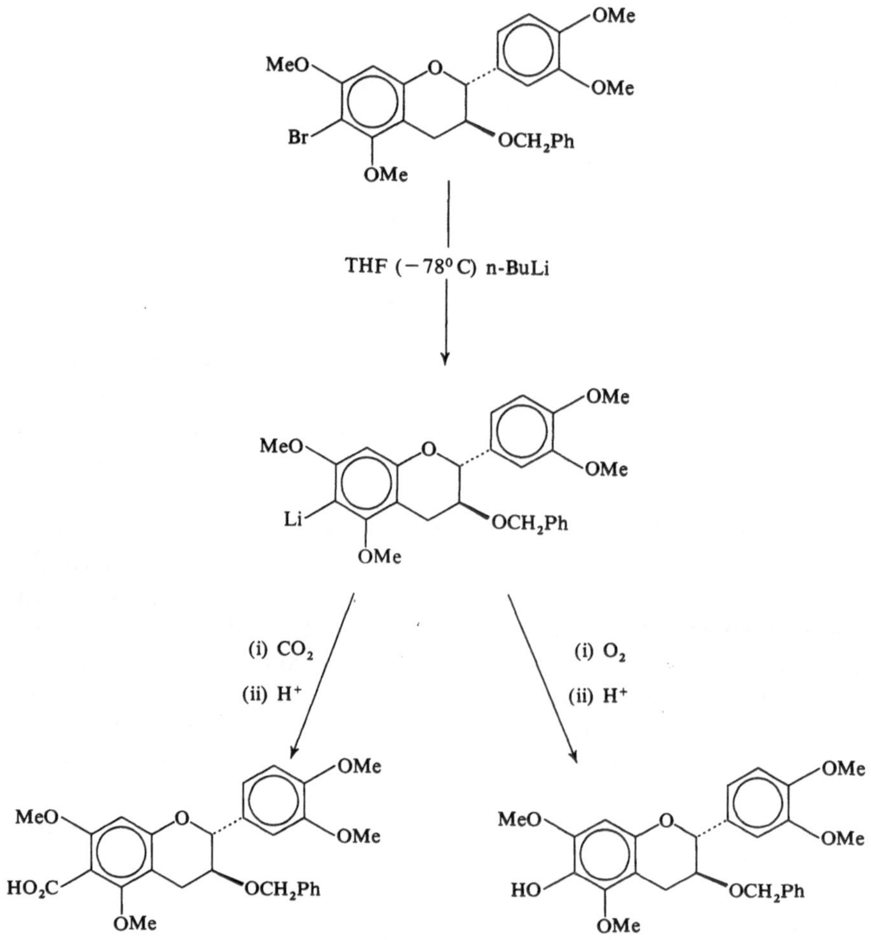

Scheme 2

By contrast, bonding to flavan-3-ols of the resorcinol-type, e. g. (−)-fisetinidol (**3**) and (+)-epifisetinidol (**4**) is regiospecific at the least hindered 6-position. Such allocation follows from singlets representing H-5 and H-8 present in the aromatic region of ^1H n.m.r. spectra of biflavanoids at high magnetic fields (> 360 MHz).

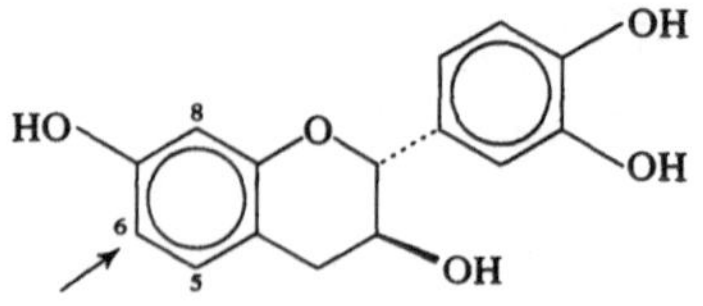

(**3**)

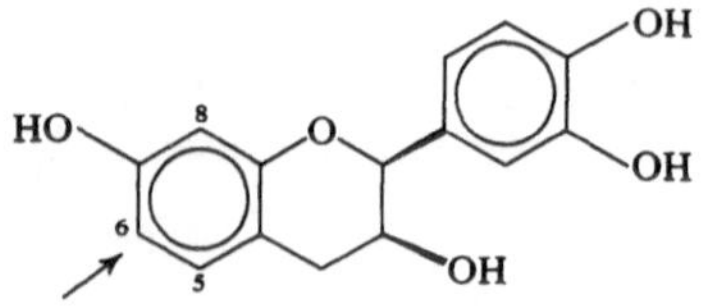

(**4**)

III. Conditions for Interflavanoid Bonding.
Stereochemical Course of the Reaction.
Chiroptical Method for Determining the Absolute
Configuration at C-4

With the knowledge that flavan-3,4-diols are freely associated* with flavan-3-ols [e. g. (+)-catechin, (−)-epicatechin, (−)-fisetinidol, (+)-epifisetinidol] in tannin mixtures comprised of these molecular entities, experimental conditions were sought which would permit formation of C-4−C-6 or 8 interflavanoid links. The conditions selected had to be sufficiently mild to avoid self-condensation of the flavan-3,4-diol, to avoid anthocyanidin formation as a side-reaction, to permit stereochemical conclusions (i. e. the reaction should be under kinetic rather than under thermodynamic control), and to allow assessments regarding the regioselectivity of the condensation.

As starting-point, the minimum requirements for the condensation of the free phenolic forms of flavan-3,4-diols (as source of carbenium ions) with the highly nucleophilic phenols, phloroglucinol and resorcinol, were examined. Promotion of the reaction by organic acids at ambient temperatures was accelerated by dilute mineral acids (e. g. 0.1 M HCl) to permit high yields over short periods (2 h) without side-effects. Thus, 2,3-*trans*-3,4-*trans*-flavan-3,4-diols yield predominantly 2,3-*trans*-3,4-*trans*-4-arylflavan-3-ols, but also the 3,4-*cis*-isomer (ratio ~2:1). By contrast to the stereoselectivity of this reaction, condensations with 2,3-*cis*-3,4-*cis*-flavan-3,4-diols give 2,3-*cis*-3,4-*trans*-4-arylflavan-3-ols stereospecifically (*7, 8*) (Scheme 3). The stereochemical course of these individual reactions follows from considerations of steric approach control factors operating around the point of attachment, and from neighbouring group participation. The expected absence of 2,3-*cis*-3,4-*cis*-4-arylflavan-3-ols from the condensation products in the latter instance is notable. These stereoisomers are, however, available photochemically from certain 4-arylflavan-3-ols where the nucleophilicity of the 4-aryl function exceeds that of the benzenoid A-ring (*9*). The fore-mentioned condensations (Scheme 3) predict the stereochemical course of *in vivo* condensations during tannin formation.

Apart from defining the ideal conditions for biomimetic condensations under kinetic control, the derivatives of these optically pure 4-arylflavan-3-ols also offered the first opportunity for fomulating a chiroptical rule which defines the absolute stereochemistry at C-4 of flavanoid units of this type and hence in biflavanoids and higher oligomers. Contrary to earlier

---

* The natural absence of leucoanthocyanidins (flavan-3,4-diols) of the delphinidin, cyanidin and pelargonidin series may be rationalized by the knowledge of their excessively high reactivity.

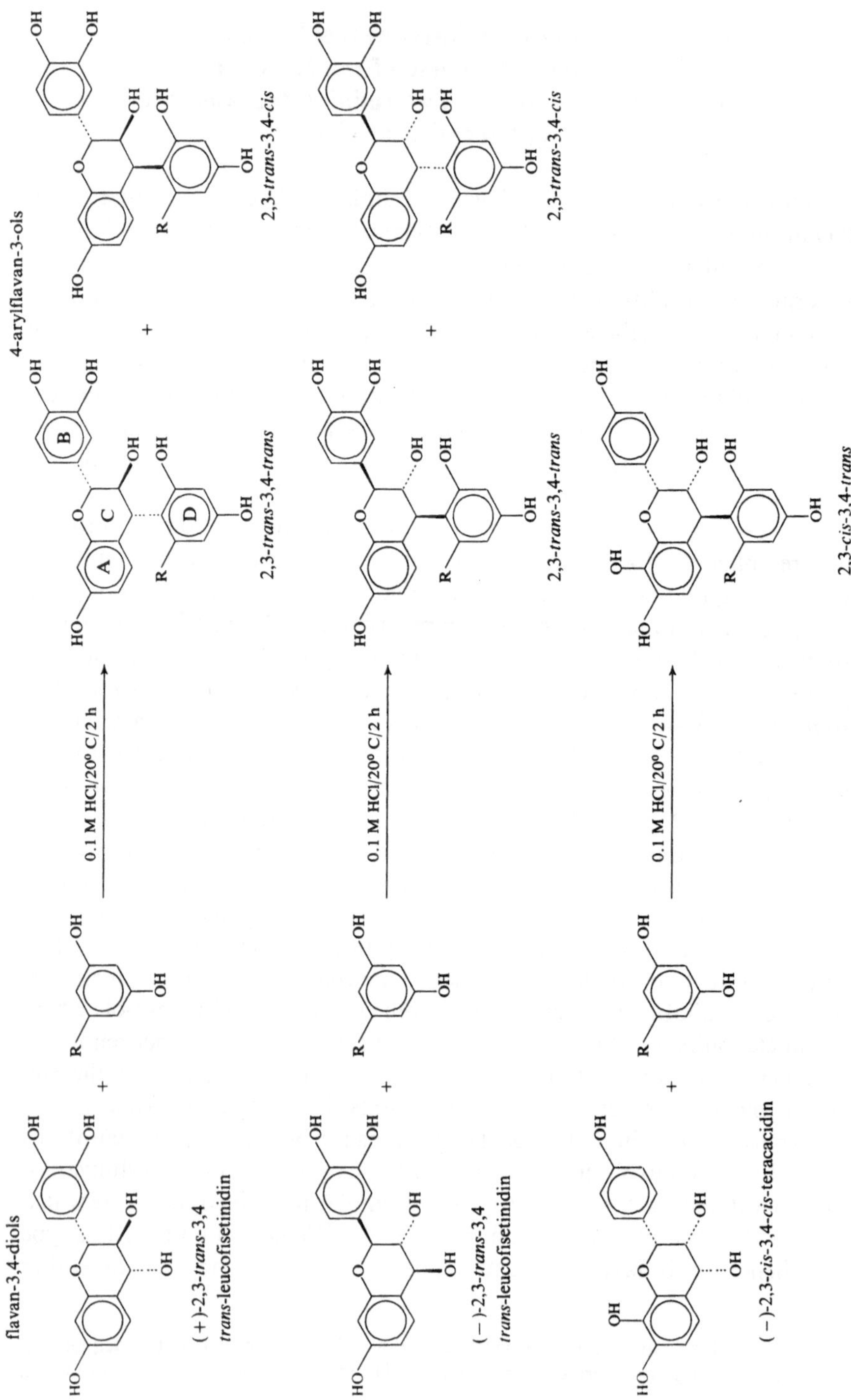
flavan-3,4-diols
4-arylflavan-3-ols
(+)-2,3-trans-3,4 trans-leucofisetinidin
(−)-2,3-trans-3,4 trans-leucofisetinidin
(−)-2,3-cis-3,4-cis-teracacidin
0.1 M HCl/20° C/2 h
0.1 M HCl/20° C/2 h
0.1 M HCl/20° C/2 h
2,3-trans-3,4-trans
2,3-trans-3,4-cis
2,3-trans-3,4-trans
2,3-trans-3,4-cis
2,3-cis-3,4-trans
A
B
C
D
R
OH
HO
O

Scheme 3

statements by HASLAM and co-workers (*10*) our examination of the sign of the high-intensity Cotton-effects which occur at low wave lenghts λ 210 – 240 nm and which are attributed mainly to chromophoric interaction at C-4 (*7, 8*) indicated that it obeys the aromatic quadrant rule for 2,3-*trans*-3,4-*trans*, 2,3-*trans*-3,4-*cis* and 2,3-*cis*-3,4-*trans* isomers (*9*). Exceptions are the 2,3-*cis*-3,4-*cis* isomers and also others, all with "abnormal" heterocyclic ring conformations or unusual conformational equilibria, as judged from ^{1}H n.m.r. coupling constants.

Thus the sign of the Cotton-effect in the fore-mentioned instances correlates with the position of the 4-aryl (or 4-flavanyl) function above

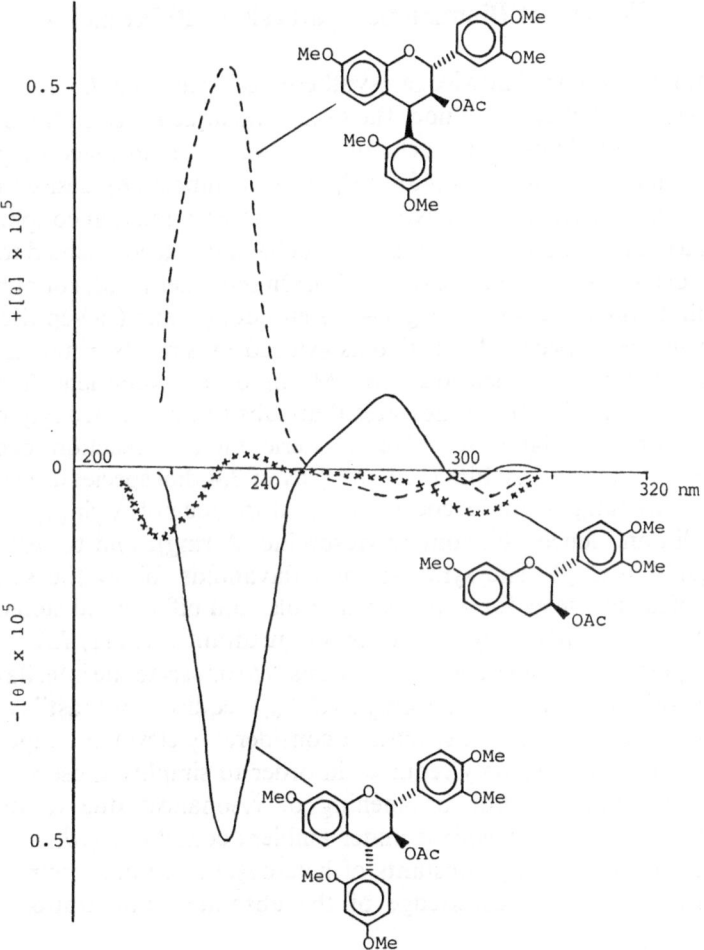

Correlation between the intensity of low wave length Cotton-effects and structure (4-arylflavan-3-ols *vs.* flavan-3-ol), and also between their sign and the stereochemistry at C-4.

(positive Cotton-effect) or below (negative) the plane of the A-ring when placed to the right in flavanoid units viewed as in Scheme 3. Correlation with the Cahn-Ingold-Prelog R and S notation at C-4 is, however, only possible in homologous series where functionalization of the A- and D-rings remains a constant factor.

Assignment of absolute configurations at C-4 is, of course, also available from ^{1}H n.m.r. spectroscopy in conjunction with the known absolute configurations at C-2 and C-3 of the flavanoid species participating in condensations; that at C-4 following from 3,4-*trans* or 3,4-*cis* relative configurations derived from coupling constants.

IV. Direct Biomimetic Synthesis of Biflavanoids

With the above knowledge, syntheses of free-phenolic forms were extended to biflavanoids. Under the same mild aqueous conditions (0.1 M HCl/20° C) couplings with nucleophilic flavan-3-ols in place of phenols occur in good overall yields (45 − 55%) to give almost any desired type of biflavanoid* (Schemes 4 − 6). Notable is the observation that coupling with (+)-catechin occurs almost exclusively at C-8, but is accompanied to a very limited extent by reaction at C-6 (*11, 12*). On the other hand, coupling with resorcinol-type flavan-3-ols, e. g. (−)-fisetinidol (**3**) and (+)-epifisetinidol (**4**), occurs regiospecifically at C-6 as evinced by singlets in the aromatic regions of ^{1}H n.m.r. spectra (360 MHz) of the products. Taken in conjunction, the foregoing indicates that substitution at strongly nucleophilic centres is highly sensitive to steric factors, reaction occurring preferentially at the least hindered site. Flavan-3,4-diol species used in these experiments range from leucocyanidin (phloroglucinol A-ring); to leuco-fisetinidin and leucorobinetinidin (resorcinol A-ring); and to teracacidin (pyrogallol A-ring). The synthesis of biflavanoids follow the same stereochemical course as for 4-arylflavan-3-ols, and biflavanoid stereochemistry at C-4 similarly obeys the aromatic quadrant rule (*11, 12*).

The presence of subsituents in positions *orto* or *peri* to the interflavanoid bond increases the activation energy ($\Delta G^{\neq}_{rot.}$) required for "fast" rotation, and recording of ^{1}H n.m.r. spectra at considerably elevated temperatures (90 − 110° C) is accordingly required in order to simplify those spectra in which duplication and/or broadening of resonances due to dynamic rotational isomerism is evident under ambient conditions (*10, 13, 14*).

Thus, from coupling constants of heterocyclic protons, from circular dichroism, and from knowledge of the absolute configuration of the

---

* Exceptions, as before, are those with 2,3-*cis*-3,4-*cis* stereochemistry of the "upper" unit (*cf. 9*).

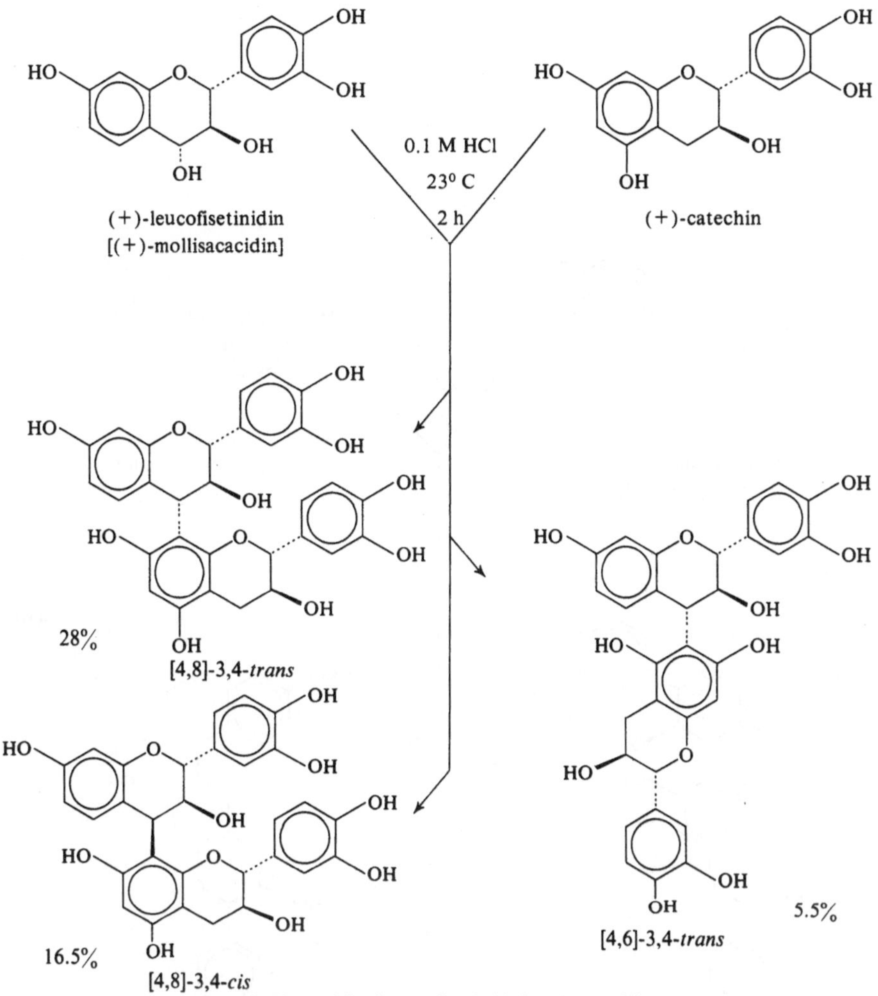

(+)-leucofisetinidin
[(+)-mollisacacidin]

0.1 M HCl
23° C
2 h

(+)-catechin

28%
[4,8]-3,4-*trans*

5.5%
[4,6]-3,4-*trans*

16.5%
[4,8]-3,4-*cis*

[4,8]- and [4,6]-(−)-fisetinidol-(+)-catechins

Scheme 4

flavanoid compounds employed in synthesis, the absolute configurations of synthetic biflavanoids and their natural counterparts could be assigned unambiguously (*11, 12*). Synthetic biflavanoids were in turn used conveniently as synthons for triflavanoids.

Biflavanoids do not in general exhibit tanning properties (*15*) (strong adsorption on collagen substrates) because of their relatively low mass and good solubility in water, but they are to be regarded as precursors of tannins considering their ubiquitous association with their own immediate precursors, i.e. flavan-3-ols and flavan-3,4-diols, and with homologues of higher mass.

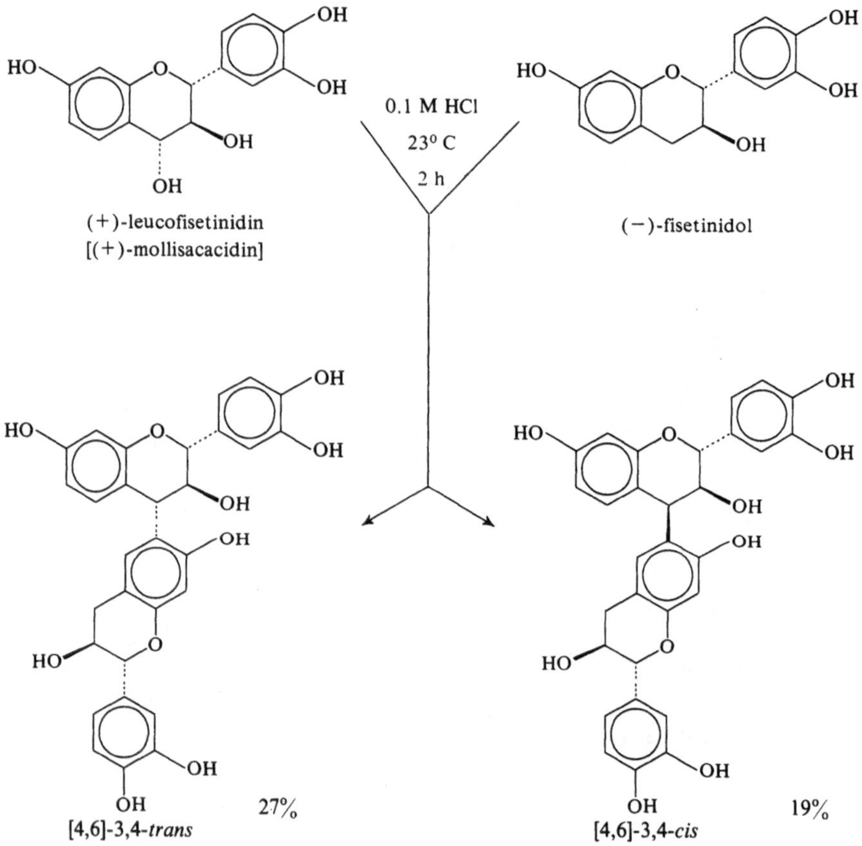

0.1 M HCl
23° C
2 h

(+)-leucofisetinidin
[(+)-mollisacacidin]

(−)-fisetinidol

[4,6]-3,4-*trans*   27%

[4,6]-3,4-*cis*   19%

[4,6]-bi-[(−)-fisetinidols]

Scheme 5

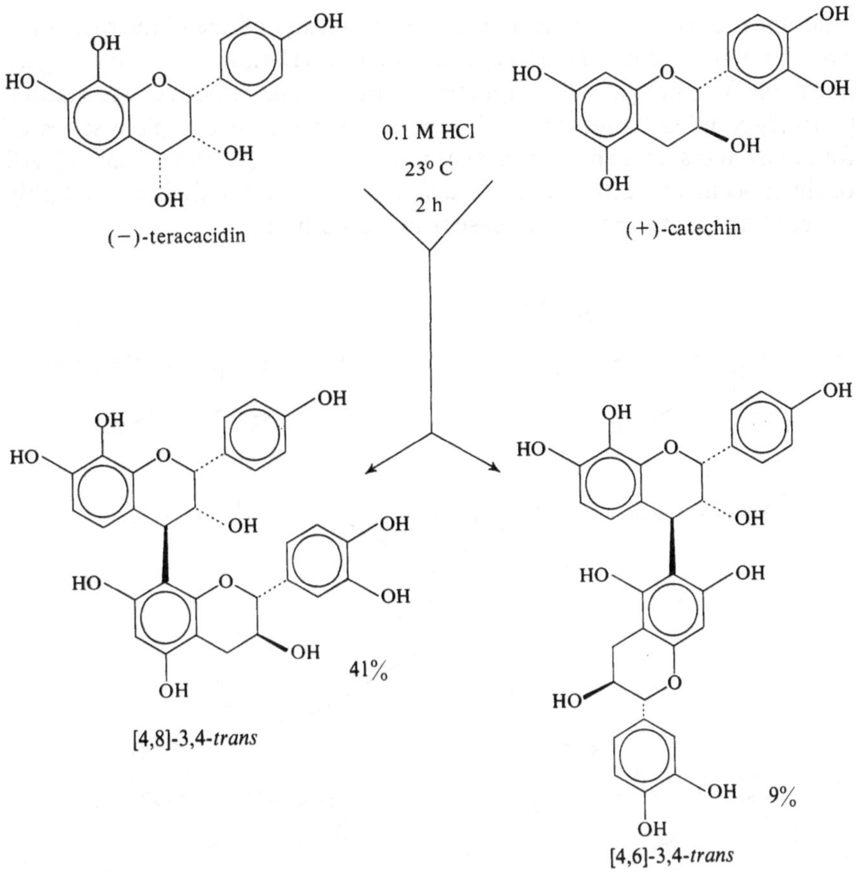

0.1 M HCl
23° C
2 h

(−)-teracacidin

(+)-catechin

41%

[4,8]-3,4-*trans*

9%

[4,6]-3,4-*trans*

[4,6]- and [4,8]-(−)-"teracacidol"-(+)-catechins

Scheme 6

V. Direct Biomimetic Synthesis of "Angular" Triflavanoids

Closer to the true tannins in terms of their relative affinity for protein substrates are the triflavanoids, compounds which constitute a greater challenge synthetically because of added structural complexity. Coupling of a flavan-3,4-diol with the terminal (lower) (+)-catechin units of [4,8]-biflavanoids under the same conditions used for biflavanoid formation, but with extended reaction times, leads to much lower overall yields of triflavanoids (~12%) presumably due to the greater degree of steric hindrance contributed by the molecular complexity of one of the reactants. The coupling occurs regiospecifically at C-6 (D-ring), the most highly nucleophilic amongst the competing centres. The products are unique [4,6:4,8]-coupled "angular" triflavanoids, of which a complete series of four diastereoisomers have been synthesized (*16, 17*) (Schemes 7 and 8). All of them occur in wattle wood *(Acacia mearnsii)*, while their more highly oxygenated counterparts are present in wattle bark.

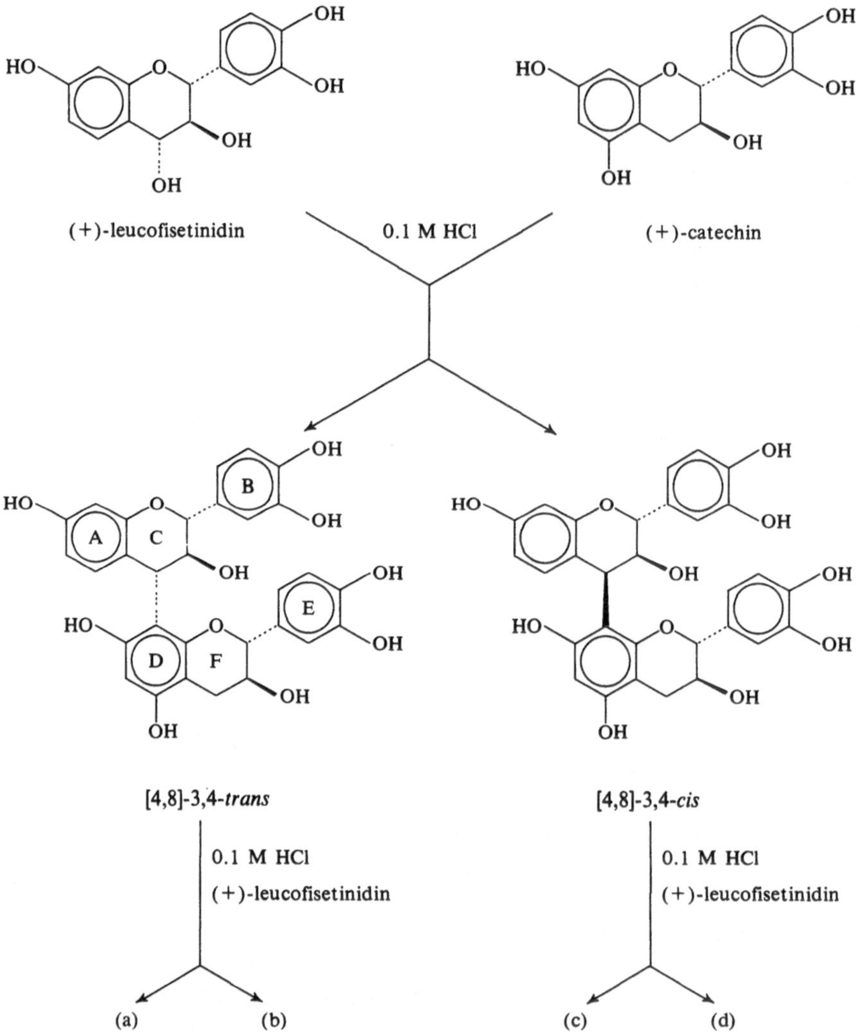

(+)-leucofisetinidin   0.1 M HCl   (+)-catechin

[4,8]-3,4-*trans*   [4,8]-3,4-*cis*

0.1 M HCl   0.1 M HCl

(+)-leucofisetinidin   (+)-leucofisetinidin

(a)   (b)   (c)   (d)

Synthesis of triflavanoids

Scheme 7

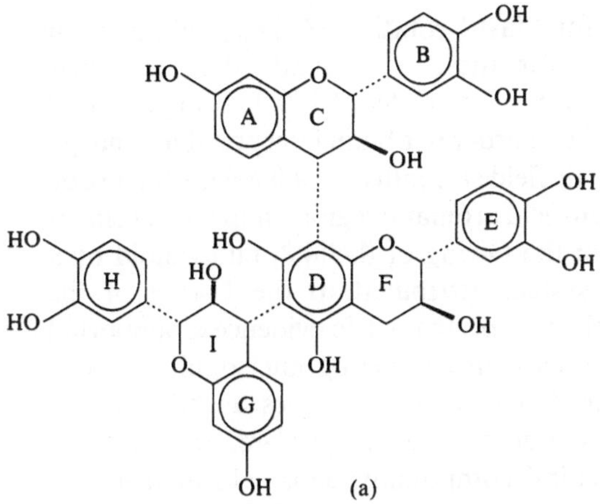

(a)
[4,6]-3,4-*trans*-[4,8]-3,4-*trans*

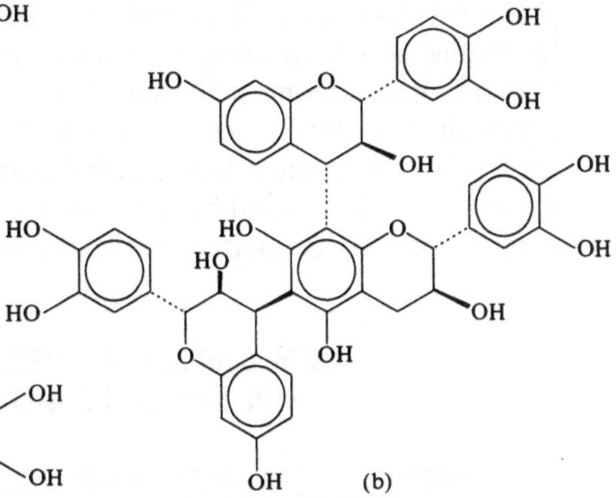

(b)
[4,6]-3,4-*cis*-[4,8]-3,4-*trans*

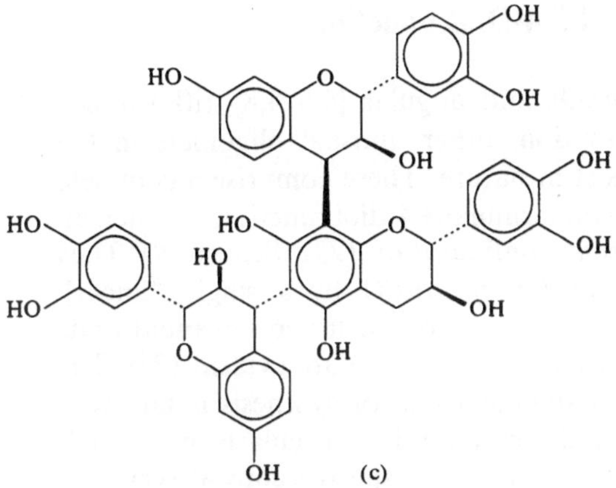

(c)
[4,6]-3,4-*trans*-[4,8]-3,4-*cis*

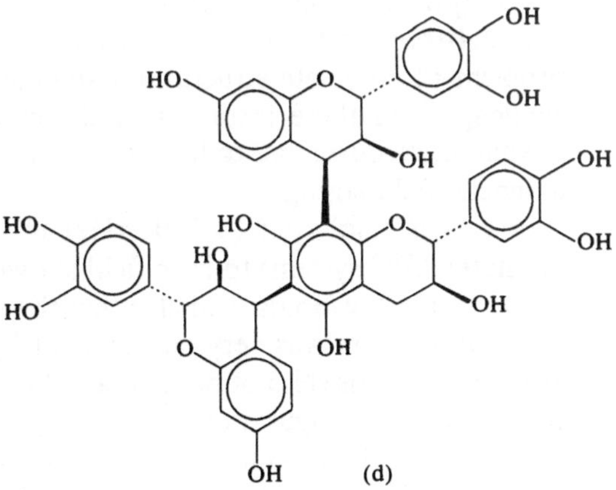

(d)
[4,6]-3,4-*cis*-[4,8]-3,4-*cis*

"Angular" [4,6:4,8]-bi-[(−)-fisetinidol]-(+)-catechins

Scheme 8

The activation energy (*18*) for "fast" rotation ($\Delta G^{\neq}_{rot.}$) about both interflavanoid bonds, is higher for these compounds than for their biflavanoid precursors; temperatures of the order of 170° C being required at 80 MHz and 200° C at 500 MHz, in order to obtain interpretable "sharp" ^{1}H n.m.r. spectra. At high magnetic field strengths (> 360 MHz) diagnostic evidence is available from the high-field aromatic region, of the replacement of the singlet allocated to H-6 of the D-ring of the [4,8]-biflavanoid by a second* high-field ABC spin system attributed to the G-ring of the [4,6:4,8]-triflavanoid. This conclusive spectrometric evidence is supported by studies of similar condensations with competing nucleophilic centres (*19*). In these unique triflavanoids (+)-catechin serves as bifunctional nucleophile in two successive condensations. As before, the absolute configurations of these highly chiral compounds, as is evident from the sequential mode of synthesis and from application of ^{1}H n.m.r. spectroscopy, is logically supported by circular dichroism (*16, 17*).

VI. Biflavanoids and a "Linear" Triflavanoid with Terminal 3,4-Diol Function

In addition to [4,8]-biflavanoids and angular [4,6:4,8]-triflavanoids which enjoy a very wide distribution, other unusual oligomers in the corresponding categories also exist in nature. These comprise a complete series of four [4,6]-biflavanoids with terminal 3,4-diol function in which all units possess the (*2R,3S*)-2,3-*trans* configuration (*20*) (Scheme 9). They are accompanied in wattle wood (*A. mearnsii*) by a single "linear" [4,6:4,6]-triflavanoid analogue (Scheme 10) in which two flavanoid units with 3,4-*trans* and one with 3,4-*cis* stereochemistry are evident (*20*). The problem of complete structural assignment and of synthesis is far more challenging in these instances, and for analysis particularly of the triflavanoid, recourse had to be taken to ^{1}H n.m.r. spectroscopy at very high magnetic field strengths.

At high resolution (> 360 MHz) the presence of two singlets (and aromatic ABC system) to high field showed that coupling was to C-6 (D-ring) in all biflavanoids, and also that the benzylic 4-proton (C-ring) at the point of junction was very heavily split by long-range coupling with two aromatic protons (H-5 of both A- and D-rings), especially for biflavanoids with 3,4-*trans* configurations in the "upper" unit. Other notable features of the hexamethyl triacetate derivatives are their adherence to the aromatic quadrant rule as regards the stereochemistry at C-4 of both flavanoid units;

---

* The A-ring derived from the [4,8]-biflavanoid already contributes a high-field ABC system.

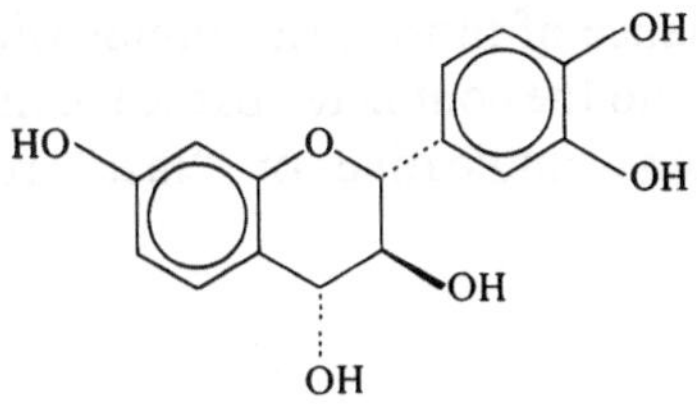

(+)-leucofisetinidin

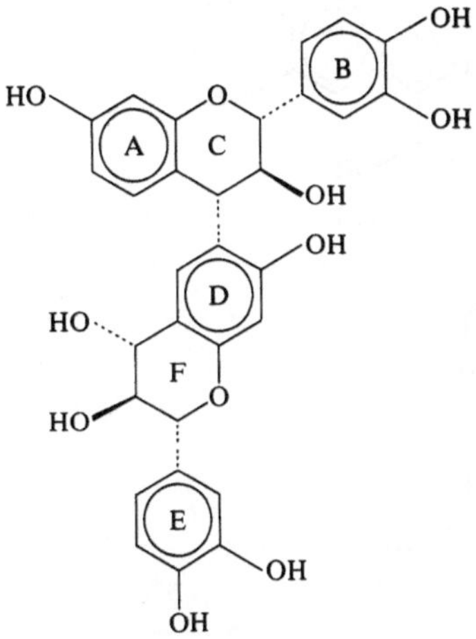

3,4-*trans*-3′,4′-*trans*

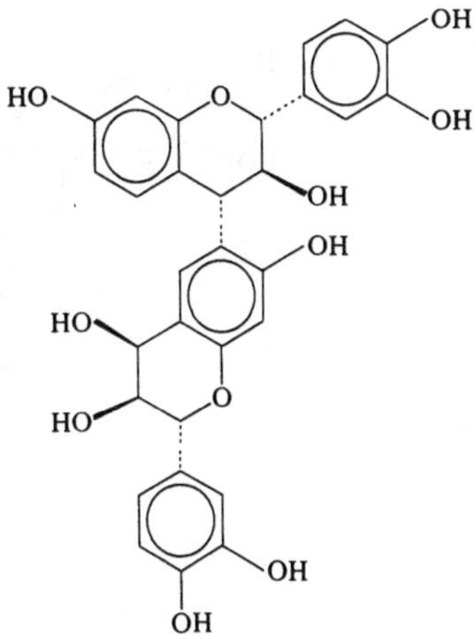

3,4-*trans*-3′,4′-*cis*

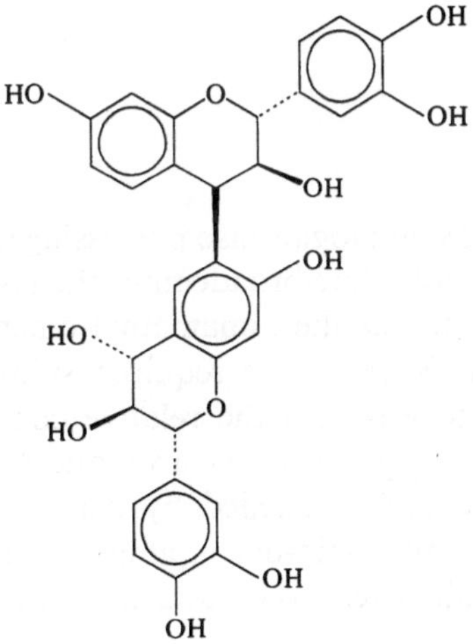

3,4-*cis*-3′,4′-*trans*

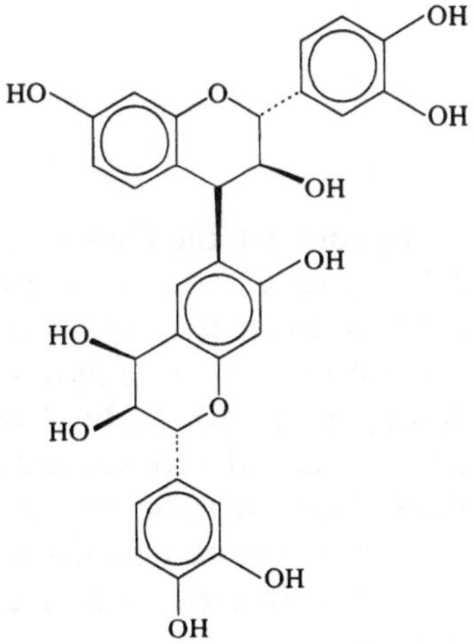

3,4-*cis*-3′,4′-*cis*

[4,6]-(−)-Fisetinidol-(+)-mollisacacidins

Scheme 9

the natural predominance of those biflavanoids with 3,4-*cis* configuration in the "upper" units; and the complete absence of line-broadening or signal duplication phenomena, indicative of "fast" rotation about the interflavanoid bond.

2,3-*trans*-3,4-*cis*:2′,3′-*trans*-3′,4′-*trans*:2″,3″-*trans*-3″,4″-*trans*

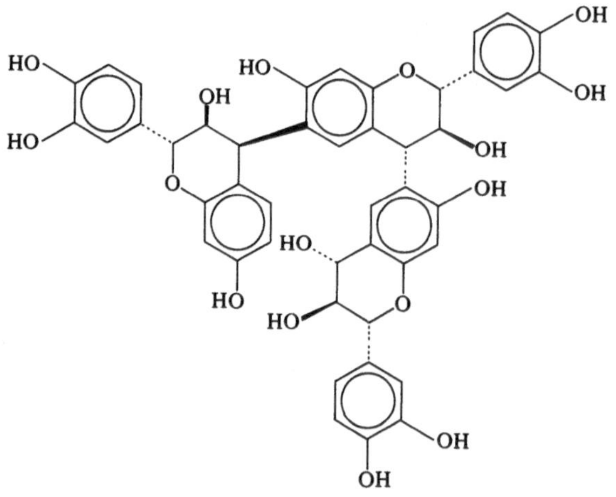

"Linear" [4,6:4,6]-bi-[(−)-fisetinidól]-(+)-mollisacacidin

Scheme 10

By contrast, the single [4,6:4,6]-triflavanoid analogue, also possessing a 3,4-*cis* configuration of the upper unit, exhibits line-broadening phenomena at ambient temperatures when examined as the nonamethyl ether tetra-acetate by ^{1}H n.m.r. spectroscopy. This derivative required spin-decoupling at 500 MHz (100°C) in order to establish the relative configurations of the constituent units; the bonding points to the aromatic A-rings (four aromatic singlets of which two are broadened by benzylic coupling); and particularly the sequence of units. Spin-decoupling at this frequency, in fact, permits assignment of all heterocyclic and aromatic protons (*20*).

The biflavanoids with 2,3-*trans*-3,4-*cis*:2′,3′-*trans*-3′,4′-*trans* and -2′,3′-*trans*-3′,4′-*cis* configurations together with the 2,3-*trans*-3,4-*cis*:2′,3′-*trans*-3′,4′-*trans*:2″,3″-*trans*-3″,4″-*trans* triflavanoid result from the *in vitro* self-condensation of the parent (2R,3S,4R)-3′,4′,7-trihydroxyflavan-3,4-

diol under carefully controlled conditions in a reaction in which the predominant products are nevertheless indefinable high condensates. The condensations taken in conjunction with known relative configurations from ^{1}H n.m.r. spectroscopy permit definition of the absolute configurations of the oligomers as indicated.

The flavan-3,4-diol resists self-condensation under those conditions (0.1 M HCl, 20°, 2 – 4 h) which catalyse its condensation with the flavan-3-ol analogue, (–)-fisetinidol. A plausible explanation for this dramatic difference lies in the inductive effect of the 4-hydroxy function of the flavan-3,4-diol (or the putative 4-carbenium ion resulting from it) as potential substrate for electrophilic attack, thus reducing the nucleophilicity of the A-ring, and inhibiting the onset of self-condensation. However, once biflavanoid formation is initiated under more drastic conditions, the upper unit in the absence of 4-oxygenation now represents a suitable substrate for accelerated electrophilic attack and the condensation runs out of control.

VII. Composition of the Metabolic Pool and Condensation Aptitudes of Tannin Precursors

With the fore-mentioned chemical principles in mind, examination of the oligomeric composition of several extracts which contain tannins is now possible in terms of stereochemistry, functionality and condensation aptitudes of their precursors or intermediates. From this it is evident that condensations follow the path of least resistance, being regulated by the stability of the carbocation arising from the flavan-3,4-diol, by the nucleophilicity of the substrate, by steric factors contributed by both electrophile and nucleophile, by the 2,3-*trans* or 2,3-*cis* stereochemistry of the electrophile, and by the relative concentrations of reactants.

Considering the simplest case first, the predominant bark metabolites of the black wattle *(A. mearnsii)* are presumably based on two electrophilic [(+)-2,3-*trans*-3,4-*trans*-leucofisetinidin and -leucorobinetinidin] and two nucleophilic species [(+)-catechin and (+)-gallocatechin] *(21)*. [4,8]-Biflavanoids resulting from these, as shown synthetically, are 2,3-*trans*-3,4-*trans*:2′,3′-*trans*-(–)-fisetinidol-(+)-catechin; its 3,4-*cis* isomer; and 2,3-*trans*-3,4-*trans* : 2′,3′-*trans*-(–)-robinetinidol-(+)-catechin and -(–)-robinetinidol-(+)-gallocatechin. Similarly the [4,6:4,8] "angular" triflavanoids present and available from these are based exclusively on the condensation of (+)-leucorobinetinidin with the last-mentioned pair of biflavanoids, giving 2,3-*trans*-3,4-*trans*:2′,3′-*trans*:2″,3″-*trans*-3″,4″-*trans*-bi-[(–)-robinetinidol]-(+)-catechin and -bi-[(–)-robinetinidol]-(+)-gallocatechin and their respective 3″,4″-*cis* diastereoisomers (Scheme 11).

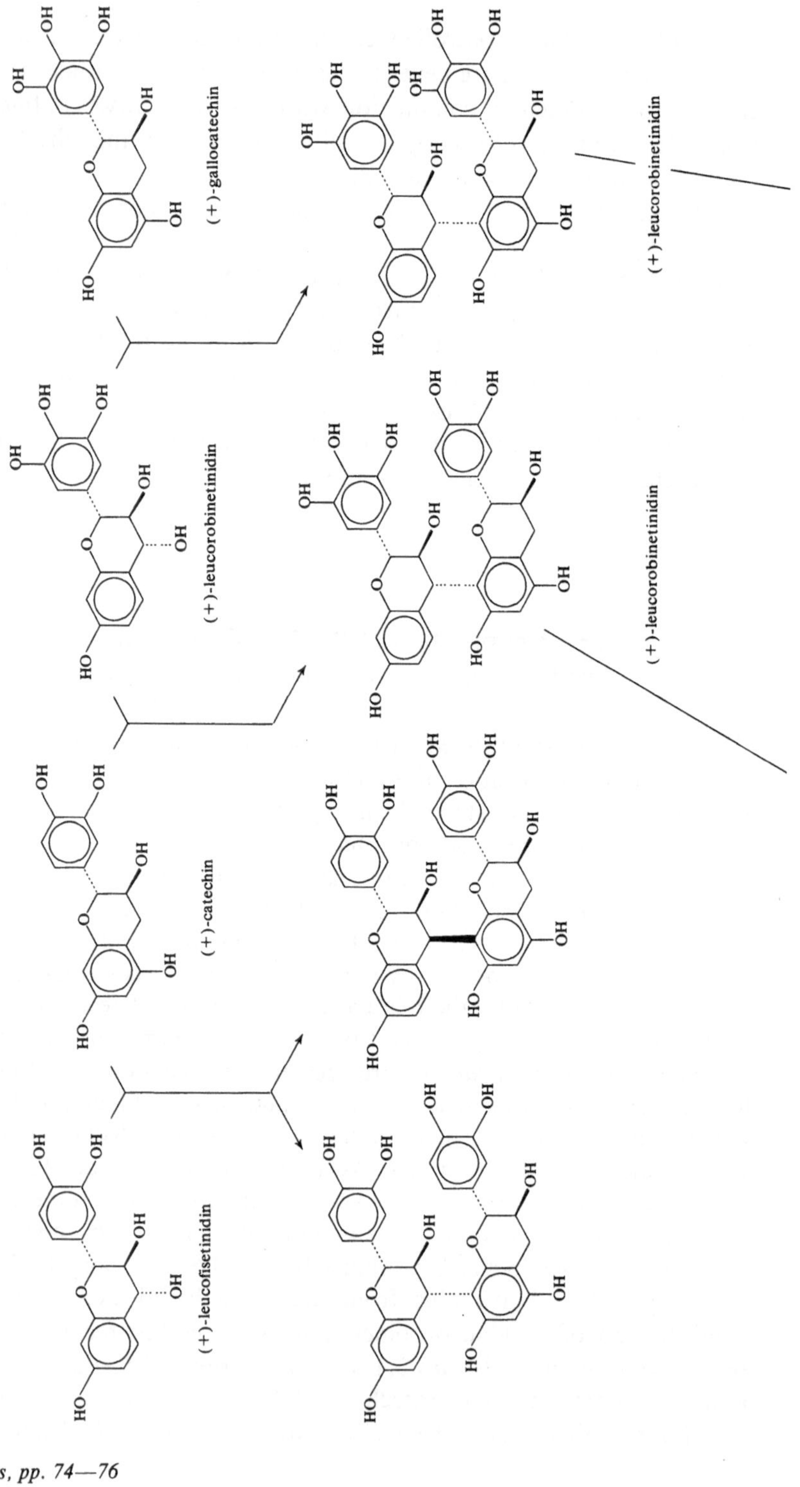

(+)-gallocatechin

(+)-leucorobinetinidin

(+)-leucorobinetinidin

(+)-leucorobinetinidin

(+)-catechin

(+)-leucofisetinidin

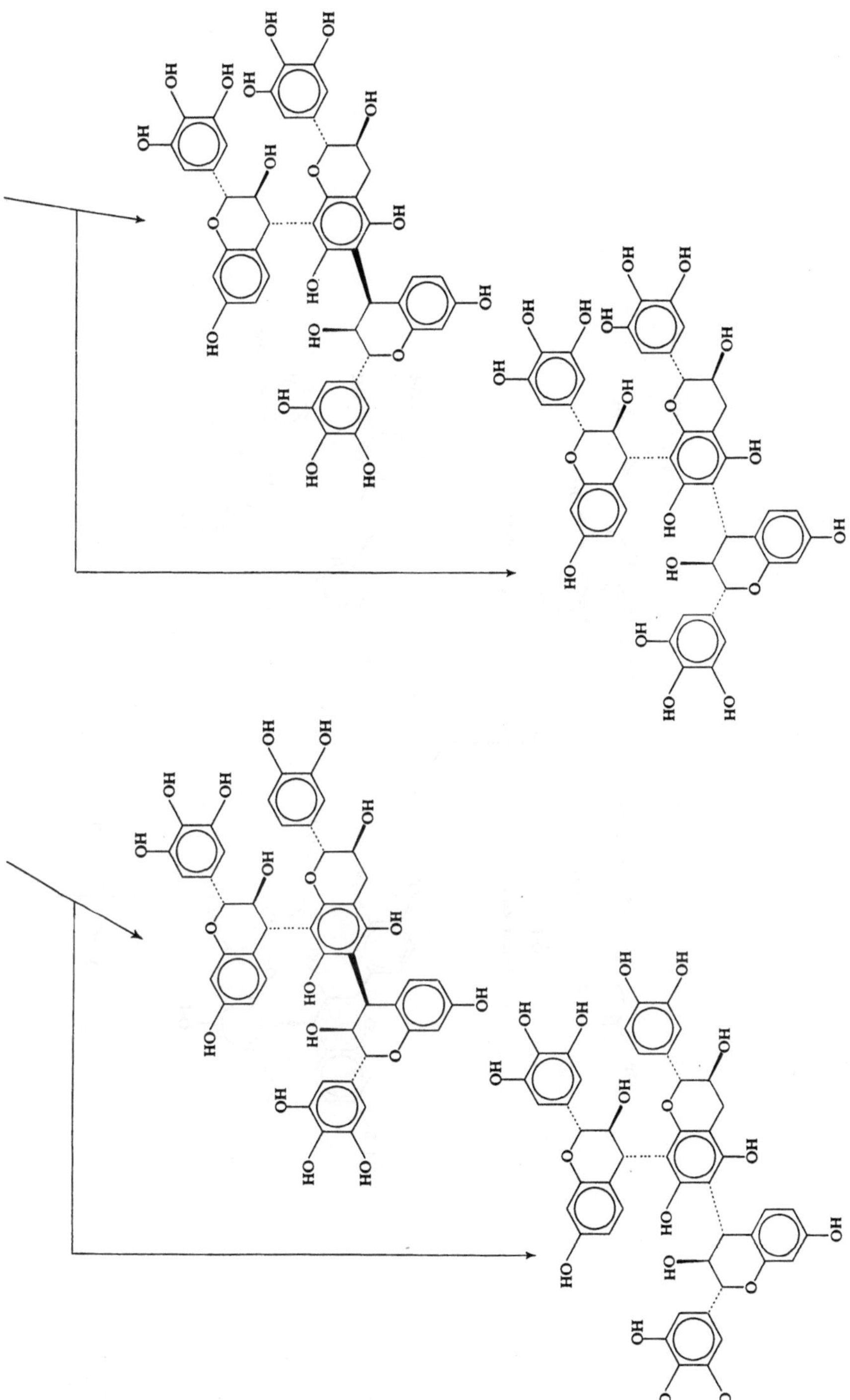

Scheme 11. Hypothetical condensation scheme for wattle bark tannins (*Acacia mearnsii*)

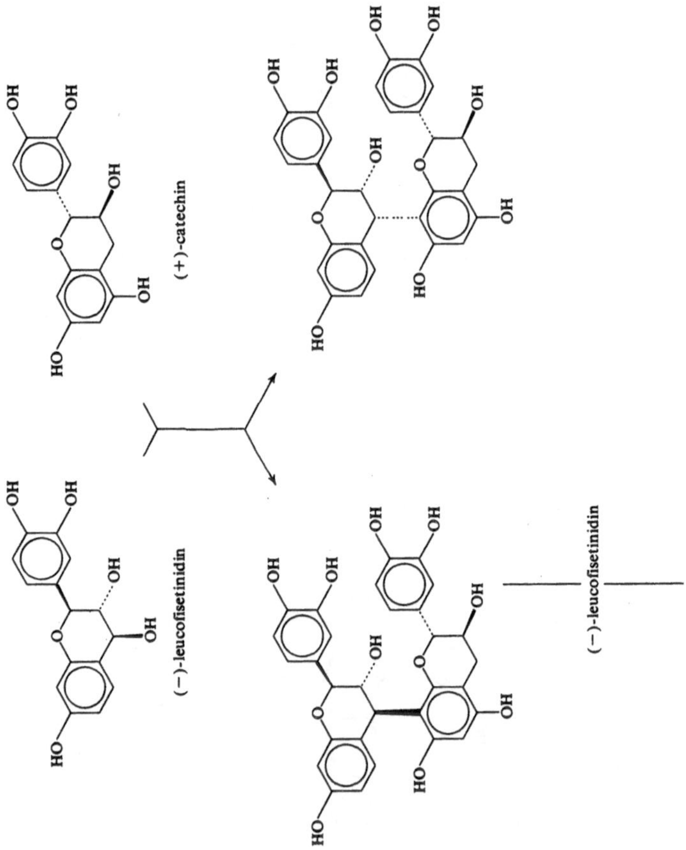
(−)-leucofisetinidin
(+)-catechin
(−)-leucofisetinidin

References, pp. 74—76

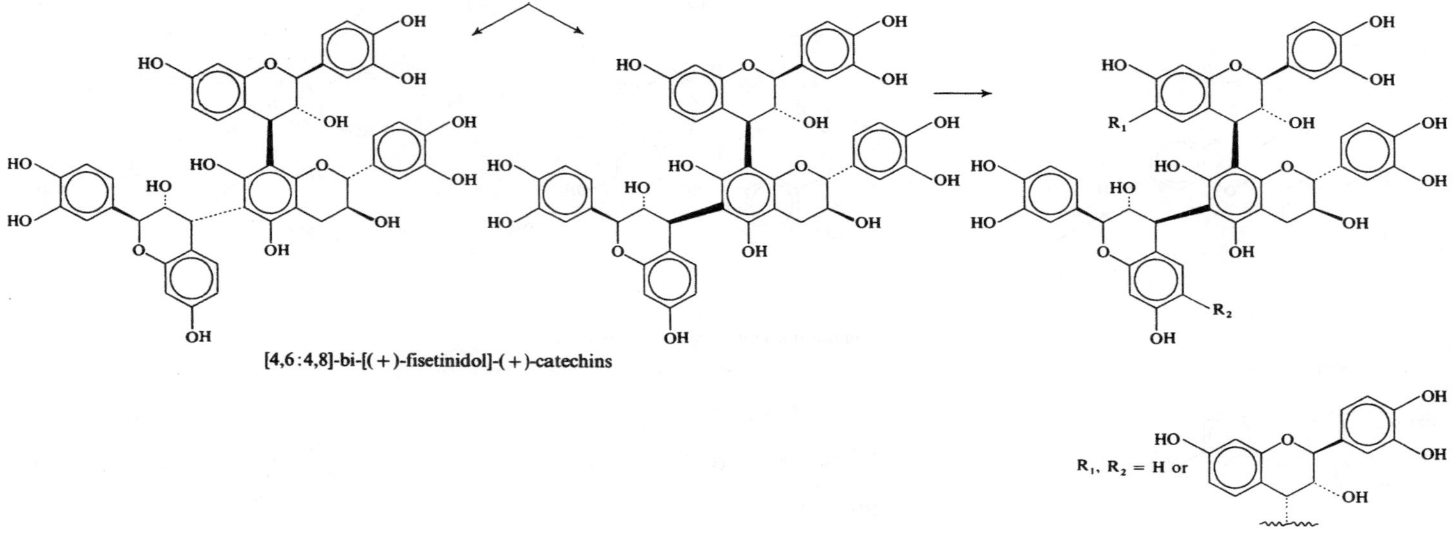

Scheme 12. Wood: *Rhus lancea; Schinopsis balansae* and *S. lorentzii*

[4,6:4,8]-bi-[(+)-fisetinidol]-(+)-catechins

[4,6:4,6:4,8]-tri-[(+)-fisetinidol]-(+)-catechin

$R_1, R_2 = H$ or

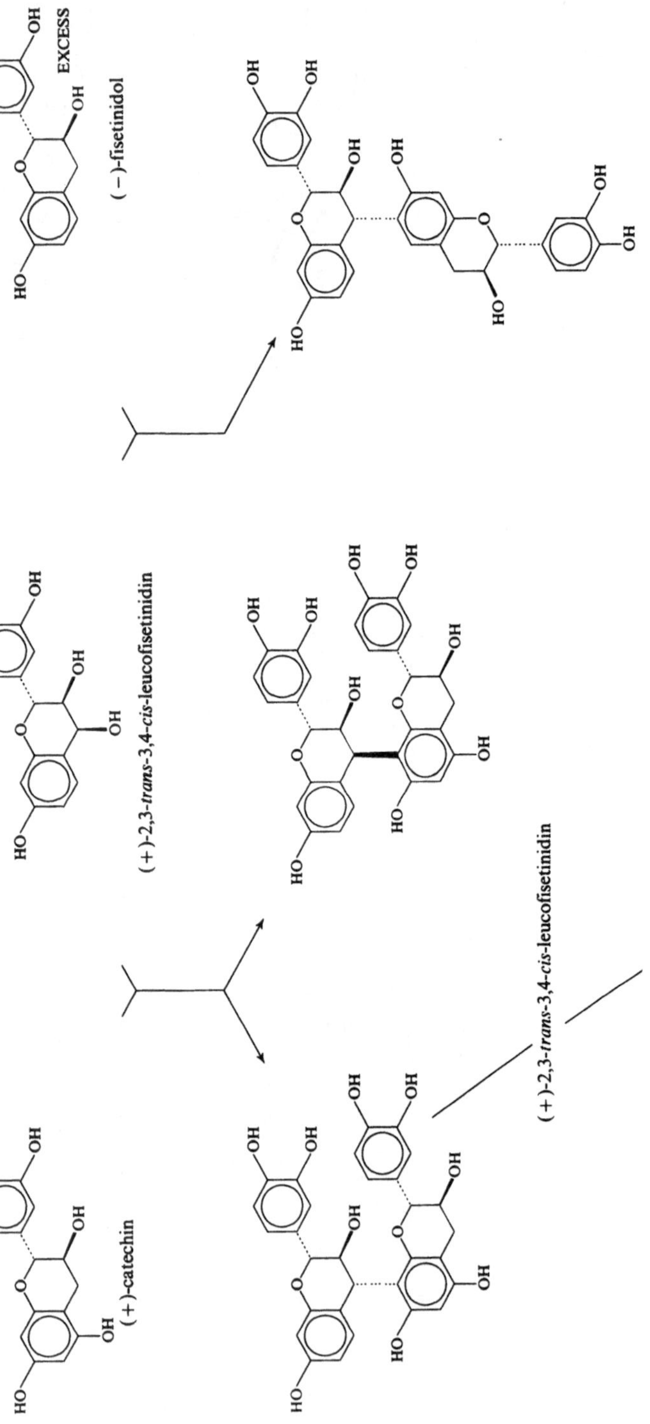

EXCESS

(−)-fisetinidol

(+)-2,3-*trans*-3,4-*cis*-leucofisetinidin

(+)-catechin

(+)-2,3-*trans*-3,4-*cis*-leucofisetinidin

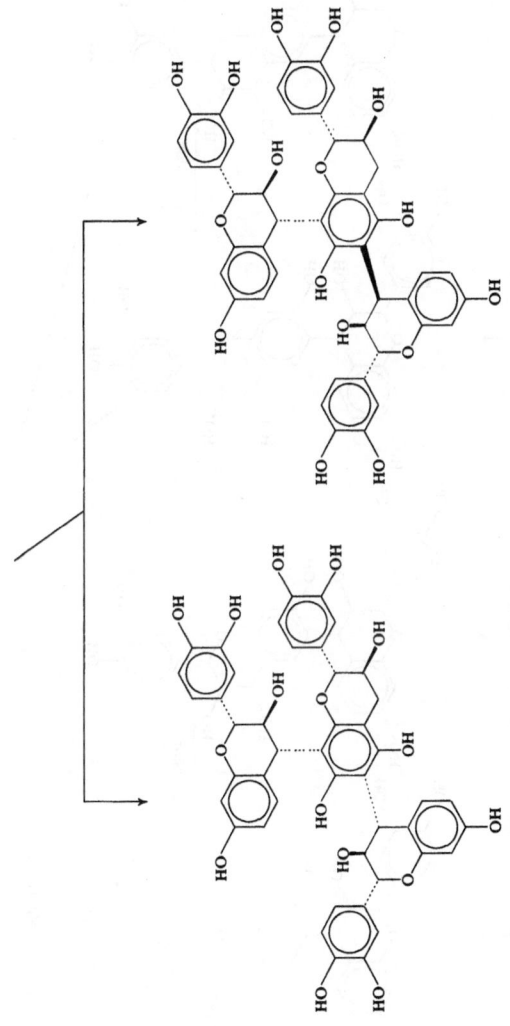

Scheme 13. Mopane wood: *Colophospermum mopane*

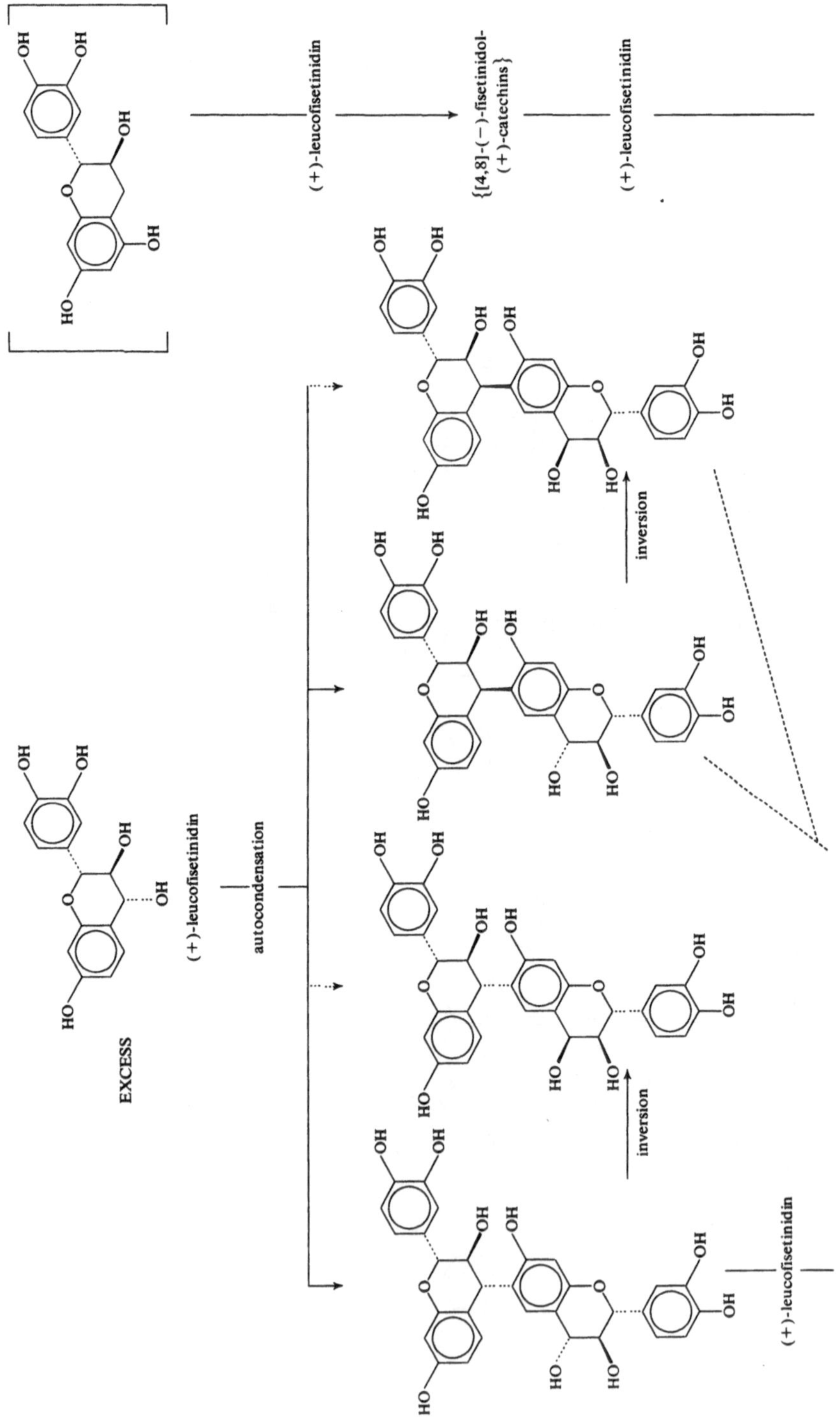

References, pp. 74—76

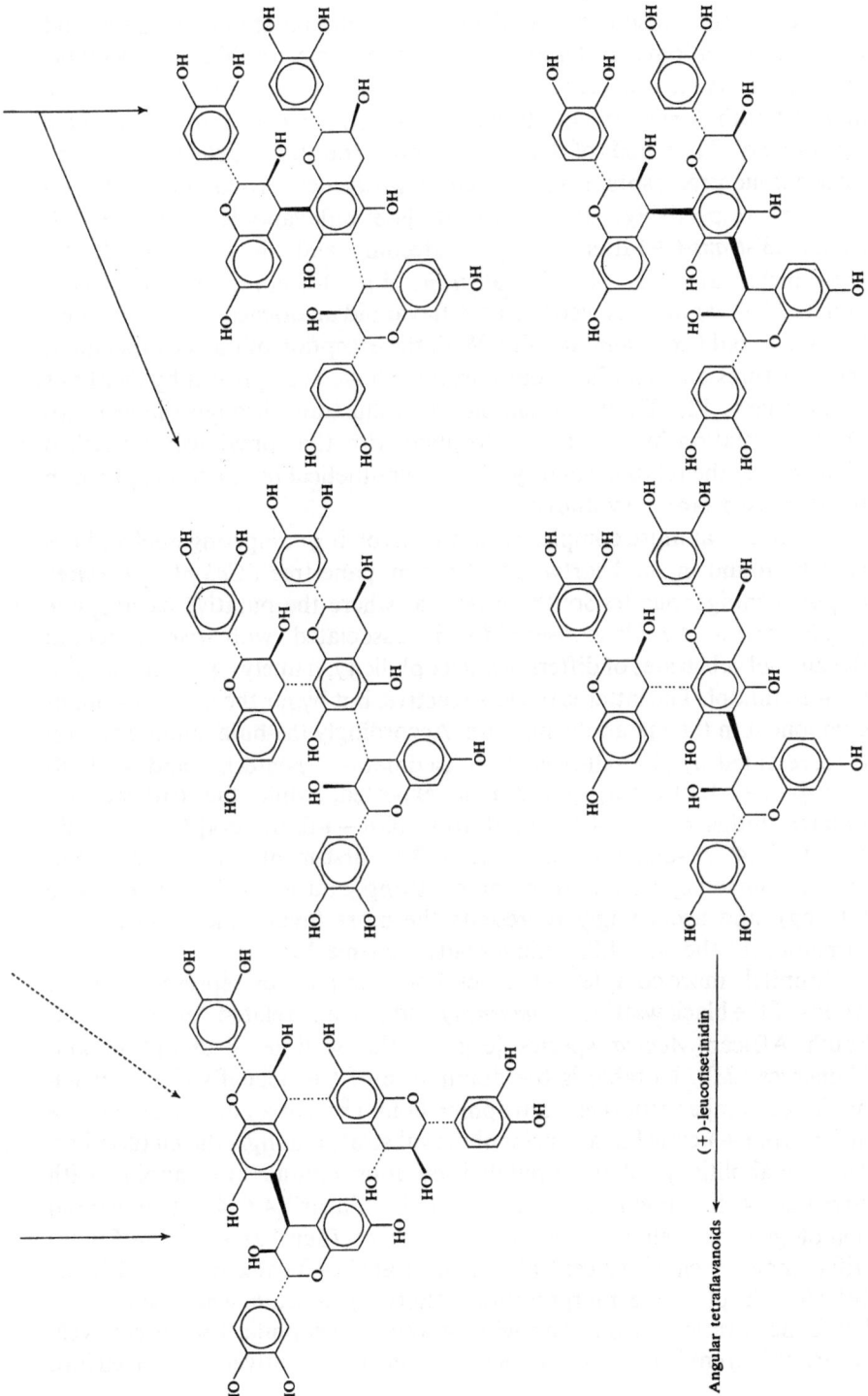

Scheme 14. Wattle wood: *Acacia mearnii*

Angular tetraflavanoids ← (+)-leucofisetinidin

Other simple instances based on compositional studies to date, and involving enantiomeric stereochemistry of the electrophile are available amongst the Anacardiaceae, i. e. *Rhus lancea* (karee) from Southern Africa and its South American counterparts *Schinopsis balansae* and *S. lorentzii* (quebracho). The woods of these trees contain the putative electrophilic and nucleophilic precursors, (−)-2,3-*trans*-3,4-*trans*-leucofisetinidin and (+)-catechin respectively; the expected [4,8]-biflavanoids, 2,3-*trans*-3,4-*trans*: 2′,3′-*trans*-(+)-fisetinidol-(+)-catechin and its 3,4-*cis* diastereoisomer; angular [4,6:4,8]-all-*trans*-bi-[(+)-fisetinidol]-(+)-catechin and its 3″,4″-*cis* diastereoisomer as triflavanoid analogues; and an angular tetraflavanoid (*ex R. lancea* only). With the exception of the tetraflavanoid the structures and absolute configurations have been proved by synthesis (*22*) (Scheme 12). These two simple cases illustrate through the enantiomorphic relationship of the electrophile with that previously described (Scheme 11) the relative validity of our hypothetical biomimetic approach to condensed tannin synthesis.

A somewhat more complex situation involving competing nucleophiles is to be found in the heartwood of the mopane tree *(Colophosphermum mopane)* indigenous to Southern Africa, where the putative electrophile (+)-2,3-*trans*-3,4-*cis*-leucofisetinidin is associated with two potential flavan-3-ol substrates of differing nucleophilicity, namely (+)-catechin and (−)-fisetinidol. The latter is the less reactive, but by far the more dominant component in the metabolic mixture. Accordingly the biflavanoid fraction is represented by [4,8]-all-*trans*-(−)-fisetinidol-(+)-catechin and [4,6]-all-*trans*-bi-[(+)-fisetinidol] (*cf. 12*) as expected, while the triflavanoids isolated hitherto are [4,6:4,8]-all-*trans*-bi-[(−)-fisetinidol]-(+)-catechin and its 3″,4″-*cis*-diastereoisomer (*16*). The first-mentioned of the biflavanoids obviously possesses the more strongly nucleophilic centre at C-6 (D-ring), and accordingly represents the more favourable substrate for formation of the [4,6:4,8]-triflavanoids (Scheme 13).

Infinitely more complex is the metabolic composition represented in the woods of the black wattle *(A. mearnsii)* and of many related Australian and South African *Acacia* species [e. g. in the sections Brunioideae and Uninerves (*23*)]. Notable is the dominance of (+)-leucofisetinidin [(+)-mollisacacidin] as putative electrophilic tannin precursor and the complete absence of (+)-catechin as nucleophilic substrate amongst the metabolites. The metabolitic pool is populated by four unique biflavanoids with terminal 3,4-diol function (Scheme 14) and a "linear" [4,6:4,6]-triflavanoid homologue, as well as a complete series of four "angular" [4,6:4,8]-triflavanoids each with a centrally incorporated (+)-catechin unit (Scheme 14). Here the plausible interpretation is that (+)-catechin, when available at low concentration in an otherwise weakly nucleophilic (but excessively electrophilic) environment is immediately and quantitatively converted into "angular" [4,6:4,8]-triflavanoids. This leaves the excess of (+)-leucofiset-inidin, with its A-ring deactivated by the inductive effect of the 4-hydroxyl, no alternative but that of condensation with the angular triflavanoids, or of self-condensation. Both processes occur.

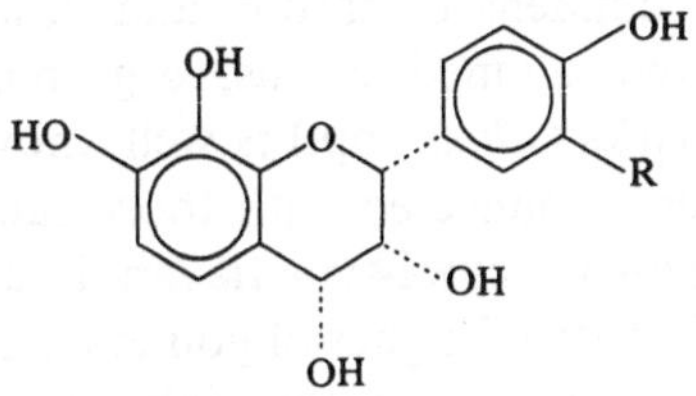

(5) R = H
(6) R = OH

Finally it is notable that where the flavan-3,4-diols (−)-teracacidin (5) and (−)-melacacidin (6) predominate in the woods of many *Acacia* spp. [e. g. in the section Plurinerves (*23*)], evidence of natural tannin formation is largely absent. This phenomenon is in line with the reduced nucleophilicity of its A-ring due to both vicinal functionalization (7,8-disubstitution) and the previously cited inductive effect of the 4-hydroxyl. These factors further limit the prospect of self-condensation, in the complete absence of suitable nucleophilic substrates [*cf.* (*24*) − (*26*)].

The above rationalizations, while providing a chemical basis for condensation phenomena (or absence thereof), in woods and barks of over a thousand species of the Leguminosae (subfamilies Mimoseae and Caesalphinieae) and Anarcardiaceae, prompt similar considerations for the formation of procyanidin and prodelphinidin tannins. Most notable is the absence of leucocyanidins and leucodelphinidins [for example (7) and (8)] from the corresponding tannin mixtures, the subject of earlier comment by CREASY and SWAIN (*1*). The high reactivity of these compounds (heightened resonance stabilization of the putative benzylcarbonium ion contributed by

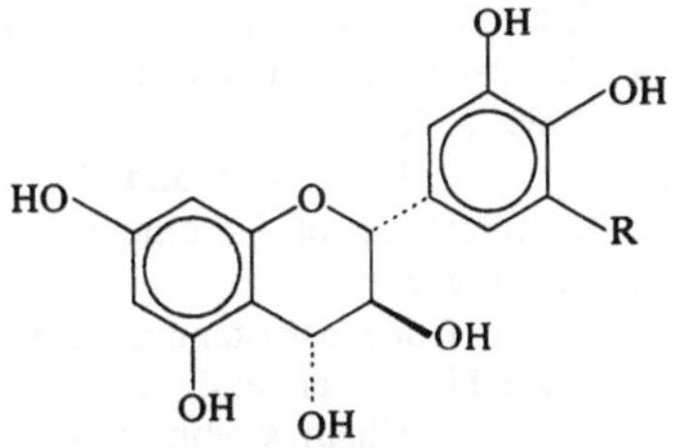

(7) R = H
(8) R = OH

the additional 5-hydroxylation of the A-ring) would lead to initiation of rapid condensation with nucleophilic (+)-catechins and (−)-epicatechins, and subsequently with the products of "dimerization", "trimerization", and of higher complexity, to terminate in high condensates. Such reactivity should predictably cause the quantitative removal of the monomeric flavan-3,4-diols. The highly condensed nature and low solubility of most procyanidin extracts re-examined by us, e. g. mangrove *(Rhizophora mucronata)* and pine barks (*Pinus* spp.) is well known.

Outstanding problems connected with those tannins which are comprised exclusively of phloroglucinol-type flavanoid units centre around (i) lack of rigid criteria of purity of higher oligomers, amongst which perhaps the most significant is assessment from high resolution (> 400 Mz) ^{1}H n.m.r. spectra recorded at elevated ($\sim 200°$ C) temperatures; (ii) lack of accurate knowledge regarding nucleophilic bonding positions and hence of the "linear" or "angular" nature of higher oligomers, although recent claims in support of the former are based on ^{13}C n.m.r. studies (*27*); and lack of unambiguous definition of the 3,4-*cis* or -*trans* stereochemistry of units present in higher oligomers and accordingly of their absolute configurations.

Condensed tannins of the prodelphinidin, procyanidin and propelargonidin series should, however, be amenable to the same direct synthetic approach as outlined in the foregoing [*cf.* biflavanoid procyanidin formation (*12*)] although 5-hydroxylation probably introduces an additional steric factor into condensation reactions, thereby reducing yields.

References

1. Creasy, L. L., and T. Swain: Structure of condensed tannins. Nature **208**, 151 (1965).

2. Geissman, T. A., and N. N. Yoshimura: Synthetic proanthocyanidin. Tetrahedron Letters **1966**, 2669.

3. Pelter, A., and P. I. Amenechi: Isoflavonoid and pterocarpinoid extractives of *Lonchocarpus laxiflorus*. J. Chem. Soc. (C) **1969**, 887; Pelter, A., P. I. Amenechi, R. Warren and S. H. Harper: The structures of two proanthocyanidins from *Julbernadia globiflora*. J. Chem. Soc. (C) **1969**, 2572.

4. Engel, D. W., M. Hattingh, H. K. L. Hundt, and D. G. Roux: X-ray structure, conformation, and absolute configuration of 8-bromotetra-*O*-methyl-(+)-catechin. J. Chem. Soc. Chem. Commun. **1978**, 695.

5. Hundt, H. K. L., and D. G. Roux: Condensed tannins: Determination of the point of linkage in "terminal" (+)-catechin units and degradative bromination of 4-flavanylflavan-3,4-diols. J. Chem. Soc. Chem. Commun. **1978**, 696.

6. — — Synthesis of condensed tannins. Part 3. Chemical shifts for determining the 6- and 8-bonding positions of "terminal" (+)-catechin units. J. Chem. Soc. Perkin I **1981**, 1227.

7. Botha, J. J., D. Ferreira, and D. G. Roux: Condensed tannins: Circular dichroism method of assessing the absolute configuration at C-4 of 4-arylflavan-3-ols and stereochemistry of their formation from flavan-3,4-diols. J. Chem. Soc. Chem. Commun. **1978**, 698.

8. Botha, J. J., D. A. Young, D. Ferreira, and D. G. Roux: Synthesis of condensed tannins. Part 1. Stereoselective and stereospecific syntheses of optically pure 4-arylflavan-3-ols, and assessment of their absolute stereochemistry at C-4 by means of circular dichroism. J. Chem. Soc. Perkin I **1981**, 1213.

9. Van der Westhuizen, J. H., D. Ferreira, and D. G. Roux: Synthesis of condensed tannins. Part 2. Synthesis by photolytic rearrangement, stereochemistry and circular dichroism of the first 2,3-*cis*-4-arylflavan-3-ols. J. Chem. Soc. Perkin I **1981**, 1220.

10. Thompson, R. S., D. Jacques, E. Haslam, and R. J. N. Tanner: Plant proanthocyanidins. Part 1. Introduction: The isolation, structure and distribution in nature of plant procyanidins. J. Chem. Soc. Perkin I **1972**, 1387.

11. Botha, J. J., D. Ferreira, and D. G. Roux: Condensed tannins: Direct synthesis, structure and absolute configuration of four biflavanoids from black wattle bark ("Mimosa") extract. J. Chem. Soc. Chem. Commun. **1978**, 700.

12. — — — Synthesis of condensed tannins. Part 4. A direct biomimetic approach to [4,6]- and [4,8]-biflavanoids. J. Chem. Soc. Perkin I **1981**, 1235.

13. Weinges, K., H. D. Marx, and K. Göritz: Die Rotationsbehinderung an der C(sp^2)-C(sp^3)-Bindung der 4-Arylsubstituierten Polymethoxyflavane. Chem. Ber. **103**, 2336 (1970).

14. Du Preez, I. C., A. C. Rowan, D. G. Roux, and J. Feeney: Hindered rotation about the sp^2-sp^3 hybridized C-C bond between flavanoid units in condensed tannins. J. Chem. Soc. Chem. Commun. **1971**, 315.

15. Sykes, R. L., and D. G. Roux: Study of the affinity of black wattle extract constituents. Part IV. Relative affinity of polyphenols for swollen chemically modified collagen. J. Soc. Leather Trades Chem. **41**, 14 (1957).

16. Botha, J. J., D. Ferreira, D. G. Roux, and W. E. Hull: Condensed tannins: Condensation mode and sequence during formation of synthetic and natural triflavanoids. J. Chem. Soc. Chem. Commun. **1979**, 510.

17. Botha, J. J., P. M. Viviers, I. C. Du Preez, D. Ferreira, D. G. Roux, and W. E. Hull: Synthesis of condensed tannins. Part 5. The first angular [4,6:4,8]-triflavanoids and their natural analogues. J. Chem. Soc. Perkin I **1982** (in press).

18. Kessler, H.: Detection of hindered rotation and inversion by NMR spectroscopy. Angew. Chem. Internat. Edn. **9**, 219 (1970).

19. Botha, J. J., P. M. Viviers, D. Ferreira, and D. G. Roux: Condensed tannins: Competing nucleophilic centres in biomimetic condensation reactions. Phytochemistry **1982** (in press).

20. Viviers, P. M., D. A. Young, J. J. Botha, D. Ferreira, D. G. Roux, and W. E. Hull: Synthesis of condensed tannins. Part 6. The sequence of units, coupling positions and absolute configuration of the first linear [4,6:4,6]-triflavanoid with terminal 3,4-diol function. J. Chem. Soc. Perkin I **1982** (in press).

21. Viviers, P. M., J. J. Botha, D. Ferreira, D. G. Roux, and H. M. Saayman: Synthesis of condensed tannins. Part 7. [4,6:4,8]-Linked angular prorobinetinidin triflavanoids from black wattle ("Mimosa") extract (submitted).

22. Viviers, P. M., J. J. Botha, D. Ferreira, and D. G. Roux: Synthesis of condensed tannins. Part 8. Oligomers of the Anacardiaceae: An enantiomeric electrophile (submitted).

23. Tindale, M. D., and D. G. Roux: An extended phytochemical survey of Australian species of *Acacia:* Chemotaxonomic and phylogenetic aspects. Phytochemistry **13**, 829 (1974).

24. Malan, E., and D. G. Roux: Flavonoids and tannins of *Acacia* species. Phytochemistry **14,** 1835 (1975).
25. Fourie, T. G., I. C. Du Preez, and D. G. Roux: 3′,4′,7,8-Tetrahydroxyflavonoids from the heartwood of *Acacia nigrescens* and their conversion products. Phytochemistry **11,** 1763 (1972).
26. Drewes, S. E., and D. G. Roux: A new flavan-3,4-diol from *Acacia auriculiformis* by paper ionophoresis. Biochem. J. **98,** 493 (1966).
27. Hemingway, R. W., L. Y. Foo, and L. J. Porter: Polymeric proanthocyanidins: Interflavanoid linkage isomerism in (epicatechin-4)-(epicatechin-4)-catechin procyanidins. J. Chem. Soc. Chem. Commun. **1981,** 320.

(Received September 4, 1981)

Streptonigrin

By St. J. Gould, School of Pharmacy, University of Connecticut, Storrs, Connecticut, U.S.A., and St. M. Weinreb, Department of Chemistry, The Pennsylvania State University, University Park, Pennsylvania, U.S.A.

Contents

I. Introduction

A collaborative effort between researchers at Chas. Pfizer, Inc. and at the Sloan-Kettering Institute led to the discovery that a filtrate from cultures of *Streptomyces flocculus* had significant anticancer and antibiotic activity. The active agent was eventually isolated, named streptonigrin, and shown to have the structure depicted in (1). During the last twenty years streptonigrin has been the subject of intensive study regarding its use as an anticancer drug, its cytotoxic mechanism of action, its laboratory synthesis, and its biosynthesis.

Two total syntheses were reported recently. Biosynthetic studies over the last four years have outlined the primary and secondary metabolism leading to streptonigrin and have revealed previously unknown biological pathways leading to formation of pyridine and quinoline rings. With

(1)

the recent recognition of the substantial enhancement of streptonigrin —
DNA binding in the presence of polyvalent metal ions, an understanding
of the mechanism of action is gaining a much firmer footing. And, as
more sophisticated protocols for cancer chemotherapy have developed,
results from clinical investigations involving streptonigrin have become
more promising. This chapter reviews each of these areas through
1980.

II. Isolation and Structure

Streptonigrin was first isolated in 1959 in the United States from a
culture of *Streptomyces flocculus* (*1*). The same compound was in-
dependently isolated in the Soviet Union a few years later from a strain
of *Actinomyces albus var. bruneomycini* (*2*) and in France from *S.
rufochromogenes* and *S. echinatus* (*3*), the antibiotic being named bruneomycin and rufochromomycin, respectively. The antibiotic forms dark
brown crystals which are slightly soluble in water, lower alcohols, ethyl
acetate or chloroform and more soluble in dioxane, pyridine, DMF or
THF. Streptonigrin has UV absorption maxima in methanol at 248 nm
(ε 38,400) and 375—380 nm (ε 17,400). It is a monobasic acid with
pKa 6.2—6.4 in 1 : 1 dioxane: water.

The structure of streptonigrin was brilliantly deduced as (**1**) by
RAO, BIEMANN, and WOODWARD in 1963 using a combination of spectral
and degradative methods (*4*). The exact elemental composition of
streptonigrin was not obtainable by standard combustion analysis, but a
high resolution mass spectrum of its derived methyl ester established
an empirical formula of $C_{25}H_{22}O_8N_4$ for the natural product. This
formula was confirmed by mass spectrometry on the hydroquinone
derivative (**2**), prepared from streptonigrin by catalytic reduction, followed by exhaustive methylation with deuteriodimethyl sulfate.

(2)

Oxidation of (1) with basic hydrogen peroxide produced tribasic streptonigric acid (3a), while the tetrabasic streptonigrinic acid (4a) was formed upon oxidation of (1) with basic permanganate. Based upon these transformations, and the empirical relationships involved, existence of an A-ring quinone or quinone-imine and a D-ring dimethoxyphenol was presupposed for streptonigrin. The postulate that streptonigrin has a quinoid structure was further supported by its color reactions with several test reagents (1).

The ^{1}H-NMR spectrum of the tetramethyl ester (4b) formed from streptonigrinic acid showed the presence of a primary amino group (2H, 7.84 δ, exchangeable with D_2O) and treatment of (4b) with nitric acid gave the desaminotetraacid (5) after ester hydrolysis. Thermal decarboxylation of (5) over soda lime produced 5-methyl-2,2'-bipyridyl (6) identical with an authentic sample.

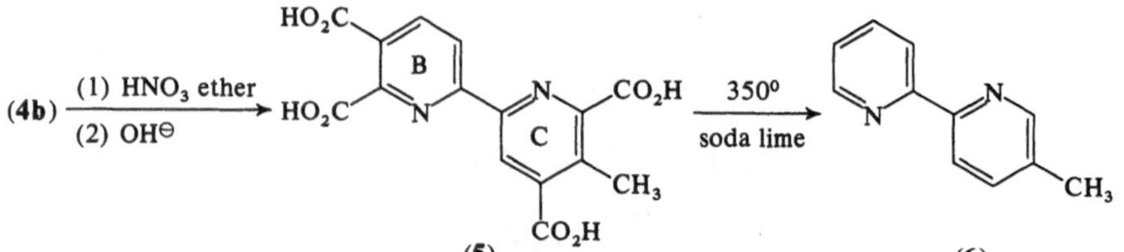

HO$_2$C — (4b) — (1) HNO$_3$ ether / (2) OH$^\ominus$ → (5) → 350° soda lime → (6)

Oxidation of streptonigrinic acid (4a) with sodium hypochlorite gave 2,3,6-pyridinetricarboxylic acid (7) which was decarboxylated to the known pyridine diacid (8), while sequential treatment of (4a) with hydrogen/ PtO$_2$/EtOH/HCl and alkaline permanganate followed by distillation from soda lime yielded 3-amino-5-methyl pyridine (9). These above reactions, coupled with the fact that streptonigrinic acid contains two adjacent aromatic hydrogens as seen by NMR, established structure (4a) for this degradation product.

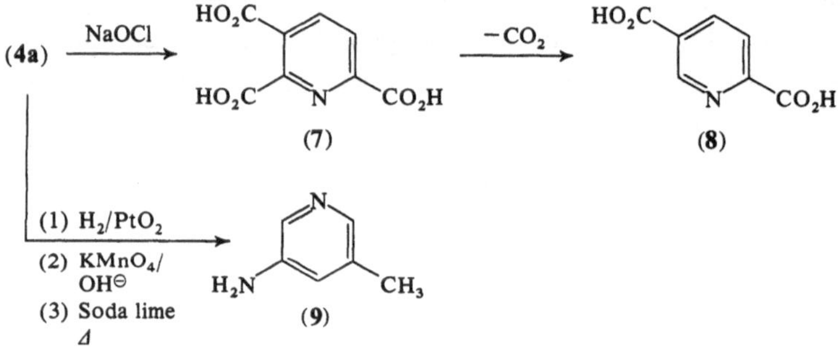

Methylation of streptonigric acid (3a) yielded methylether ester (3b), which upon basic ester hydrolysis gave triacid (10). Oxidation of (10) with hot basic permanganate afforded the known 2,3,4-trimethoxybenzoic acid (11). The exact position of the phenolic group in the D-ring of streptonigrin was cleverly determined by conversion of the trideuterio-methyl ether (12), prepared from streptonigric acid, to the undeuterated benzofuran (13) upon treatment with nitric acid.

Methylation of streptonigrin gave a non-acidic compound containing only *two* new O-methyl groups [i.e. (14)], thus establishing the A-ring of the antibiotic as an aminoquinone rather than as a hydroxyquinone imine. Condensation of this derivative with hydroxylamine followed by dithionite reduction yielded a o-diamino compound (15). Treatment of (15) with biacetyl gave a quinoxaline (16), which was oxidized with permanganate to yield acid (17). Basic ester hydrolysis and thermal decarboxylation then gave (18). The NMR spectrum of (18) was con-

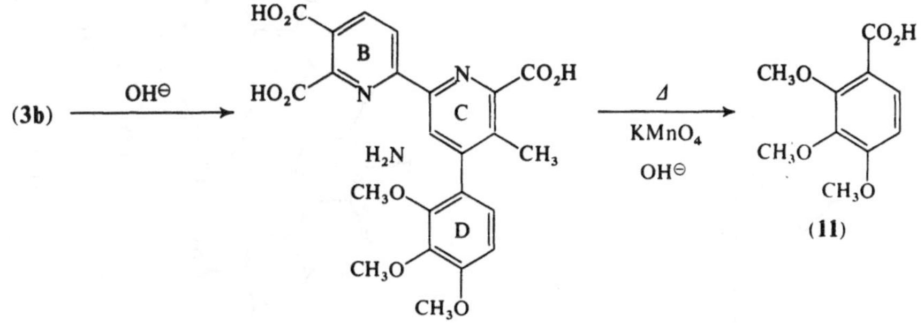

(10)

(11)

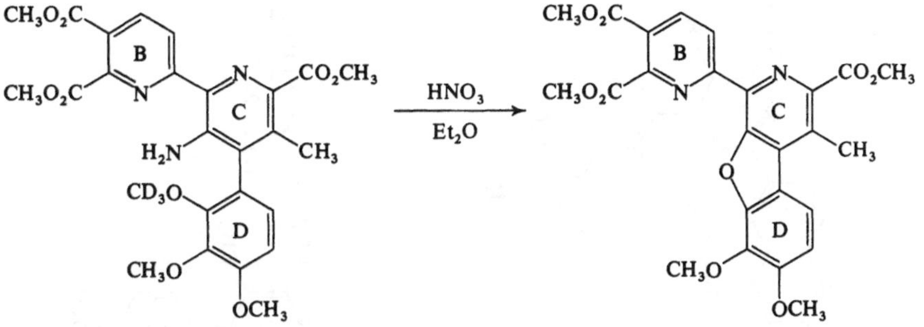

(12)   (13)

sistent with the arrangement of aromatic hydrogens shown, and supports the complete structure indicated in (1) for streptonigrin.

The proton NMR spectrum of (1) (not previously reported) is shown in Table 1. (5) and is in complete accord with this structure. The carbon-13 NMR spectrum of streptonigrin was originally reported by LOWN and BEGLEITER (6). During biosynthetic studies a number of resonances in this spectrum were reassigned (7) and this ^{13}C-NMR data is listed in Table 2.

Table 1. '*H NMR Spectrum of Streptonigrin in d_6-DMSO at 270 MHz*

Chemical Shift (integral)	Multiplicy (J, Hz)	Assignment
2.18 (3)	s	3'-CH$_3$
3.76 (3)	s	OCH$_3$
3.82 (3)	s	OCH$_3$
3.86 (3)	s	OCH$_3$
6.71 (1)	d (10.1)	11', 12'
6.73 (1)	d (10.1)	
8.36 (1)	d (8.4)	3
9.01 (1)	d (8.4)	4

* Chemical shifts in p.p.m. from internal TMS.

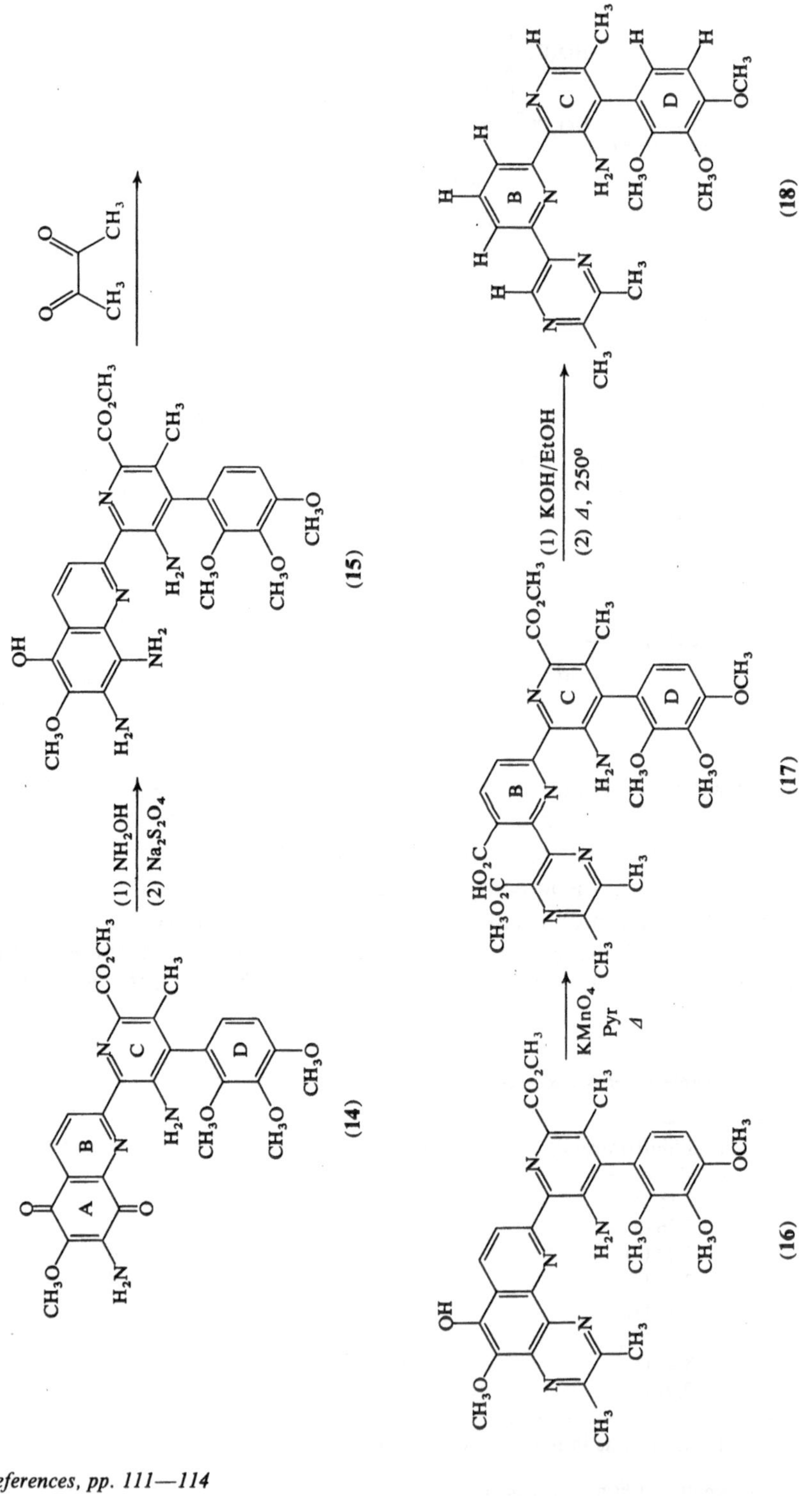
(1) NH2OH
(2) Na2S2O4
KMnO4
Pyr
Δ
(1) KOH/EtOH
(2) Δ, 250°
(14)
(15)
(16)
(17)
(18)

Table 2. ^{13}C *Chemical Shifts*
of Streptonigrin in p.p.m. from
TMS in d_6-DMSO at 67.88 MHz

Chemical Shift	Assignment
180.165	8
175.850	5
166.842	CO_2H
159.748	10'
153.059	6
147.989	8'
145.615	5'
143.997	8a
141.354	7
136.931	2 or 6'
136.149	3'
135.717	6' or 2
134.422	2'
133.910	4'
133.287	3
129.487	9'
126.601	**4a**
125.927	4
124.578	**12'**
114.788	7'
104.458	11'
60.253	OCH_3
59.632	OCH_3
55.614	OCH_3
16.856	CCH_3

Table 3. *^1H NMR Spectrum of Lavendamycin in d_6-DMSO at 100 MHz** (9)

Chemical Shift (integral)	Multiplicity (J, Hz)	Assignment
3.08 (3)	s	CH_3
5.94 (1)	s	6
7.42 (1)	dd (8.6, 6)	11'
7.40 (2)	bs	NH_2
7.69 (2)	d (6)	9', 10'
8.34 (1)	d (8.5)	12'
8.42 (1)	(d) (8)	3
8.95 (1)	(d) (8)	4

* Chemical shifts in p.p.m. from internal TMS.

A single crystal X-ray diffraction analysis reported by CHIU and
LIPSCOMB in 1975 (8) confirmed the original structure assignment for
streptonigrin (1). Of interest is the fact that in the crystal the A, B,

and C rings of the antibiotic are nearly coplanar and the D ring is almost perpendicular to the plane of the other three rings. The ABC-coplanarity is due to a hydrogen bond between one hydrogen of the C-ring amino group and the quinoline B-ring nitrogen atom. It is not presently known whether streptonigrin has this same conformation in solution.

(19)

Table 4. ^{13}C Chemical Shifts of
Lavendamycin in p.p.m. from TMS
in TFA/CD$_2$Cl$_2$ (1/9) at 25 MHz (9)

Chemical Shift (multiplicity)	Assignment
178.3	5
178.3	8
153.1 (s)	7
150.6 (s)	
147.6 (s)	
145.7 (s)	
145.5 (s)	
138.7 (s)	
138.2 (s)	
138.1 (s)	
134.7 (d)	3
134.7 (s)	
134.5 (s)	
129.4 (s)	
126.8 (d)	4
126.1 (d)	10'
126.1 (d)	11'
125.5 (d)	12'
122.2 (s)	7'
114.7 (d)	9'
17.4 (q)	CH$_3$

A new metabolite which is closely related to streptonigrin has recently been isolated from *Streptomyces lavendulae* by DOYLE (9) and has been named lavendamycin. This compound is a dark red solid which decomposes above 300°, and has an empirical formula of $C_{22}H_{14}N_4O_4$. Structure (19) has been proposed for lavendamycin, based upon a careful analysis of its spectral data. The metabolite has ultraviolet absorption maxima in methanol at 234, 246 and 391 nm. Table 3 lists its proton NMR absorptions at 100 MHz along with peak assignments for structure (19), and Table 4 contains the ^{13}C-NMR data obtained, with assignments made to date.

III. Structure Activity Relationships and Mechanism of Action

Streptonigrin has not proved readily amenable to chemical modification and relatively few derivatives of the antibiotic have been reported. Isopropylazastreptonigrin (20) and streptonigrin methyl ester have been prepared by Pfizer chemists. The former compound has no anticancer activity (10) while the latter compound is weakly efficient due to partial *in vivo* hydrolysis to streptonigrin (11). Inactive compounds also resulted from replacement of the 7-NH$_2$ by either –OH or –OMe (12), while 6-demethylstreptonigrin (13) was still slightly active. ROSAZZA (14) has now obtained the amide derivative (21) of streptonigrin and orsellinic acid by a high yield microbial synthesis using a strain of *Streptomyces griseus*. This compound shows significant *in vivo* activity against a particularly refractive mouse mammary tumor used in the National Cancer Institute solid tumor screen *(vide infra)*.

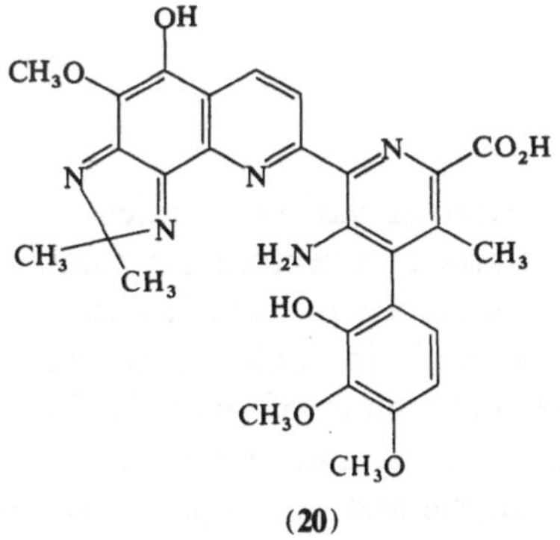

(20)

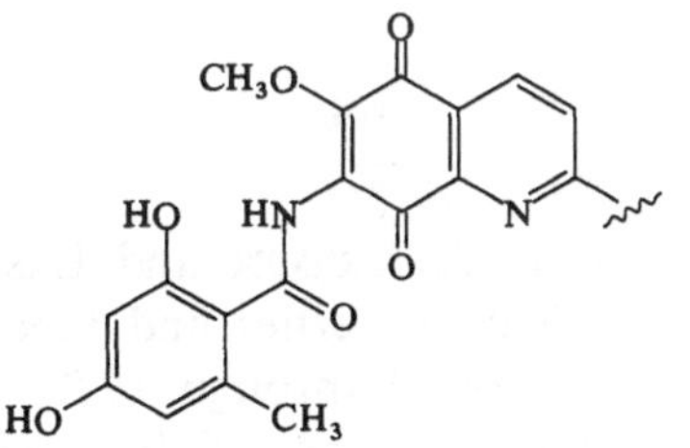

(21)

In recent years a considerable number of streptonigrin analogs have been synthesized which are primarily on variations of the 5,8-quinoline quinone structure with different substituents at C-2, C-6, and C-7

(*15, 16,* and *17*). Most of these compounds have only been examined *in vitro* during studies related to determining the mechanism of action of streptonigrin. A few of the compounds have been tested for antibiotic activity (*18*) and for *in vitro* cytotoxicity (*12*). Compound (**22**) was the most promising antibiotic, being twice as active as streptonigrin against *Bacillus subtilis*. Kende (*17*) has prepared compound (**23**), which proved inactive against KB cells. Rao (*12*) has suggested partial structure (**24**) as the minimum necessary for activity, although this postulate may have to be modified in view of Rosazza's recent work (*14*). However, *in vivo* hydrolysis of the orsellinamide (**21**), regenerating streptonigrin, could well explain the activity of this compound.

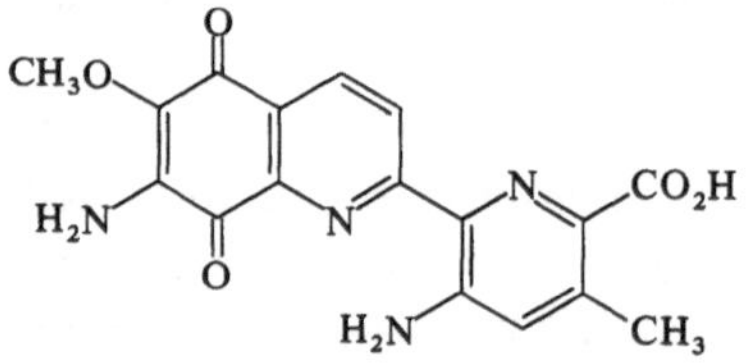

(**22**)

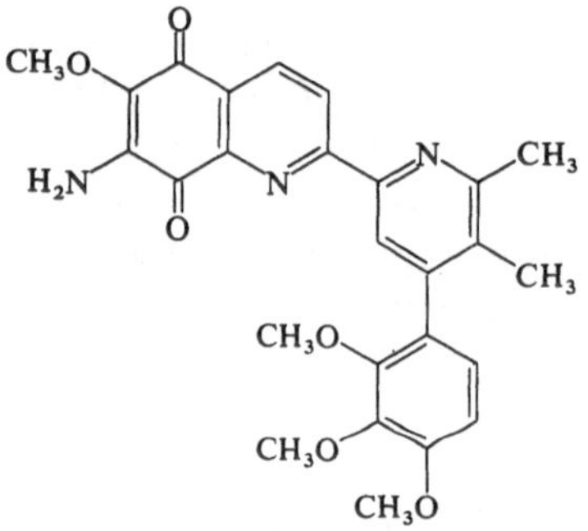

(**23**)

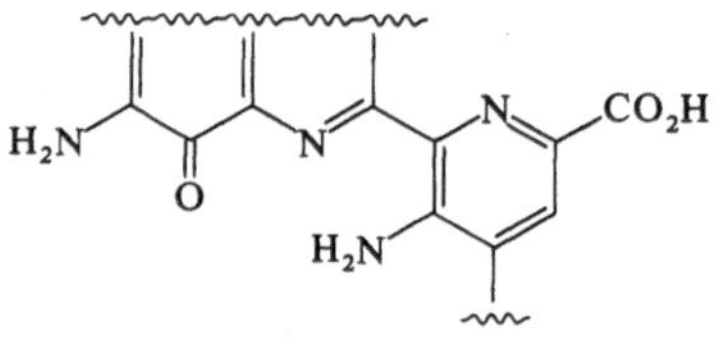

(**24**)

In 1975, Kremer and Laszio (*19*) reviewed various studies of the biological properties and biochemical effects of streptonigrin that had been reported through 1970. The preponderance of data indicated at that time that the cellular toxicity of streptonigrin could be due to the depletion of NADPH and NADH, the uncoupling of oxidative phosphorylation, and/or the formation of single strand breaks in DNA caused by radicals generated from streptonigrin and/or oxygen. A variety of data were noted which suggested that in bacteria at low concentrations of streptonigrin the lethal sequence involved reduction of streptonigrin by NADPH and NADH followed by a rapid autoxidation which bypassed oxidative phosphorylation and led to depletion of cellular

ATP. However, cultures of *E. coli* exposed to higher concentrations of streptonigrin exhibited an initial first-order decline in viability, implying that a single hit per cell was lethal and pointed to the bacterial chromosome as the site of action.

The relevance of these observations to the lethal effect of streptonigrin *in vivo* on human tumors has not been established. Virtually all work reported during the past decade has focused on the *in vitro* interaction of streptonigrin with DNA and on the formation and identity of the putative radical(s) responsible for the single strand breaks.

Early studies on streptonigrin/DNA interactions yielded conflicting claims for irreversible binding. MIZUNO and GILBOE (20) reported irreversible binding of ^{3}H-streptonigrin and DNA to form a complex that was stable to both dialysis and gel filtration. When the complex was hydrolyzed with either deoxyribonuclease 1 or with snake venom phosphodiesterase 1, more radioactivity was associated with the dCMP fraction than with any of the other three nucleosides. These results are in sharp contrast to those of Russian workers (21) who found that the ^{3}H-streptonigrin/DNA complex dissociated during sephadex gel permeation chromatography. In addition, WHITE (22) demonstrated that, unlike the situation with the quinone anticancer agent mitomycin C, sodium borohydride reduction of ^{3}H-streptonigrin in the presence of DNA did not result in a covalent streptonigrin-DNA bond.

Recent studies of streptonigrin/DNA binding in the presence of polyvalent metal ions present a more coherent picture of the requirements for strong complexes and may indicate the presence of factors that had been unrecognized in the earlier reports. Very tight binding of streptonigrin to DNA has been found to occur in the presence of metals such as Zn^{++} (23), where a 1 : 7 : 25 complex of streptonigrin : Zn : DNA was stable to dialysis and gel permeation chromatography. Significant interactions of Cu^{++}, Zn^{++}, Mn^{++}, Cd^{++}, and Fe^{+++} but not of Ca^{++} or Mg^{++} with streptonigrin (24, 25) and with various model 5,8-quinoline quinones (26) have now been observed. Cupric or ferrous ions were shown to promote single strand DNA breaks while Zn^{++} increased the lethality of reduced streptonigrin towards *E. coli* by two orders of magnitude (25).

HAJDU (27) has presented spectroscopic and electrochemical evidence for the formation of 1 : 1 complexes of streptonigrin with Cu^{++} and Zn^{++}, in each case with the release of one equivalent of proton. The pKa of streptonigrin was lowered 2.3 and 3.3 pKa units in the presence of one equivalent of Zn^{++} and Cu^{++}, respectively. In acetonitrile solution, reduction of streptonigrin was accelerated in the presence of Cu^{++} but was inhibited by Zn^{++}. HAJDU suggested that in acetonitrile structure

(25) represents the complex activated toward reduction while (26) represents the inhibited complex. In the latter case, metal-ion-assisted tautomerization to an aza-substituted o-quinoid structure (27) (28) may have been responsible for the decreased reducibility. Such tautomerism in A-ring analogs of streptonigrin has been suggested previously (29).

The ^{13}C NMR spectrum of the Zn^{++} complex of streptonigrin in DMSO-d_6 showed distinct downfield chemical shifts for the COOH, C-2, C-3, C-4, and C-4a resonances (28). The structural significance of these shifts is not clear.

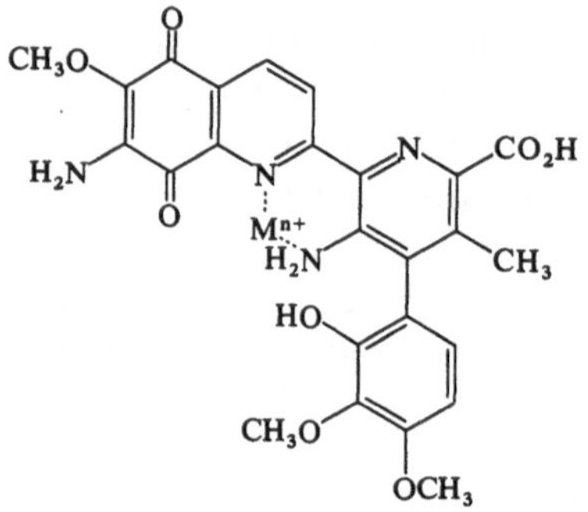

(25)

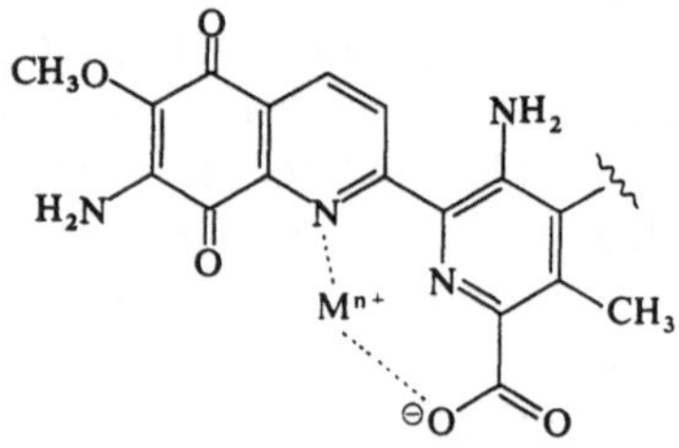

(26)

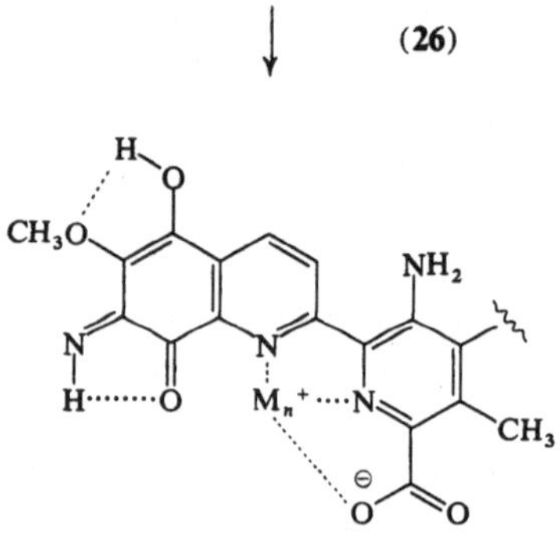

(27)

The nature of streptonigrin-metal ion complexes is strongly dependent on solvent and pH (28). In contrast to acetonitrile solutions, aqueous solutions of streptonigrin containing Zn^{++} or Cu^{++} are more easily reduced than solutions in the absence of the metal ion. These aqueous complexes exhibit UV spectra quite different from those in acetonitrile, and the structures of all these complexes are still very much in doubt. WHITE (25) has suggested a structure similar to (26) to explain the need for a free carboxylic acid for streptonigrin activity, but an alternate explanation would be that the carboxylate is needed for biological transport.

Various authors (20, 24) have pointed out the similarity of DNA degradation due to streptonigrin and that due to γ- and X-irradiation. Irradiation is generally believed to generate destructive hydroxyl radicals from water. The formation of streptonigrin semiquinone (28) both *in vivo* and *in vitro* was detected by ESR spectroscopy (*30—33*). Oxygen consumption was also demonstrated (*26*). Thus either (28) or a reduced form of oxygen, generated from the autoxidation of reduced streptonigrin, could be the active cytotoxic agent. Most work to date has focused on determining which reactive forms of oxygen possibly are involved in DNA degradation.

(28)

WHITE (*22*) found evidence showing that hydrogen peroxide could not be responsible for the lethality of streptonigrin. Later, LOWN (*24, 26, 31*) provided strong evidence for the involvement of superoxide and hydroxyl radical in streptonigrin mediated DNA degradation. They found that such degradation was completely inhibited by superoxide dismutase and by catalase. More recently, N-*t*-butyl-α-phenylnitrone (BPN) was used as a spin-trapping agent for semiquinones generated by NaBH$_4$ reduction of mitomycins C and B, and streptonigrin in pH 7.0 phosphate buffer (*32*). The resulting nitroxyl radical derived from BPN was clearly observed by ESR spectroscopy, and its formation was selectively inhibited by catalase and superoxide dismutase. Its formation was also strongly suppressed by EDTA. LOWN has no proposed the following redox scheme to account for the generation of hydroxyl radicals by antitumor antibiotics (A) such as streptonigrin:

$$A + 2H \longrightarrow AH_2$$

$$AH_2 + O_2 \longrightarrow AH \cdot + HO_2 \cdot$$

$$HO_2 + \longrightarrow H^+ + O_2^{\overline{\cdot}}$$

$$2H^+ + 2O_2^{\overline{\cdot}} \xrightarrow{\text{superoxide dismutase}} H_2O_2 + O_2$$

$$H_2O_2 \xrightarrow{\text{catalase}} 2H_2O + O_2$$

$$M(II) + H_2O_2 \cdot \longrightarrow M(III) + OH \cdot + OH^-$$

$$M(III) + O_2^{\overline{\cdot}} \longrightarrow M(II) + O_2$$

A slightly different view of streptonigrin lethality has been offered by Bachur (33). In studying the enzymatic generation of semiquinones from some fifteen quinone anticancer antibiotics, Bachur found streptonigrin to be one of the most effective substrates for NADPH cytochrome P-450 reductase. Based upon the fact that the biochemical reduction of vitamin K and coenzyme Q generate superoxide and hydroxyl radicals and that these radicals are detoxified by enzymes such as superoxide dismutase and catalase, it was suggested as an alternative that streptonigrin and other antitumor antibiotics, such as the mitomycins and the anthracyclinones, are initially reduced enzymatically to free radicals which first bind to DNA. These radicals then react directly with DNA "or produce reactive oxygen species within the DNA complex which could damage the DNA" (33). However, lowered levels of superoxide dismutase in tumor cells has been reported (34) and Lown (24) has shown that streptonigrin inactivates this enzyme.

It will likely prove very difficult to differentiate between such fine distinctions. However, there is now general agreement that DNA degradation requires reduction of streptonigrin to a semiquinone and involves molecular oxygen, metal ions and free radicals. With the recognition of the profound influence of trace metals upon the binding of streptonigrin to DNA one may hope that a definitive picture will emerge during the coming years.

The clinical use of streptonigrin has been under investigation for more than a decade. Early preclinical trials showed streptonigrin to have good antitumor activity in a variety of *in vivo* and *in vitro* systems. This early work culminated in a Phase III clinical trial in which streptonigrin appeared to be as effective as chlorabucil against lymproliferative diseases but also demonstrated a greater degree of toxicity (35).

Reports of the use of streptonigrin in combination with other chemotherapeutic agents in clinical trials have appeared during the last few years. In the United States, these combinations have included vincristine and prednisone, and sometimes also bleomycin, for treating lymphosarcoma and reticulum cell sarcoma (36, 37). Similar combination chemotherapy has been used in Europe and has indicated possible benefits for treatment of primary malignant melanoma (38) and non-Hodgkin lymphoma in children (39).

The microbial transformation product (21) is apparently the first chemical agent found effective against the CD8F1 spontaneous mouse mammary tumor (14). Reductions in tumor size of 62, 74 and 90% were obtained in mice at doses of 100, 200 and 400 mg/kg. These dose levels were well-tolerated: of ten animals tested at each level, only one — at the highest dose — died during the study. Such results are very promising, but still quite preliminary.

IV. Biosynthesis

The biosynthesis of streptonigrin is now known in considerable detail, and the antibiotic can be viewed as being constructed *via* a convergent pathway involving the two major units shown by the dotted line in the structure below. The four methyl groups of streptonigrin are unexceptionally derived from methionine (*40, 41*). However, previously unknown pathways are involved in the formation of both the quinoline and pyridine portions of (**1**) (*41, 42*).

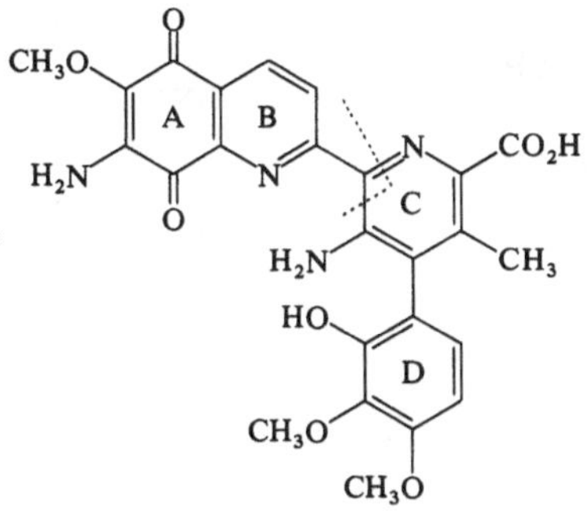

Incorporations of β-^{14}C-tryptophan (**29a**) and β-^{13}C-tryptophan (**29b**) quickly established that this amino acid is the biological precursor to the C and D rings (*41*). Kuhn-Roth oxidation of labeled (**1**) derived from (**29a**), followed by Schmidt degradation of the resulting acetic acid, located all of the radioactivity at C-3' of (**1**). This result was later confirmed by ^{13}C NMR analysis of streptonigrin derived from (**29b**). Other experiments showed that while methionine had labeled the four methyls of (**1**) equally, C-3 of serine — the major precursor to the biological one-carbon pool — labeled the three O-methyls almost exclusively (*41*).

The pathway for formation of the CD-system shown in Scheme 1 was proposed to account for these findings. The unusual pyridine C-ring of streptonigrin with five substituents could be viewed as derived *via* a β-carboline intermediate (**30**) formed as shown from a carboxylic acid AB ring precursor and β-methyltryptophan (**31**). Ring cleavage of a hydroxylated β-carboline intermediate (**32**) would generate the streptonigrin fully substituted pyridine C-ring. Early C-methylation of (**29**) and late O-methylations — perhaps of (**32**) — would account for the observed 3-^{14}C-serine labeling considering that time is needed for metabolism to produce ^{14}CH$_3$-L-methionine.

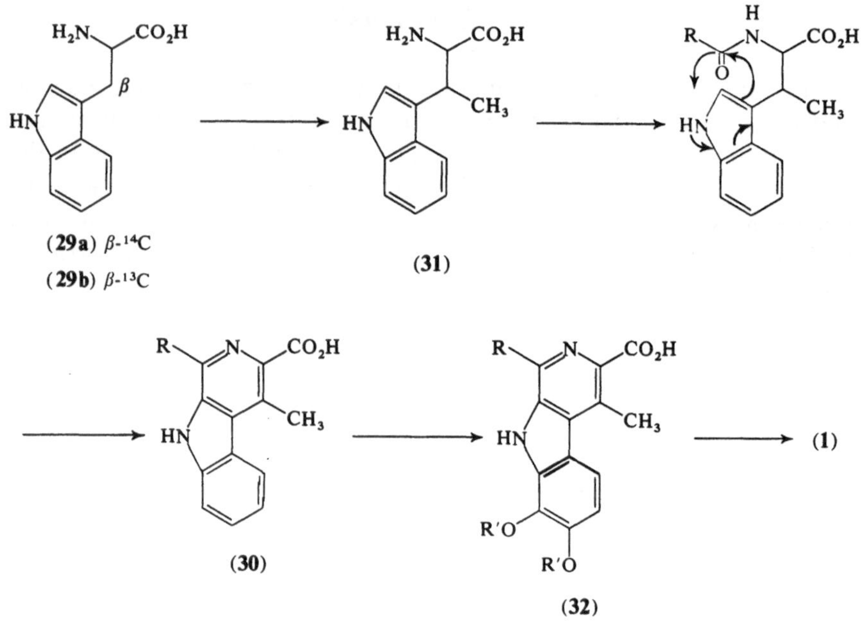

Scheme 1

Support for this proposal has been obtained by the actual isolation of (31) from *S. flocculus* and its reincoporation into (1) (*41, 44*). Additionally, the ^{13}C NMR spectrum of (1) derived from $^{13}CH_3$-L-methionine added to the fermentation at an early stage showed somewhat greater enrichment of the C-methyl compared to the O-methyls, and this enrichment was reversed when the methionine was introduced much later in the fermentation (*41*). (Table 5).

Table 5. *Incorporation of* [$^{13}CH_3$]-*L-Methionine at Different Stages of the Fermentation*

Carbon atom	Chem. shift, ppm from Me$_4$Si[a]	Single pulse S/11'[b]	Double pulse S/11'[b]
3'-CH$_3$	17.8	1.9	0.8
6'-OCH$_3$	56.1	0.8	2.4
9'-OCH$_3$	60.1	1.3	2.2
10'-OCH$_3$	60.5	1.3	2.3
11'	105.1		

[a] Spectra were run in pyridine-d$_5$.
[b] Peak intensities of the enriched methyl signals were normalized to signals in an unenriched spectrum by using the C-11' signal as an internal standard in each spectrum.

The recently characterized antibiotic lavendamycin (**19**) retains the β-carboline of the putative intermediate (**30**) (*43*). It has not yet been established whether (**19**) is an intermediate in the biosynthesis of (**1**) or a shunt metabolite.

Incorporation of $2\text{-}^{13}C,1\text{-}^{15}N$-tryptophan (**29c**) and examination of the ^{13}C NMR spectrum of the derived (**1**) has revealed the direction of β-carboline cleavage (*42, 44*). The presence and location of the ^{15}N label in (**1**) was determined from its spin-coupling ($J_{CN} = 14.7$ Hz) to the enriching ^{13}C at C-5′, demonstrating that it was the original $N_b\text{-}C_{7a}$ bond of tryptophan which had been broken. Such a cleavage reaction of an indole or a β-carboline, either chemically or biochemically, is unprecedented.

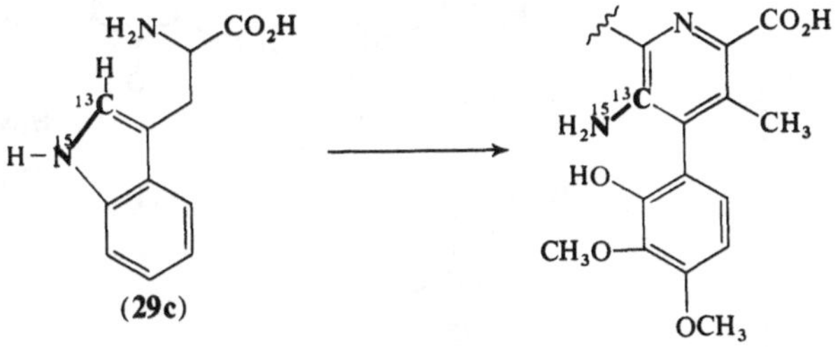

(**29c**)

SPEEDIE (*45*) has obtained a cell-free preparation from *S. flocculus* which catalyzes the formation of (**31**) from L-tryptophan and S-adenosyl-L methionine. The crude enzyme has been purified 2-fold by ammonium sulfate fractionation, and preliminary results with this preparation after dialysis indicated that pyridoxal phosphate is not required, but may cause some stimulation of enzyme activity. At this stage of purification tryptophan transaminase activity was also present, and it has not yet been possible to determine whether the true methylase substrate is an activated tryptophan, or indole pyruvic acid (**33**), as has been demonstrated to be the case in the biosynthesis of indolmycin (**34**) (*46*).

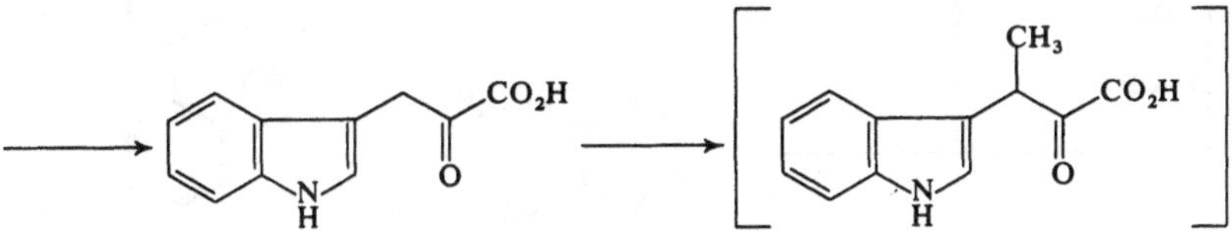

(**33**)

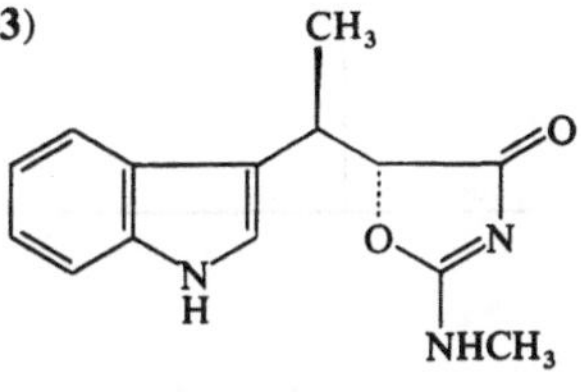

(**34**)

Additional feedings using β-^{14}C,7a-^{14}C-tryptophan, ^{14}COOH-anthra-
nilic acid, U-^{14}C-shikimic acid, and 2-^{14}C-pyruvic acid failed to implicate
any known pathway in the formation of the quinoline portion of
streptonigrin (41, 42). Suspicions that a new pathway for quinoline
biosynthesis was involved have recently been confirmed and the basic
features of the biosynthesis of this portion of streptonigrin have been
determined by analysis of the ^{13}C NMR spectrum of the antibiotic
derived from feeding UL-^{13}C$_6$-D-glucose (35) to S. flocculus (7, 47).

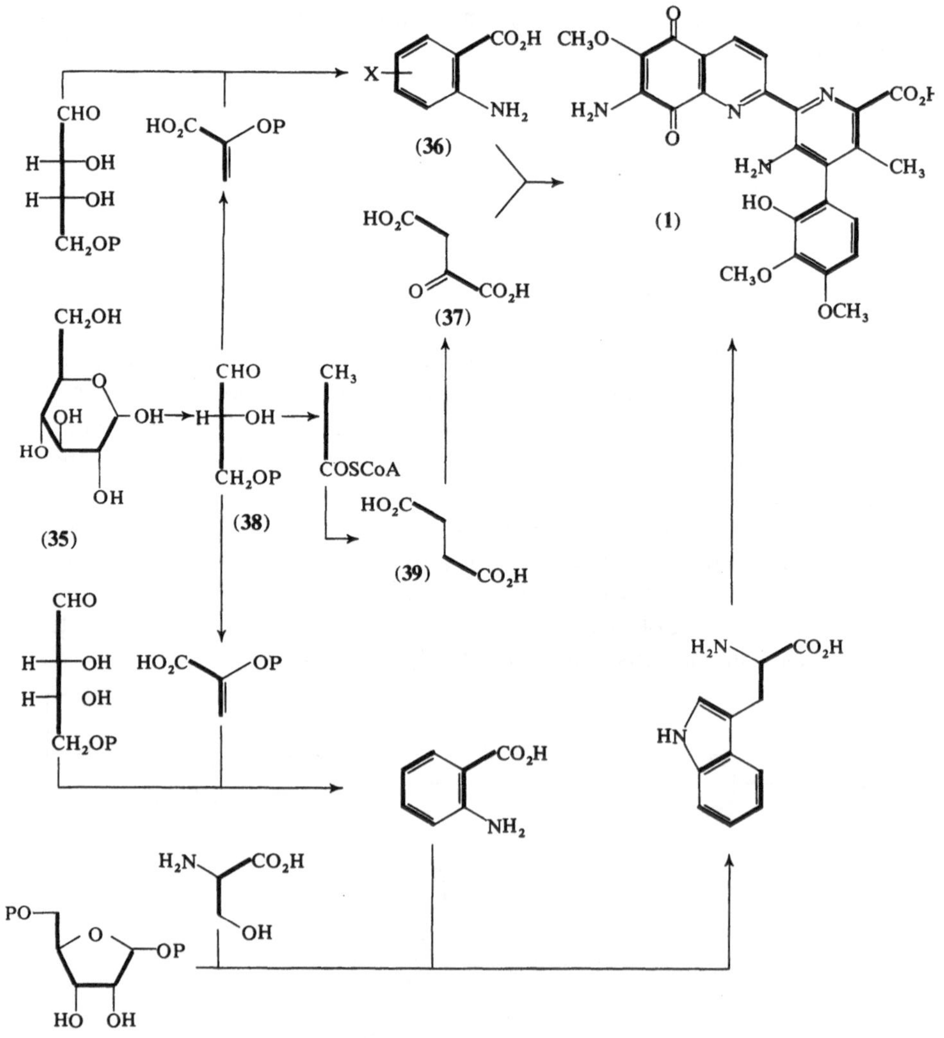

Scheme 2

With glucose (35) as the sole carbon source intermediates in all active metabolic pathways were labeled. Metabolic intermediates derived from (35) and retaining one or more ^{13}C-^{13}C bonds intact can be recognized by the homonuclear spincoupling(s). The size of each such intermediate subsequently utilized in the formation of (1) was determined by the multiplicities of the ^{13}C signals and by matching the observed coupling constants.

The validity of this novel approach for revealing biosynthetic pathways was substantiated by the ease in identifying the well established biosynthesis of tryptophan from the labeling pattern of the CD-system of (35)-derived streptonigrin (1). Both this and the labeling pattern of the AB-rings are outlined in Scheme 2.

The labeling pattern of the non-tryptophan derived portion of (1) is most easily explained by a modified shikimate pathway leading to a substituted anthranilic acid (36) which condenses with a four-carbon diacid [e. g. oxaloacetic acid (37)] derived via the citric acid cycle. The anthranilate carboxyl group is presumably lost in cyclization and aromatization to the quinoline. Since glycolysis would have converted (35) to 1,2-$^{13}C_2$-acetyl CoA (38), its conversion to (37) via the intermediacy of symmetrical succinic acid (39) would then yield the labeling pattern observed for C-6', C-2-3, C-4 of the labeled (1).

In view of the structure of lavendamycin (19), it is likely that intermediate (36) is 4-aminoanthranilic acid and both (1) and (19) contain a m-phenylenediamine (C_6N_2) unit. Since the carbocyclic ring of nybomycin (40) has been shown to be derived via a shikimate pathway (48), it, too, may now be viewed as biogenetically part of the same family. The ionophore A-16239 (41) (49) also has a m-phenylenediamine grouping, but in this case it is part of an isomeric 6-aminoanthranilic acid.

(40)   (41)

V. Synthetic Studies

The combination of unique structural features and high degree of functionalization in streptonigrin have presented a formidable challenge to the synthetic chemist. During the past 15 years a number of research groups have actively worked in this area and have described various synthetic approaches to streptonigrin and/or streptonigrin analogs. These studies have recently culminated in two successful total syntheses of the antibiotic.

In general, preliminary synthetic work dealt with (1) developing methods for constructing a quinoline-quinone AB-ring system having the desired arrangement of A-ring substituents and (2) finding means to build the fully substituted pyridine C-ring. The following discussion will first cover the basic methodology reported for synthesis of some appropriate streptonigrin quinoline-quinone systems and then will turn to the assorted methods developed for synthesis of the requisite penta-substituted pyridines and polycyclic systems. Finally, the aforementioned streptonigrin total syntheses will be described.

In an early series of papers (50, 51) Kametani et al. attempted to prepare some quinoline-quinones having methoxyl and primary amino substituents in the A-ring. However, these workers were only able to make the demethyl compound (43) via the route outlined in Scheme 3. A key step here involved use of a classical Skraup synthesis for formation of the intermediate quinoline (42).

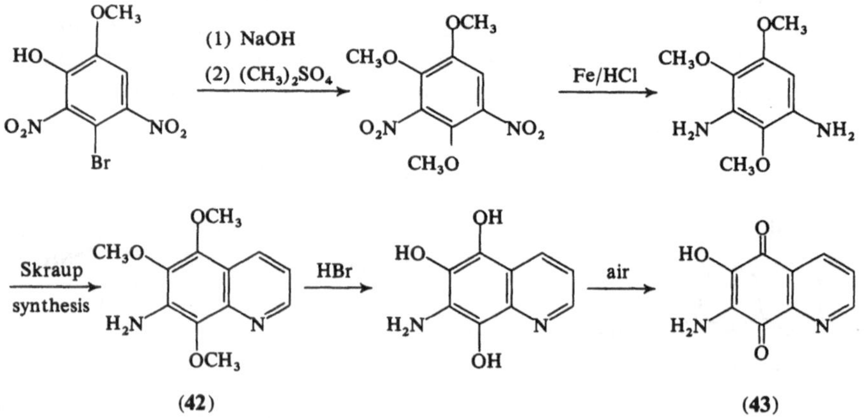

Scheme 3

A more successful and versatile route to the properly substituted AB-ring system of streptonigrin was described by Liao, Nyberg and Cheng (29, 52) (Scheme 4). Quinoline (44), a previously known compound prepared *via* a Skraup synthesis from 2-nitroanisidine, was nitrated to give (**45**), and subsequently converted as shown to methoxyquinone (**46**). This compound could be cleanly brominated to yield bromoquinone (**47**). Attempted replacement of the bromine atom by ammonia did not give the desired aminoquinone (**49**), but rather gave a product derived from displacement of the C-6 methoxyl group, a not surprising result based

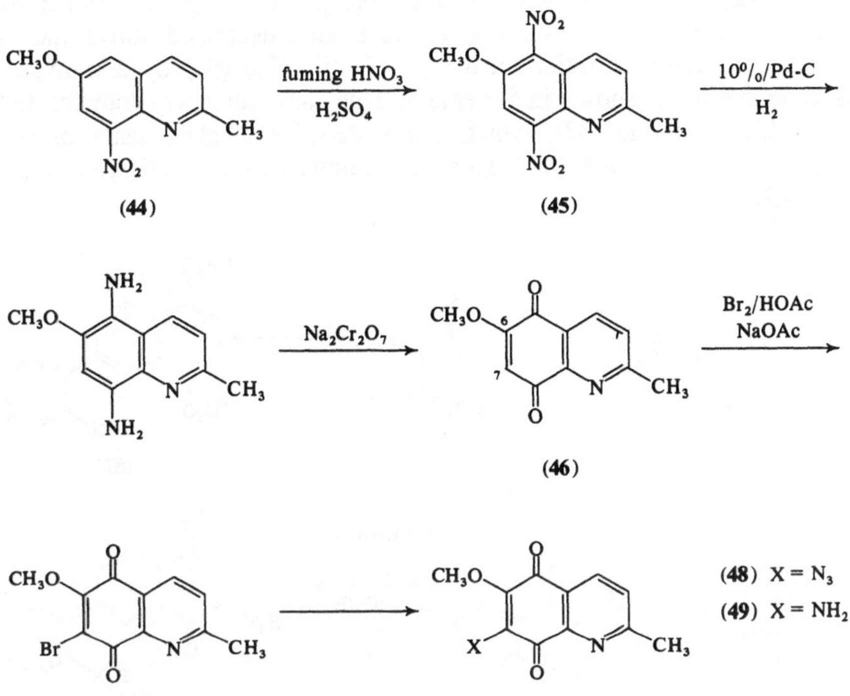

Scheme 4

upon known quinoline-quinone chemistry. Treatment of bromoquinone (**47**) with azide, however, gave the desired azidoquinone (**48**). An interesting explanation was offered by these authors for the product difference in the ammonia *vs.* azide displacements. It was suggested that azide ion initially adds to bromoquinone (**47**) at the C-6 methoxyl-bearing carbon to produce (**50**) which rearranges *via* triazoline (**51**) to the observed 7-azidoquinone (**48**). The strategy developed here for elaboration of the A-ring substituents was later applied by others, and was an important feature of both streptonigrin total syntheses.

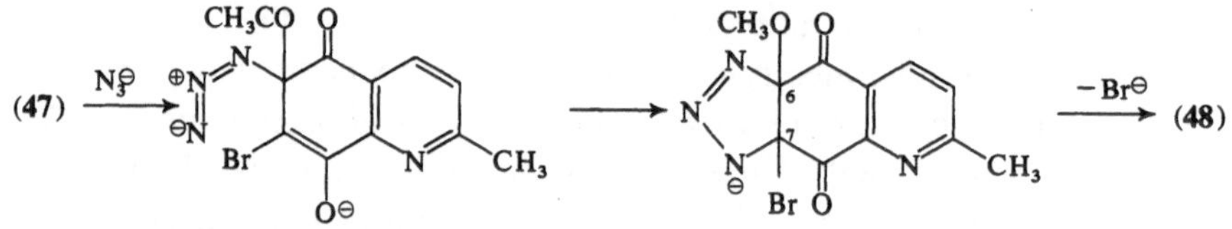

(50)   (51)

LOWN and SIM (53) have prepared several streptonigrin analogues for a study of structure-activity relationships. The approach taken by these workers was quite closely related to that described above and in one series involved initial formation of quinoline (52) by a modified Skraup synthesis, shown in Scheme 5. This compound was transformed to dichloroquinone (53), which upon stepwise displacements of the chlorines by methoxide and by azide, followed by catalytic reduction, gave (54).

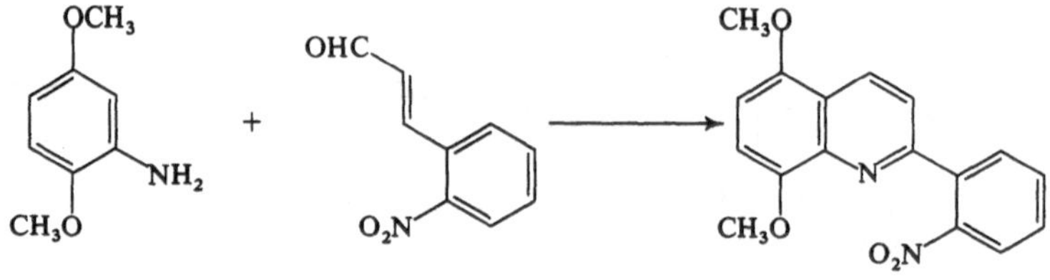

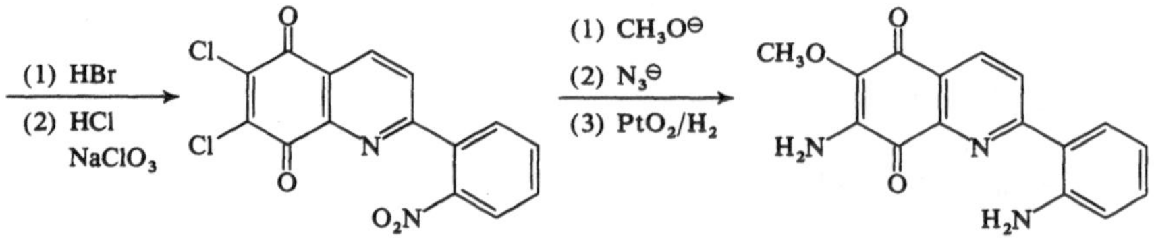

Scheme 5

Several streptonigrin quinoline-quinone analogs have also been synthesized by RAO (16) (Scheme 6) using a modified Friedländer synthesis for construction of the quinoline system. The compound prepared which structurally is most like the natural product is the pyridylquinoline-quinone (59). Chalcone (55), prepared from 3,5-dimethoxybenzaldehyde and 2-acetylpyridine, was nitrated to give (56). Reductive cyclization of (56) with sodium dithionite led to quinoline (57) which upon treatment with nitric acid/sulfuric acid yielded nitroquinone (58) in 50% yield along with 40% of a product of straightforward mononitration. Reduction of the nitro group of (58) and O-methylation with diazomethane afforded (59).

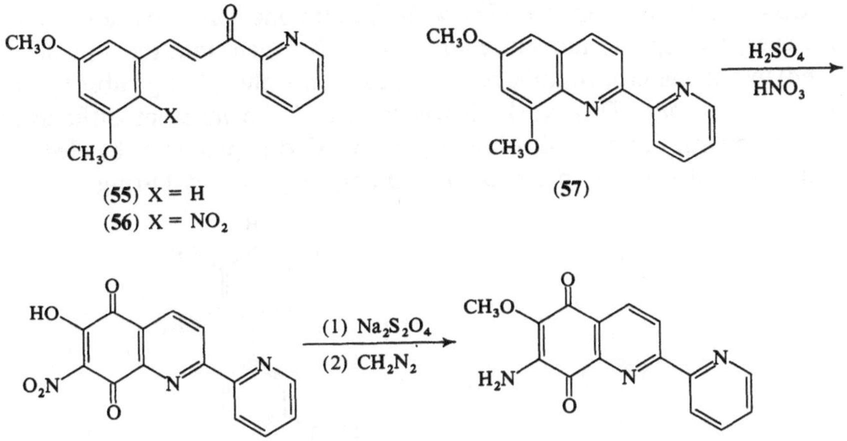

(55) X = H
(56) X = NO$_2$
(57)
(58)
(59)

Scheme 6

HIBINO and WEINREB (*54*) have reported approaches to synthesis of the AB-quinoline-quinone system of (**1**) (Scheme 7). In a classical Friedländer synthesis, the known aminobenzaldehyde (**60**) was combined with 2-acetylpyridine to afford quinoline (**61**) after removal of the sulfonyl protecting group. Fremy's salt oxidation of (**61**) gave methoxyquinone (**62**) which was readily converted to the compound (**59**) previously prepared by RAO (*16*).

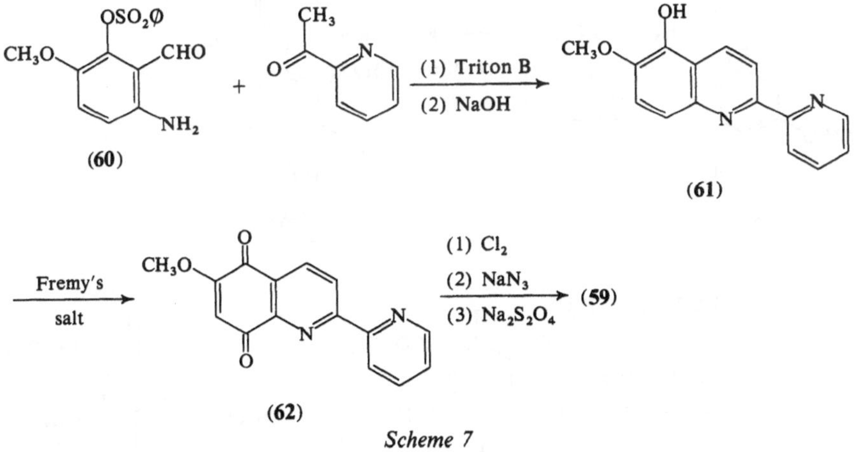

(60)
(61)
(62)

Scheme 7

In a series of publications by KAMETANI and coworkers (*55—62*) a number of routes to systems related to the pyridine C-ring of streptonigrin were described. Most of these routes wer based upon classical methods for pyridine or 2-pyridone ring formation.

Condensation of nitrile (**63**) with 2-butanone gave pyridone (**64**). However, attempts to introduce an amino substituent into the 3-position of (**64**) by various electrophilic reactions gave only phenyl ring-substitution products (*55, 58*). This work clearly indicated the inherent difficulties associated with functionalization of pyridine C-ring precursor by electrophilic chemistry in the presence of a highly oxygenated D-ring.

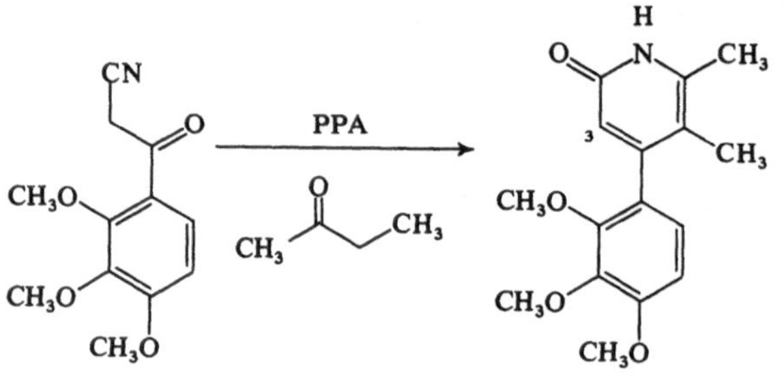

(63)   (64)

A far more successful sequence (*62*) which produced a fully substituted pyridone (**66**) involved ethoxide promoted condensation of (**65**) with cyanoacetamide. Similarly, (**65**) could be combined with (**67**) to give, after methylation, dihydropyridone (**68**) which on dehydrogenation yielded the fully substituted pyridone (**69**).

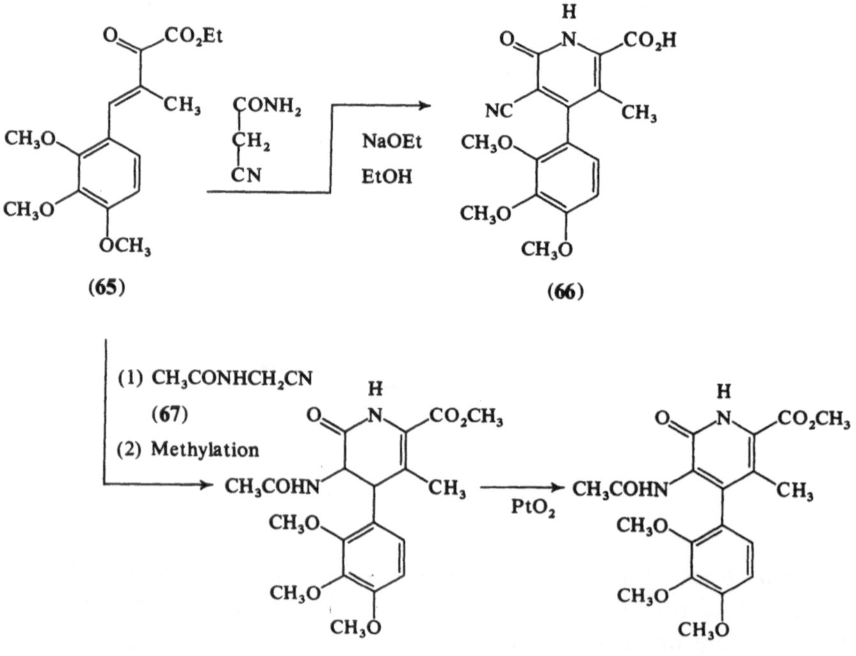

(65)   (66)

(68)   (69)

Both KAMETANI and coworkers (*61*) and LIAO *et al.* (*63*) have described synthesis of pyridone (**72**) which now has the D-ring oxygen functionality differentiated. The former research group began with α,β-unsaturated ketone (**70**); the latter group prepared and utilized diketone (**71**) (Scheme 8), and described subsequent conversion of product (**72**) to chloro-pyridine (**73**).

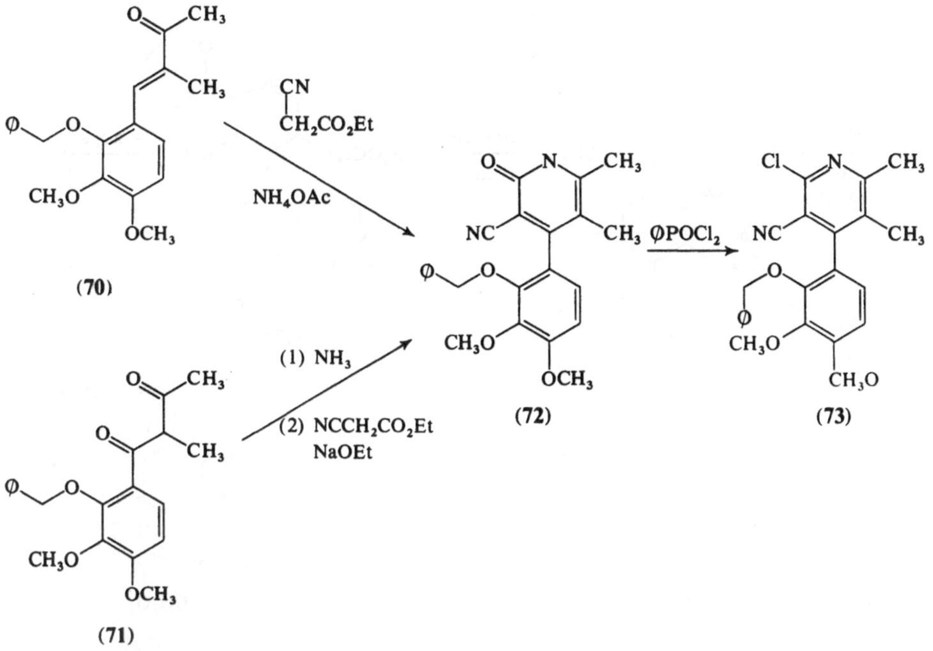

Scheme 8

The LIAO group (*63*) further transformed (**73**) to the streptonigrin CD-ring analog (**78**) (Scheme 9). Catalytic hydrogenation of (**73**) led to removal of both the chlorine atom and the benzyl group to afford (**74**). After reprotection of the phenol, it was possible to oxidize the 2-methyl group selectively with selenium dioxide to the corresponding aldehyde, which was protected as the acetal (**75**). Hydrolysis of (**75**) to amide (**76**), and subsequent Hoffmann rearrangement afforded the desired aminopyridine derivative (**77**), which was converted in a few simple steps to (**78**).

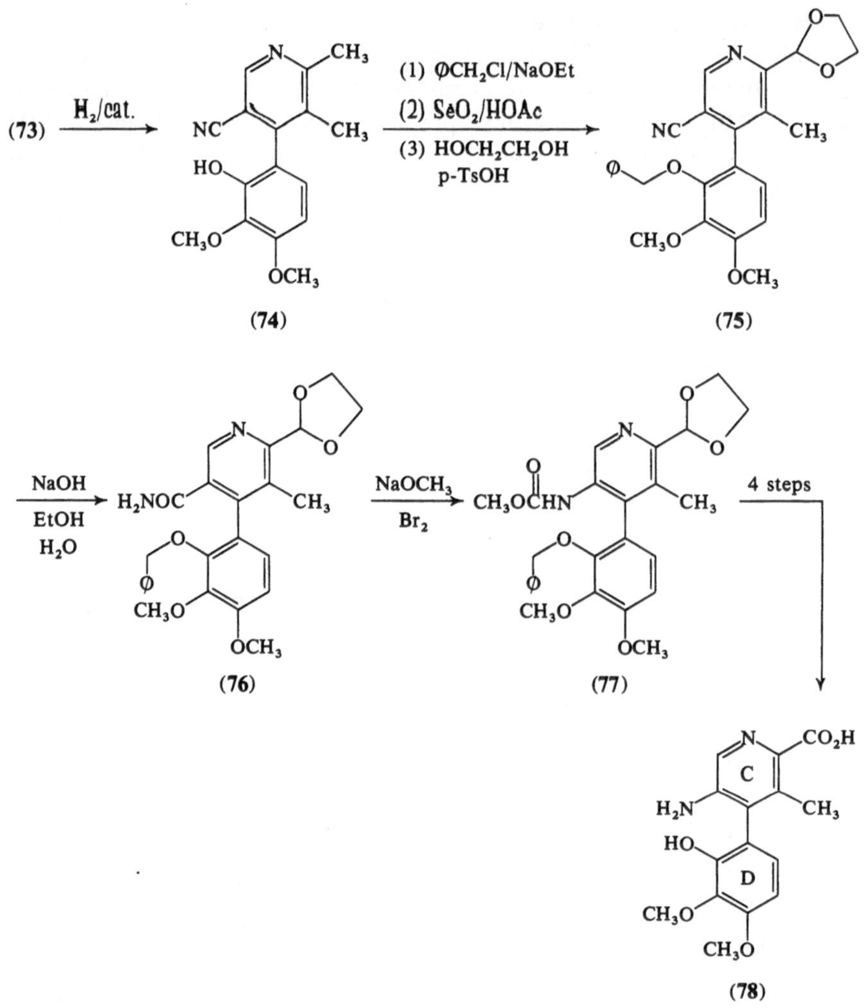

Scheme 9

A second and more efficient synthesis of (78) has been reported (64) which uses an Ullmann coupling (Scheme 10) to construct the CD-biaryl system. This mixed coupling between readily available compounds (79) and (80) could be effected in fair yield to produce (81). Compound (78) was prepared in three steps from this intermediate.

Several publications by Rao and coworkers (18, 65, 66) have reported syntheses of various streptonigrin analogs. A key pyridine intermediate used in this work was the system (82), lacking the streptonigrin D-ring, which was prepared via the straightforward sequence shown in Scheme 11.

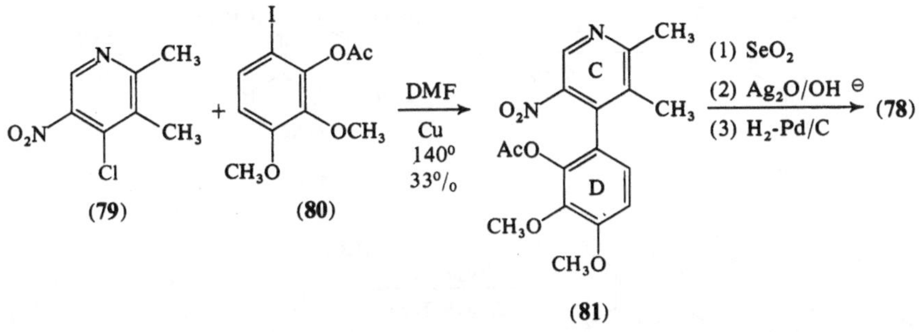

Scheme 10

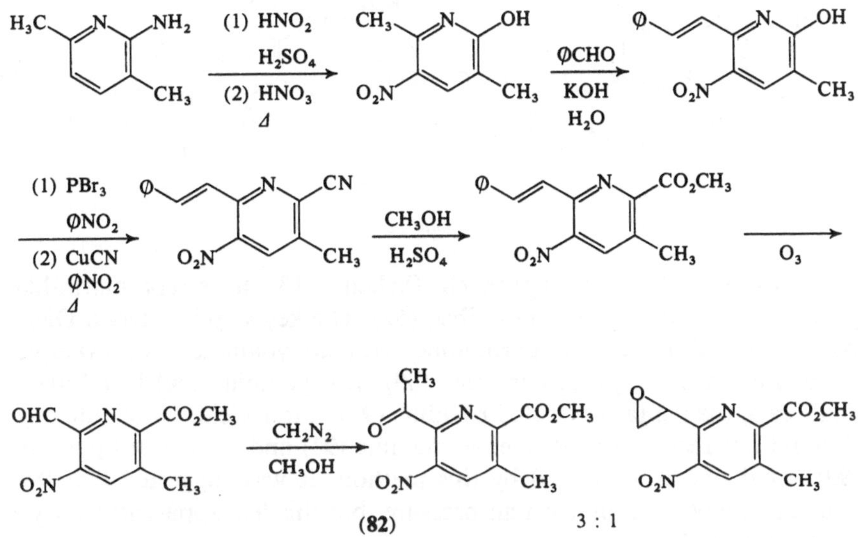

Scheme 11

Acetylpyridine (82) has been used to synthesize some tricyclic ABC-ring streptonigrin analogs, in particular compound (88) (*18, 66*) (Scheme 12). One of the more interesting of several reported routes to (88) began with commercially available acid (83) which was readily converted to nitroaldehyde (84). Condensation of (84) with the acetylpyridine (82) gave chalcone (85), and hydrosulfite reduction led to quinoline (86). Fremy's salt oxidation of (86) yielded quinoline-quinone (87), and the A-ring of this molecule was then properly elaborated using previously established methodology to give tricyclic compound (88).

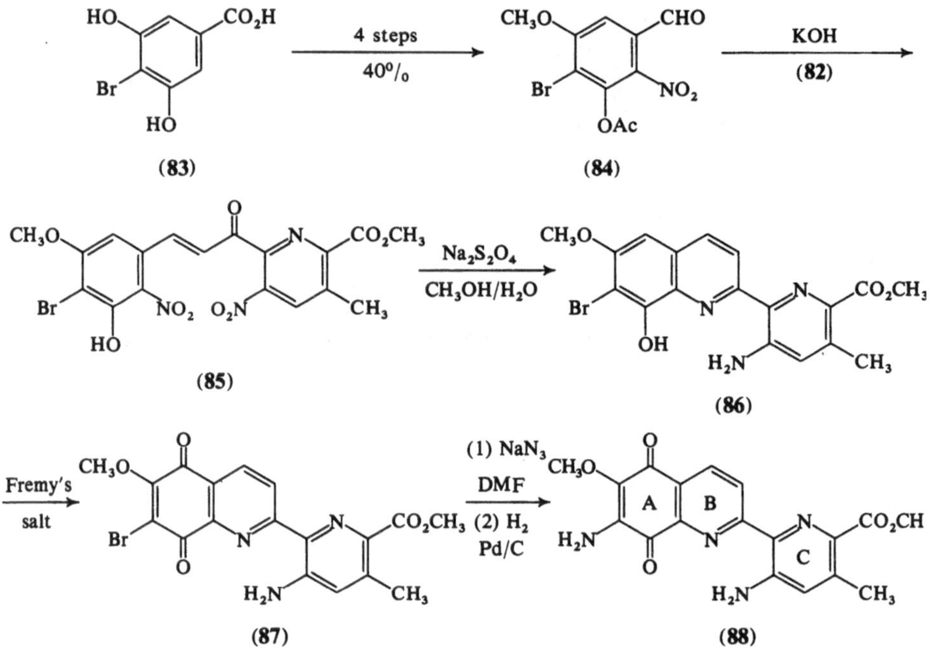

Scheme 12

A rather interesting approach (Scheme 13) to streptonigrin-like pyridines has recently been described (*67*). The key step involves a Diels-Alder cycloaddition of a pyrimidine with an ynamine. For example, condensation of dicyanopyrimidine (**89**) with ynamine (**90**) in refluxing THF gave the pentasubstituted pyridine (**92**) in fair yield. Unfortunately, (**89**) did not react with the phenyl ynamine (**91**) and the desired pyridine (**93**) could not be produced by this method. It was suggested that (**89**) and (**91**) might react under high pressure, but this has apparently not yet been attempted.

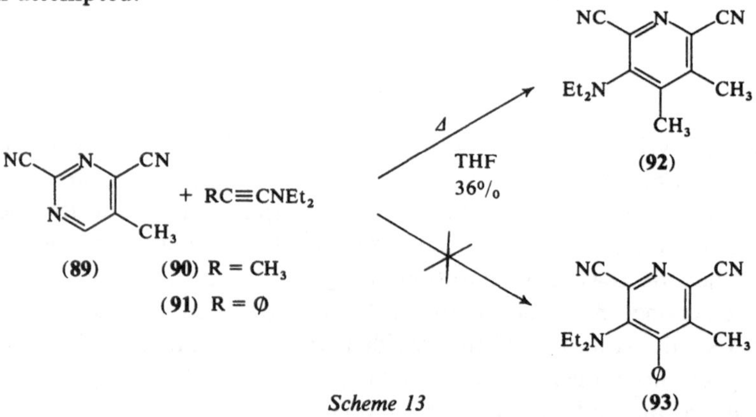

Scheme 13

KAMETANI, OGASAWARA and KOZUKA (56) were the first to synthesize a tetracyclic system related to streptonigrin. In a Hantzsch-type cyclization (Scheme 14) enamino-nitrile (94) was combined with unsaturated ketone (95) to give an intermediate dihydropyridine which could be oxidized to tetracyclic pyridine (96). Using a strategy for C-ring amino group introduction seen in some previous studies, the nitrile group of (96) was first converted to the corresponding amide, which successfully underwent a Hoffmann rearrangement to afford aminopyridine (97).

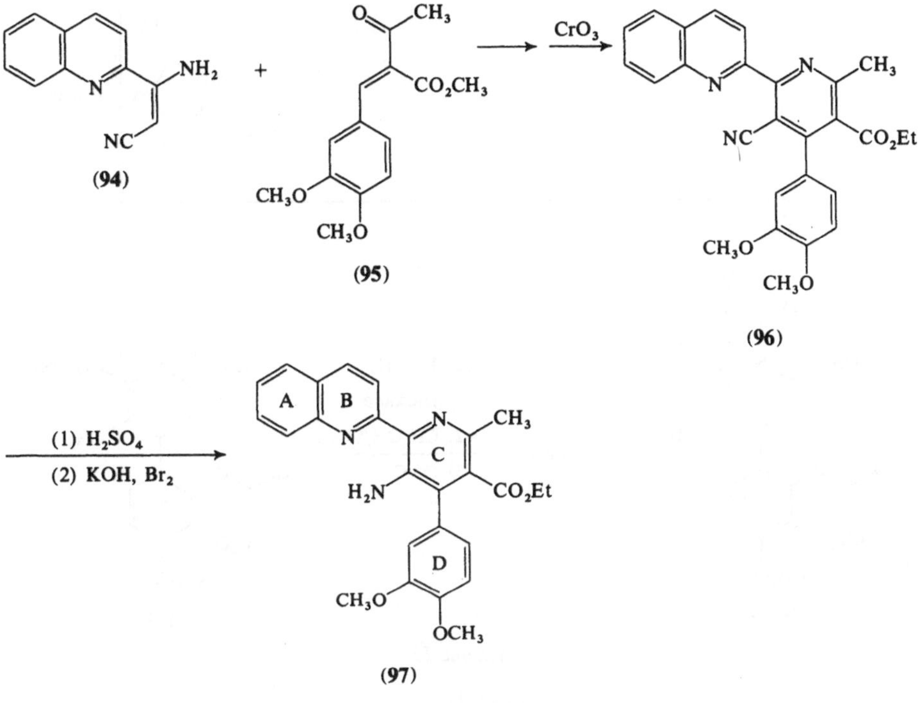

Scheme 14

After extensive preliminary studies (68, 69), WEINREB and coworkers completed the first total synthesis of streptonigrin in 1980 ref. (70). The central strategy in their approach involved the use of an imino Diels-Alder reaction for construction of the CD framework. The quinoline system was ultimately synthesized by a new modification of the Friedländer condensation after complete elaboration of all of the necessary pyridine substituents.

Readily available aldehyde (98) could be converted to α,β-unsaturated aldehyde (99) in three simple steps. This compound on treatment with ethylidene triphenylphosphorane at low temperature, followed by n-butyl

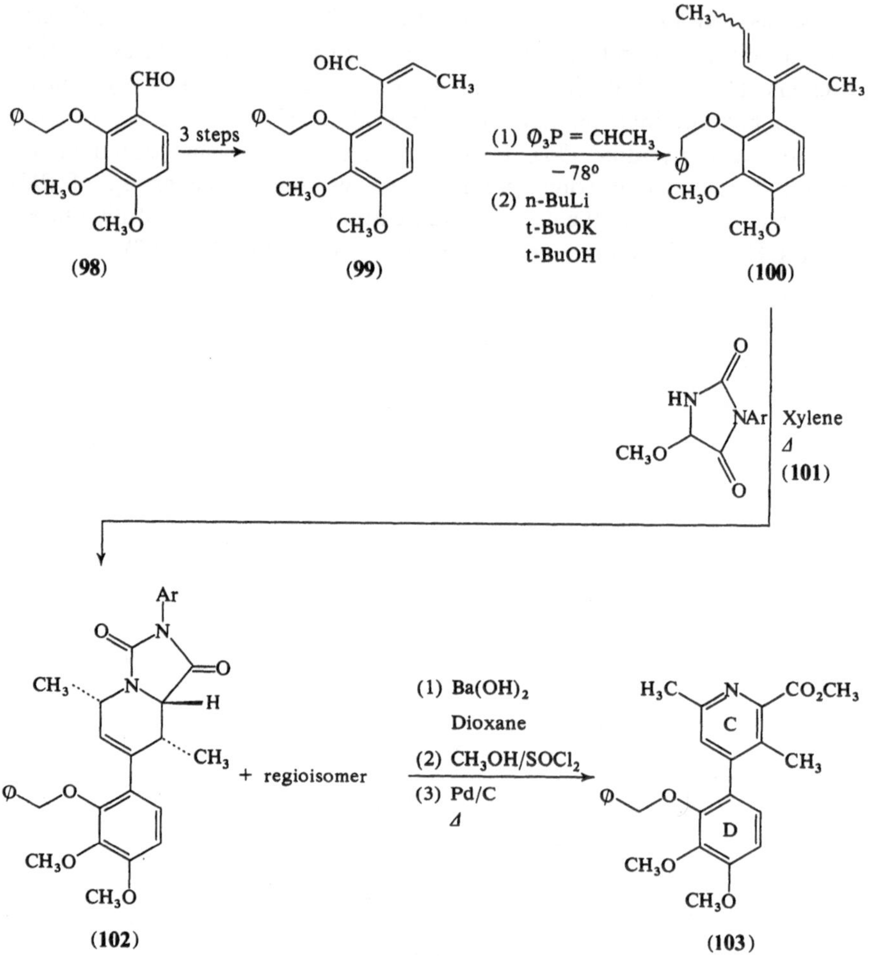

Scheme 15

lithium and potassium *t*-butoxide (Schlosser procedure) gave diene (**100**) as an inseparable 2.5/1 mixture of *trans/cis* isomers. The mixture of dienes reacted with methoxyhydantoin (**101**) in refluxing xylene to afford adduct (**102**) as the major product, along with some of the corresponding Diels-Alder regioisomer. Without separation, this mixture of adducts was transformed in three steps to the key tetrasubstituted pyridine (**103**) (Scheme 15).

The next stage of the synthesis was concerned with introduction of the missing amino substituent into the vacant position in the pyridine ring of (**103**). Since direct electrophilic substitution of this pyridine ring in the presence of the highly oxygenated D-ring did not appear feasible (*cf. 55, 58*), a more circuitous route was necessary. Thus, pyridine (**103**) was oxidized with m-chloroperoxybenzoic acid to the corresponding N-oxide, which underwent a Polonovsky rearrangement on heating in acetic anhydride to afford acetate (**104**). This compound was converted as shown to quaternary salt (**105**) (Scheme 16), which on treatment with potassium t-butoxide under very carefully controlled conditions, followed by hydrolysis of the initial [2,3]-sigmatropic rearrangement product, yielded aldehyde (**106**). N-Oxide acid (**107**) was then prepared from (**106**), and the Yamada modification of the Curtius rearrangement served to produce amine (**108**).

(**103**) $\xrightarrow[\substack{(2)\ Ac_2O \\ \Delta}]{(1)\ m\text{-CPBA}}$ (**104**) $\xrightarrow[\substack{(2)\ SOCl_2 \\ (3)\ \text{pyrrolidine-}CN,\ DMSO}]{(1)\ K_2CO_3/CH_3OH}$ (**105**)

(**104**) $\xrightarrow[\substack{(1)\ t\text{-BuOK} \\ DMSO,\ THF \\ -12^\circ \\ (2)\ (CO_2H)_2 \\ H_2O}]{}$ (**106**) $\xrightarrow[\substack{(1)\ KMnO_4 \\ (2)\ m\text{-CPBA}}]{}$ (**107**) $\xrightarrow[\substack{(1)\ (\varnothing O)_2PON_3 \\ NEt_3 \\ (2)\ H_2O}]{}$

(**108**) $\xrightarrow[\substack{(1)\ Ac_2O,\ \Delta \\ (2)\ K_2CO_3/CH_3OH \\ (3)\ MnO_2/CHCl_3}]{}$ (**109**) $\xrightarrow[\substack{(1)\ CH_2PO(OCH_3)_2 \\ THF/HMPA \\ (2)\ MnO_2/CHCl_3}]{}$

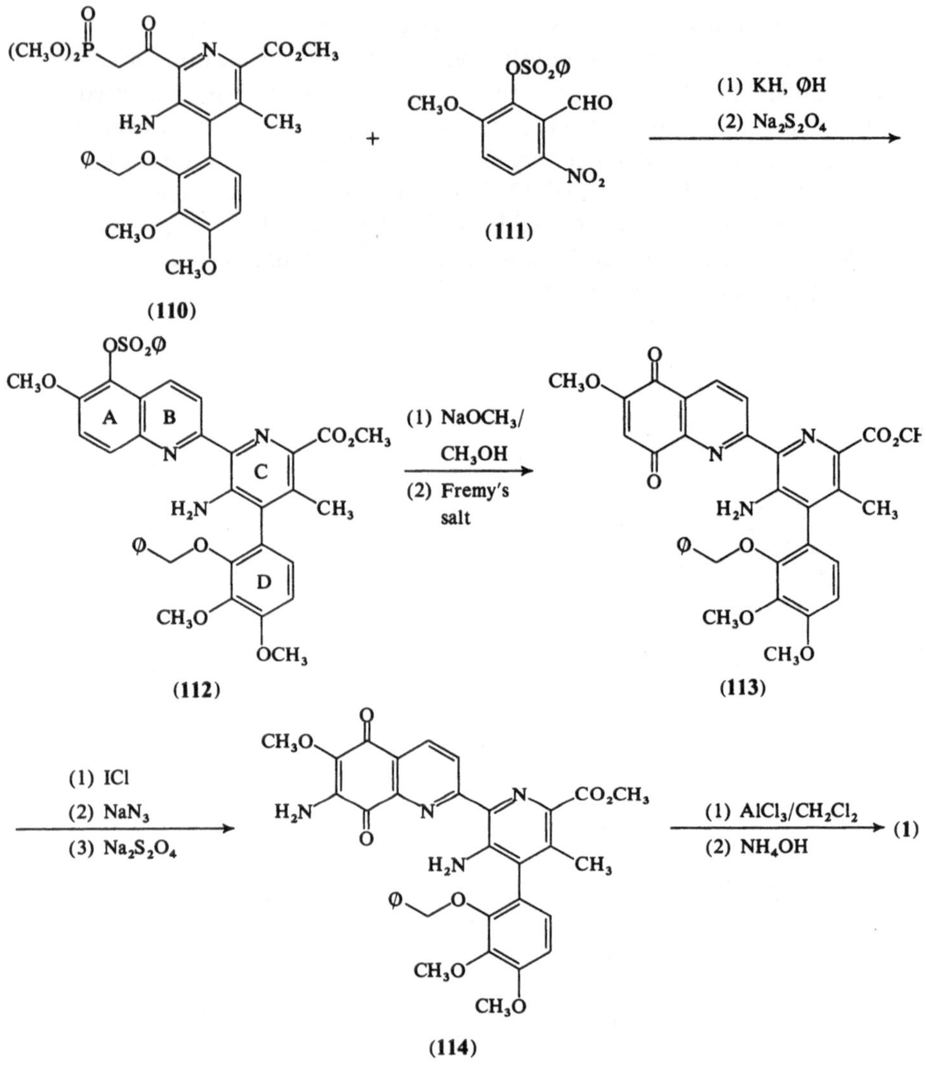

(110)   +   **(111)**   (1) KH, ØH   (2) Na₂S₂O₄

(1) NaOCH₃/CH₃OH   (2) Fremy's salt

(112)   **(113)**

(1) ICl   (2) NaN₃   (3) Na₂S₂O₄

(1) AlCl₃/CH₂Cl₂   (2) NH₄OH   **(1)**

(114)

Scheme 16

The next sequence of steps was directed towards generating a "handle" for attachment of this pyridine CD system to a quinoline AB framework. Compound (**108**) was therefore transformed to β-ketophosphonate (**110**) in five additional steps *via* aldehyde (**109**). Condensation of the anion derived from (**110**) with nitroaldehyde (**111**) gave an intermediate chalcone, which on reductive cyclization with sodium hydrosulfite afforded the tetracyclic system (**112**).

Functionality in the A-ring of (**112**) was properly elaborated using methodology developed in model studies (*54*). Cleavage of the sulfonate protecting group of (112) to the corresponding phenol, and Fremy's salt oxidation led to quinoline-quinone (113). The A-ring amino substituent was introduced by the three step sequence shown, affording the amino-quinone (114). Removal of the O-benzyl group of (114) was achieved with aluminum chloride to give streptonigrin methyl ester, which was hydrolyzed with ammonium hydroxide to material identical with the natural product (1).

Recently another total synthesis of streptonigrin has been completed by KENDE, LORAH and BOATMAN (71) based upon some preliminary model studies (17). As in the first total synthesis described above, an appropriately substituted pyridine CD-fragment was initially prepared, and the quinoline system was annulated onto this piece at a latter stage using a modified Friedländer approach.

Ketoenamine (115), which had previously been used in studies by LAO, WITTEK and CHENG (cf. Scheme 8) was condensed with ethyl acetoacetate to produce acylpyridone (116) (Scheme 17). Reduction of the acyl group to the alcohol and further treatment with ØPOCl₂ simultaneously dehydrated this alcohol to the vinyl group and converted the pyridone to the chloropyridine (117). Cuprous cyanide served to replace the chlorine with a cyano group giving (118) which was transformed to the methyl ketone (119) with methylmagnesium bromide.

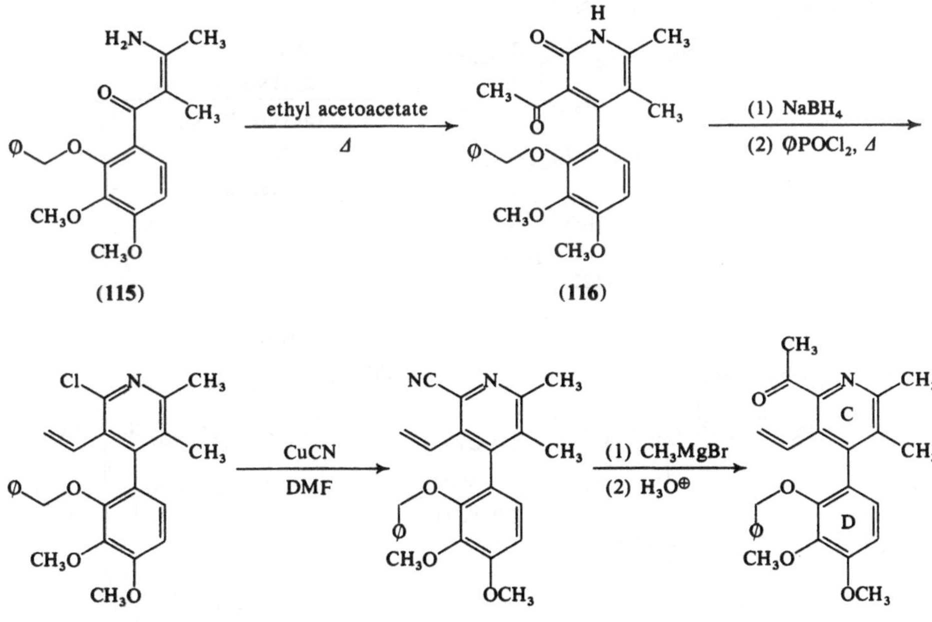

Scheme 17

The precursor **(121)** for the A-ring was prepared in three steps from aldehyde **(120)**. This compound was condensed with ketone **(119)** in a Borsche modification of the Friedländer reaction to give tetra-cyclic quinoline **(122)** (Scheme 18). Selective cleavage of the A-ring p-methoxybenzyl protecting group allowed further transformation to nitro compound **(123)**.

The next series of reactions served to establish the necessary C-ring functionality. Cleavage of the vinyl group of **(123)** led to acid **(124)**. Selenium dioxide was found to oxidize the 2-methyl group of **(124)** giving an aldehyde which was further oxidized to the acid, and selectively esterified to provide ester acid **(125)**. The Yamada modification of the Curtius rearrangement was used to prepare amine **(126)** from acid **(125)**. Finally, the A-ring nitro group of **(126)** was reduced to the amine with sodium hydrosulfite, and Fremy's salt oxidation of this material gave methoxyquinone **(113)** identical with that prepared by the Weinreb group.

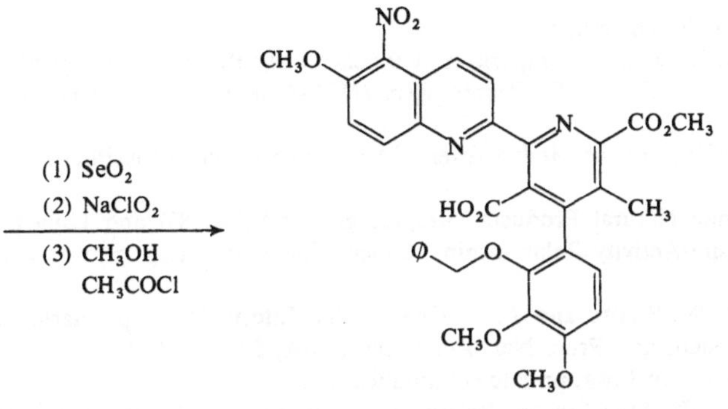

Scheme 18

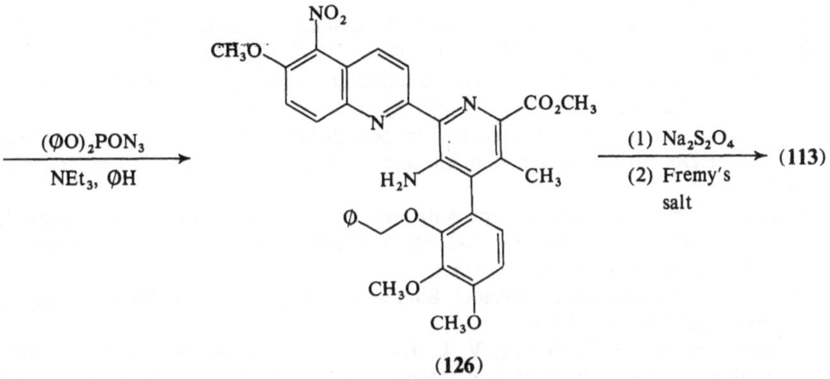

References

1. RAO, K. V., and W. P. CULLEN: Streptonigrin, An Antitumor Substance I. Isolation and Characterization. Antibiot. Annu. 950 (1959—1960).
2. KUDRINA, E. S., O. L. OLKHOVATOVA, L. I. MURAV'EVA, and G. F. GAUZE: Systematic Position and Variation of the Organism Producing Bruneomycin, an Antitumor Antibiotic. Antibiotiki 11, 400 (1966). BRAZHNIKOVA, M. G., V. I. PONOMARENKO, I. N. KOVSHAROVA, E. B. KRUGLYAK, and V. V. PROSHLYAKOVA: Study on Bruneomycin Produced by Act. albus var. Bruneomycini and its Identification with Streptonigrin. Antibiotiki 13, 99 (1968).
3. Société des usines chimiques Rhône-Poulenc, Brit. Pat. 872, 261, July 5, 1961; Chem. Abstr. 55, p25158a (1961).
4. RAO, K. V., K. BIEMANN, and R. B. WOODWARD: The Structure of Streptonigrin. J. Amer. Chem. Soc. 85, 2532 (1963).
5. GOULD, S. J.: Unpublished results.
6. LOWN, J. W., and A. BEGLEITER: Studies Relating to Aziridine Antitumor Antibiotics. Part II. ^{13}C and ^{1}H Nuclear Magnetic Resonance Spectra of Mitomycin C and Structurally Related Streptonigrin. Canad. J. Chem. 52, 2331 (1974).
7. GOULD, S. J., and D. E. CANE: Unpublished results.
8. CHIU, Y.-Y., and W. N. LIPSCOMB: Molecular and Crystal Structure of Streptonigrin. J. Amer. Chem. Soc. 97, 2525 (1975).

9. DOYLE, T. W.: Unpublished results.
10. KREMER, W. B., and J. LASZLO: Comparison of Biochemical Effects of Isopropylidine Azastreptonigrin (NSC-62709) with Streptonigrin (NSC-45383). Cancer Chemother. Rep. **51**, 19 (1967).
11. — — Biochemical Effects of the Methyl Ester of Streptonigrin. Biochem. Pharmacol. **15**, 1111 (1966).
12. RAO, K. V.: Quinone Natural Products. Streptonigrin (NSC-45383) and Lapachol (NSC-11905) Structure-Activity Relationships. Cancer Chemother. Rep. part 4, **4**, 11 (1974).
13. COHEN, M. M., M. W. SHAW, and A. P. CRAIG: The Effects of Streptonigrin on Cultured Human Leucocytes. Proc. Nat. Acad. Sci. (USA) **50**, 16 (1963).
14. ROSAZZA, J.: University of Iowa, private communication.
15. LOWN, J. W., and S.-K. SIM: Studies Related to Antitumor Antibiotics. Part VII. Synthesis of Streptonigrin Analogs and Their Single Strand Scission of DNA. Canad. J. Chem. **54**, 2563 (1976).
16. RAO, K. V.: Streptonigrin and Related Compounds, I. Some 2-Phenyl- and 2,2-Pyridyiquinoline-5,8-diones. J. Heterocycl. Chem. **12**, 725 (1975).
17. KENDE, A. S., and P. C. NAEGELY: Total Synthesis of the Streptonigrin Quinone Carbon Framework. Tetrahedron Letters 4775 (1978).
18. RAO, K. V.: Streptonigrin and Related Compounds III. Synthesis and Microbiological Activity of Destrioxyphenylstreptonigrin and Analogs. J. Heterocycl. Chem. **14**, 653 (1977).
19. KREMER, W. B., and J. LASZLO: In: Antineoplastic and Immunosupressive Agents II, Handb. Exp. Pharm. 38/2, pp. 633 641. eds. A. C. SARTORELLI and D. G. JOHNS. Berlin-Heidelberg-New York: Springer. 1975.
20. MIZUNO, N. S., and D. P. GILBOE: Binding of Streptonigrin to DNA. Biochem. Biophys. Acta **224**, 319 (1970).
21. DUDNIK, YU. V., G. G. GAUZE, V. L. KARPOV, L. I. KOZMYAN, and E. PADRON: Interaction in vitro of Bruneomycin (Streptonigrin) with DNA. Antibiotiki. **18**, 968 (1973); Chem. Abstr. **80**; 105000e.
22. WHITE, H. L., and J. R. WHITE: Lethal Action and Metabolic Effects of Streptonigrin in Escherichia coli. Mol. Pharmacol. **4**, 549 (1968).
23. RAO, K. V.: Interaction of Streptonigrin with Metals and with DNA. J. Pharm. Sci. **68**, 853 (1979).
24. CONE, R., S. K. HASAN, J. W. LOWN, and A. R. MORGAN: The Mechanism of the Degradation of DNA by Streptonigrin. Canad. J. Biochem. **54**, 219 (1976).
25. WHITE, J. R.: Streptonigrin-Transition Metal Complexes: Binding to DNA and Biological Activity. Biochem. Biophys. Res. Comm. **77**, 387 (1977).
26. LOWN, J. W., and S.-K. SIM: Studies Related to Antitumor Antibiotics. Part VIII. Cleavage of DNA by Streptonigrin Analogs and the Relationship to Antineoplastic Activity. Canad. J. Biochem. **54**, 446 (1976).
27. HAJDU, J., and E. C. ARMSTRONG: Interaction of Metal Ions with Streptonigrin. 1. Formation of Copper (II) and Zinc (II) Complexes of the Antitumor Antibiotic. J. Amer. Chem. Soc. **103**, 232 (1981).
28. HAJDU, J.: Private communication.
29. LIAO, T. K., W. H. NYBERG, and C. C. CHENG: Synthetic Studies of the Antitumor Antibiotic Streptonigrin. I. Synthesis of the A-B Ring Portion of Streptonigrin. J. Heterocycl. Chem. **13**, 1063 (1976).
30. ISHIZU, K., H. H. DEARMAN, M. T. HUANG, and J. R. WHITE: Electron Paramagnetic Resonance Observations on Biogenic Semiquinone and 5-Methyl Phenazinium Radicals. Biochem. Biophys. Acta **165**, 283 (1968).
31. WHITE, J. R., and H. H. DEARMAN: Generation of Free Radicals from Phenazine

Methosulfate, Streptonigrin, and Rubiflavin in Bacterial Suspensions. Proc. Nat. Acad. Sci. (USA) **54**, 887 (1965).

32. LOWN, J. W., S.-K. SIM, and H.-H. CHEN: Hydroxyl Radical Production by Free and DNA-Bound Aminoquinone Antibiotics and its Role in DNA Degradation. Electron-Spin Resonance Detection of Hydroxyl Radicals by Spin Trapping. Canad. J. Biochem. **56**, 1042 (1978).

33. BACHUR, N. R., S. L. GORDON, M. V. GEE, and H. KON: NADPH Cytochrome P-450 Reductase Activation of Quinone Anticancer Agents to Free Radicals. Proc. Nat. Acad. Sci. (USA) **76**, 954 (1979).

34. See reference *32*, note 27.

35. KUANG, D. T., R. M. WHITTINGTON, H. H. SPENSER, and M. E. PATNO: Comparison of Chlorambucil and Streptonigrin (NSC 45383) in the Treatment of Chronic Lymphocytic Leukemia. Cancer **23**, 597 (1969).

36. NISSEN, N. I., T. PAJAK, O. GLIDEWELL, H. BLOM, M. FLAHERTY, D. HAYES, R. MCINTYRE, and J. F. HOLLAND: Overview of Four Clinical Studies of Chemotherapy for Stage III and Stage IV Non-Hodgkin's Lymphomas by the Cancer and Leukemia Group B. Cancer Treatment Reports **61**, 1097 (1977).

37. FORCIER, R. J., O. R. MCINTYRE, N. I. NISSEN, T. F. PAJAK, O. GLIDEWELL, and J. F. HOLLAND: Combination Chemotherapy of Non-Hodgkin Lymphoma. Med. and Pediatr. Oncol. **4**, 351 (1978).

38. BANZET, P., C. JACQUILLAT, J. CIVATTE, A. PUISSANT, J. MARAL, C. CHASTANG, L. ISRAEL, S. BELAICH, J. C. JOURDAIN, M. WEIL, and G. AUCLERC: Adjuvant Chemotherapy in the Management of Primary Malignant Melanoma. Cancer **41**, 1240 (1978).

39. GOUT-LEMERLE, M., C. RODARY, and D. SARRAZIN: Arch. Fr. Pediatr. **33**, 527 (1976).

40. KARPOV, V. L., and L. G. ROMANOVA: Carbon and Tritium-Labeled Bruneomycin Obtained by Biosynthetic Method. Antibiotiki **17**, 419 (1972).

41. GOULD, S. J., and C. C. CHANG: Streptonigrin Biosynthesis. 3. Determination of the Primary Precursors to the 4-Phenylpicolinic Acid Portion. J. Amer. Chem. Soc. **102**, 1702 (1980).

42. GOULD, S. J., C. C. CHANG, D. S. DARLING, J. D. ROBERTS, and M. SQUILLACOTE: Streptonigrin Biosynthesis. 4. Details of the Tryptophan Metabolism. J. Amer. Chem. Soc. **102**, 1707 (1980).

43. DOYLE, T. W., and S. J. GOULD: Unpublished results.

44. GOULD, S. J., and C. C. CHANG: Studies of Nitrogen Metabolism Using ^{13}C NMR Spectroscopy. 1. Streptonigrin Biosynthesis. J. Amer. Chem. Soc. **100**, 1624 (1978).

45. SPEEDIE, M. K.: University of Maryland, private communication.

46. HORNEMANN, U., L. H. HURLEY, M. K. SPEEDIE, and H. G. FLOSS: The Biosynthesis of Indolmycin. J. Amer. Chem. Soc. **93**, 3028 (1971).

47. This technique was first used to study the sesquiterpene antibiotic pentalenolactone: CANE, D E., T. ROSSI, and J. P. PACHLATKO: The Biosynthesis of Pentalenolactone. Tetrahedron Letters, 3639 (1979).

48. NADZAN, A. M., and K. L. RINEHART, JR.: Nybomycin. 8. Biosynthetic Origin of the Central Ring Carbons Studied by ^{13}C-Labeled Substrates. J. Amer. Chem. Soc. **98**, 5012 (1976).

49. ZMIJEWSKI, M. J.: Biosynthesis of Antibiotic A23187. Incorporation of Precursors into A23187. J. Antibiotics **33**, 447 (1980).

50. KAMETANI, T., and K. OGASAWARA: Streptonigrin and Related Compounds. I. Syntheses of 5,6,8-Trimethoxy-7-dimethylaminoquinoline and 7-Amino-6-hydroxy-5,8-quinoline-dione. Yakugaku Zasshi **85**, 985 (1965).

51. — — Streptonigrin and Related Compounds II. Syntheses of 7-Aminoquinoline Derivatives from Hexachlorocyclohexane. Yakugaku Zasshi **86**, 55 (1966).

52. Liao, T. K., W. H. Nybert, and C. C. Cheng: Synthesis of 7-Amino-6-methoxy-5,8-quinolinedione. Angew. Chem. Int. Ed. **6**, 82 (1967).

53. Lown, J. W., and S.-K. Sim: Studies Related to Antitumor Antibiotics. Part VII. Synthesis of Streptonigrin Analogues and Their Single Strand Scission of DNA. Canad. J. Chem. **54**, 2563 (1976).

54. Hibino, S., and S. M. Weinreb: Synthetic Approaches to the Quinolinequinone System of Streptonigrin. J. Organ. Chem. (USA) **42**, 232 (1977).

55. Kametani, T., K. Ogasawara, and M. Shio: Streptonigrin and Related Compounds. III. Syntheses of 4-Phenylpyridine Derivatives. Yakugaku Zasshi **86**, 809 (1966).

56. Kametani, T., K. Ogasawara, and A. Kozuka: Streptonigrin and Related Compounds. IV. Syntheses of 4-(3,4-Methylenedioxyphenyl)- and 4-(3,4-Dimethoxyphenyl)-3-cyano-5-ethoxycarbonyl-6-methyl-2-quinolylpyridine. Yakugaku Zasshi **86**, 815 (1966).

57. Kametani, T., K. Ogasawara, A. Kozuka, and M. Shio: Streptonigrin and Related Compounds. V. Syntheses of the Compounds having Streptonigrin-type Structure. Yakugaku Zasshi **87**, 254 (1967).

58. Kametani, T., K. Ogasawara, M. Shio, and A. Kozuka: Streptonigrin and Related Compounds. VI. The NMR Spectra of 4-(2,3,4-Trimethoxyphenyl)-2,3-dimethylpyridine Derivatives. Yakugaku Zasshi **87**, 260 (1967).

59. Kametani, T., K. Ogasawara, A. Kozuka, and K. Nyu: Streptonigrin and Related Compounds. VII. Synthesis of Streptonigrin Nucleus by the Formation of Pyridine Ring. Yakugaku Zasshi **87**, 1189 (1967).

60. Kametani, T., K. Ogasawara, and A. Kozuka: Streptonigrin and Related Compounds. VIII. Hydrolysis and Hoffmann Reaction of Ethyl 3-Cyano-4-(3,4-dimethoxyphenyl)-6-methyl-2-quinolyl-5-pyridinecarboxylate. Yakugaku Zasshi **87**, 1195 (1967).

61. Kametani, T., A. Kozuka, and S. Tanaka: Streptonigrin and Related Compounds. IX. Syntheses of 4-Phenyl-2,3-dimethylpyridone Derivatives. Yakugaku Zasshi **90**, 1574 (1970).

62. Kametani, T., S. Tanaka, and A. Kozuka: Syntheses of Streptonigrin and Related Compounds. X. A Synthesis of Methyl 3-Acetamido-1,2-dihydro-4-(2,3,4-trimethoxyphenyl)-5-methyl-2-oxo-6-pyridinecarboxylate. Yakugaku Zasshi **91**, 1068 (1971).

63. Liao, T. K., P. J. Wittek, and C. C. Cheng: Synthetic Studies of the Antitumor Antibiotic Streptonigrin. II. Synthesis of the C-D Ring Portion of Streptonigrin. J. Heterocycl. Chem. **13**, 1283 (1976).

64. Wittek, P. J., T. K. Liao, and C. C. Cheng: Synthetic Studies of the Antitumor Antibiotic Streptonigrin. 3. Synthesis of the C-D Ring of Streptonigrin by an Unsymmetrical Ullmann Reaction. J. Organ. Chem. (USA) **44**, 870 (1979).

65. Rao, K. V., and P. Venkateswarlu: Streptonigrin and Related Compounds. II. Synthesis of the C-Ring Precursors. J. Heterocycl. Chem. **12**, 731 (1975).

66. Rao, K. V., and H.-S. Kuo: Streptonigrin and Related Compounds. IV. Precursors for the A-Ring. J. Heterocycl. Chem. **16**, 1241 (1979).

67. Martin, J. C.: Synthesis of Pyridines from Dicyanopyrimidines. A Diels-Alder Approach to the C-Ring of Streptonigrin. J. Heterocycl. Chem. **17**, 1111 (1980).

68. Kim, D., and S. M. Weinreb: A Diels-Alder Approach to the Pyridine C-Ring of Streptonigrin. J. Organ. Chem. (USA) **43**, 121 (1978).

69. — — Elaboration of the Pyridine C-Ring Functionality in a Streptonigrin Precursor. J. Organ. Chem. (USA) **43**, 125 (1978).

70. Basha, F. Z., S. Hibino, D. Kim, W. E. Pye, T.-T. Wu, and S. M. Weinreb: Total Synthesis of Streptonigrin. J. Amer. Chem. Soc. **102**, 3962 (1980).

71. Kende, A. S., D. P. Lorah, and R. J. Boatman: A New and Efficient Total Synthesis of Streptonigrin. J. Amer. Chem. Soc. **103**, 1271 (1981).

(Received March 19, 1981)

The Pyrrolizidine Alkaloids

By D. J. ROBINS, Department of Chemistry,
University of Glasgow, Scotland

Contents

I. Introduction

Previous chapters in this series by WARREN appeared in 1955 (*310*) and 1966 (*311*). Since then an authoritative book by BULL, CULVENOR, and DICK, on the chemistry and pharmacology of pyrrolizidine alkaloids has been published (*60*). Comprehensive reviews on the alkaloids (*312*), their chemotaxonomic significance (*77*), and general pyrrolizidine chemistry (*243*) are available. A series of Annual Reports including pyrrolizidine alkaloids was introduced in 1971 (*245*). The present review follows the format of previous articles in this series and covers the literature up to early 1981.

The past fifteen years have seen a tremendous growth in interest in pyrrolizidine alkaloids. Their economic importance is now more widely appreciated because of the increasing numbers of deaths to livestock attributable to consumption of plants containing pyrrolizidine alkaloids. Many cases of human liver disease are now also known to be due to ingestion of pyrrolizidine alkaloids (Section VII). Thus study of the chemistry and pharmacology of these compounds has been intensified. Improved methods for detection and chromatographic separation (chiefly high performance liquid chromatography) of alkaloid mixtures have been developed (Section IV). Many new alkaloids have been isolated and structures assigned. The highly desirable practice of hydrolysing or hydrogenolysing the alkaloid to its constituent basic (necine) and acidic (necic acid) moieties, and identifying both portions is not now carried out by all workers, sometimes due to the small amounts of material available. There has been an increasing reliance on spectral data — chiefly ^{1}H nuclear magnetic resonance (n.m.r.) spectrometry (although [13]C n.m.r. spectrometry will gain in importance), and mass spectrometry. X-ray diffraction analyses of a number of pyrrolizidine alkaloids have been carried out and have provided useful information not only on stereochemical details, but also on the conformations of these alkaloids. This may help in understanding the mode of action and biosynthesis of the alkaloids. Progress is at last being made in determining the biosynthetic pathways to the necine bases and necic acids (Section VI).

One of the main features of this review is the inclusion of three comprehensive Tables (Section VIII). All of the plant genera from 13 plant families known to contain pyrrolizidine alkaloids are listed in Table 1. Up to 1955, pyrrolizidine alkaloids were known to be present in 70 plant species (*310*); by 1966, the number had risen to 130 (*311*); during the last 15 years 240 more species have been shown to contain pyrrolizidine alkaloids. These are listed with alkaloid content in Table 2. Likewise, there were about 50 known pyrrolizidine alkaloids in 1955, 100 by 1966, and now the total stands in excess of 200. The structures of these alkaloids appear in Table 3 classified according to the necine base they contain. It should be noted, that in many cases the stereochemistry of the acidic and base portions has not yet been established.

In the following two sections, new material relating to the synthesis and stereochemistry of the constituent necines and necic acids will be discussed.

II. The Necine Bases

1. Structure and Stereochemistry

a) 1-Hydroxymethylpyrrolizidines

All four stereoisomeric forms of 1-hydroxymethylpyrrolizidine have been found as part of pyrrolizidine alkaloids. The final member of the series was discovered by BRANDANGE *et al.* (*53, 56*). Phalaenopsine Is (Table 3 F) is an ester derivative of (−)-isoretronecanol (**1**), present in *Phalaenopsis equestris*. Both forms of supinidine also occur naturally. An ester derivative of (+)-supinidine (**2**), cynaustine (Table 3 H) has been found in *Cynoglossum australe* by CULVENOR and SMITH. (*91*).

(1)   (2)

b) Pyrrolizidine Diols

A new base, petasinecine (**5**), has been found as part of two ester alkaloids, petasinine and petasinoside (Table 3 J), in *Petasites japonicus* by YAMADA *et al.* (*322*). The key feature in the identification of this base was the mass spectrum which suggested a 1-hydroxymethylpyrrolizidine structure. Furthermore, the prominent fragment ions at m/e 98 and 83 are indicative of an unsubstituted ring A, thereby fixing the position of the secondary hydroxy group at C-2 since petasinecine is not a carbinolamine. The relative configuration of petasinecine was apparent, since it had been synthesised previously as a racemate by ADAMS *et al.* (*7*), as outlined in Scheme 1. Addition of hydrogen to the sterically less hindered face of the enolic double bond of (**3**) gave the saturated ester (**4**), which yielded petasinecine (**5**) on reduction. The assignment of the relative stereochemistry of (**5**) was supported by ^{1}H n.m.r. spectroscopic data and the conversion of (**5**) into ($\pm$)-heliotridane (**6**) (*2*).

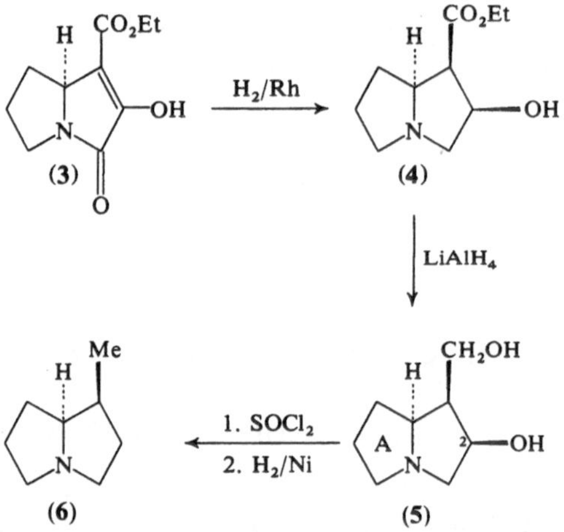

Scheme 1

Macronecine (**8**) is a stereoisomer of petasinecine and has been isolated from only one alkaloid, macrophylline (Table 3 I). AASEN and CULVENOR (*2*) synthesised macronecine from the same ester (**3**) used in the route to petasinecine (Scheme 2). Reduction of (**3**) with zinc and acetic acid gave (**7**) as the major product, from which the racemic pyrrolizidine diol was formed on reduction. Resolution gave the (+)-isomer which was identical with natural (+)-macronecine. The absolute configuration of macronecine follows from the work of DANILOVA and UTKIN (*94*) who degraded macronecine to laburnine (**9**).

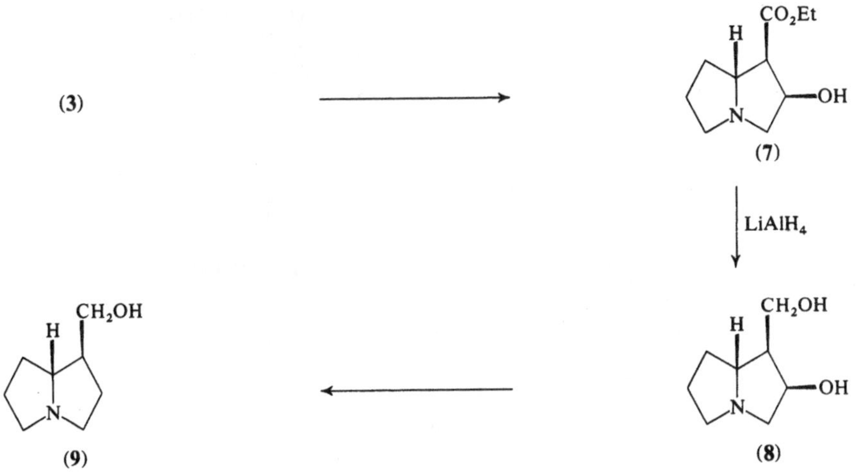

$$CO_2Et$$

(7)

(3) $\longrightarrow$

$\downarrow$ LiAlH$_4$

(9)

(8)

Scheme 2

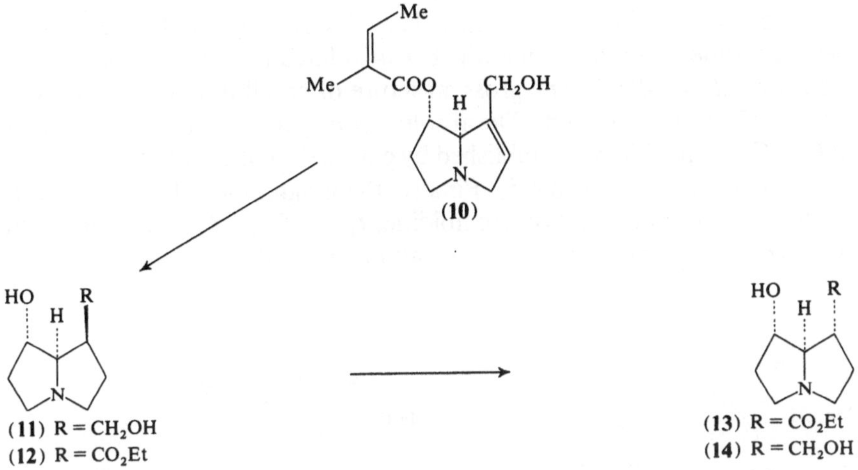

(10)

(11) R = CH$_2$OH
(12) R = CO$_2$Et

$\longrightarrow$

(13) R = CO$_2$Et
(14) R = CH$_2$OH

Scheme 3

The stereochemistry of hastanecine and turneforcidine has been established by CULVENOR *et al.* (*3*). 7-Angelylheliotridine (**10**) was converted into dihydroxyheliotridane (**11**) of known absolute configuration by reduction and hydrolysis steps (Scheme 3). Oxidation and esterification of (**11**) gave the ester (**12**) which was epimerised at C-1 with sodium ethoxide to give the more stable isomer (**13**). Reduction of (**13**) with lithium aluminium hydride gave the enantiomer (**14**) of hastanecine, thus establishing the relative and absolute configuration of hastanecine. The absolute configuration of turneforcidine (**15**) was determined by AASEN and CULVENOR (*1*) when they prepared (**15**) from the enantiomer (**14**) of hastanecine as outlined in Scheme 4.

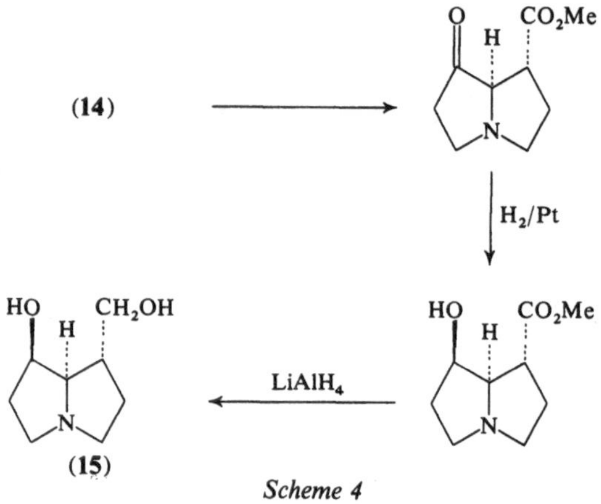

(14) ⟶

Scheme 4

c) Pyrrolizidine Triols

Two new pyrrolizidine triols have been discovered. Croalbinecine is the base constituent of croalbidine which was isolated by SAWHNEY et al. (266) from Crotalaria albida. The gross structure of croalbinecine was deduced from its ^{1}H n.m.r. spectrum. The absolute configuration of the substituents at C-1, C-7, and C-8, was established by conversion of croalbidine (**16**) into turneforcidine (**15**) (Scheme 5). Finally, the magnitudes of the ^{1}H n.m.r. spectral coupling constants in croalbidine, $J_{1\beta,2} = J_{2,3\beta} = 8$ Hz, are consistent with trans-disposed hydrogens in an almost diaxial arrangement.

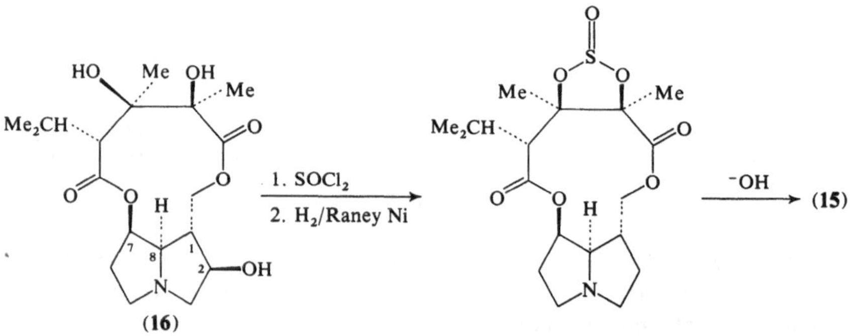

Scheme 5

Eight esters of crotanecine (**17**) have been characterised (Table 3 S) by CULVENOR et al. (22, 93). The similarity of the ^{1}H n.m.r. and mass spectra of crotanecine to those of retronecine (**18**) led to formulation of the gross structure, while the relative stereochemistry at the three chiral centres was deduced to be as in (**17**) from the ^{1}H n.m.r. spectral coupling constants: $J_{5\alpha,6\alpha} = 6.5$ Hz, $J_{5\beta,6\alpha} = 9.5$ Hz, and $J_{6\alpha,7\alpha} = J_{7\alpha,8\alpha} = 3.5$ Hz.

(17) R = OH
(18) R = H

d) Dihydropyrrolizines

As part of their extensive chemotaxonomic studies on the Compositae, BOHLMANN *et al.* (*39—45*) have recently discovered several new types of pyrrolizidine alkaloids with a dihydropyrrolizine nucleus in *Senecio* species. These are all neutral compounds. The senampelines, which occur as isomeric mixtures that have not yet been separated, contain the triol (19) of unidentified stereochemistry (Table 3 T). Another series of compounds contains the related oxidised dihydropyrrolizinone (20) which occurs naturally with either 7 α or 7 β stereochemistry as part of the macrocyclic diester system (Table 3 U).

(19)   (20)

2. Synthesis of Necine Bases

A great many syntheses of necine bases have been reported. These have been aimed mainly at the relatively easy synthetic targets of the saturated 1-hydroxymethylpyrrolizidines. More attention should be directed towards the synthesis of the unsaturated necines, particularly those with more than one hydroxy group. Synthetic routes to the necine bases have been recently reviewed by ROBINS (*243*) and they are also discussed annually (*245*). Only a few selected examples will be described here.

The first synthesis of the 1-hydroxymethylpyrrolizidines in optically active form has been reported by ROBINS and SAKDARAT (*249*) and is outlined in Scheme 6. 1,3-Dipolar cycloaddition of ethyl propiolate to the *NO*-diformyl derivative (21) of natural (−)-4-hydroxyl-L-proline took place in a regiospecific manner to give the (+)-ester (22), which was deformylated with ammonia to give (23). In the next key step, catalytic hydrogenation of (23) afforded a single optically active ester (24) in good yield. The absolute stereochemistry of this ester was established by converting it into (+)-isoretronecanol (26) by treatment with thionyl chloride, catalytic hydrogenation [to give (25)] and reduction with lithium aluminium hydride. (+)-Laburnine (9) was prepared from the intermediate (25) by epimerisation of the ester function to the more stable isomer, followed by hydride reduction. A useful method for converting saturated pyrrolizidine esters into their 1,2-didehydro-analogues has been described by ROBINS and SAKDARAT (250). A phenylseleno-group was introduced α to the ester function in (25), and then thermal elimination of the derived selenoxide gave (+)-supinidine (2). To complete the synthesis of all the stereoisomeric forms of the 1-hydroxymethylpyrrolizidines, the chiral centre in (23) was epimerised by S_N2 displacement of the tosyl derivative with formate anion. The epimer of (22) thus produced, was converted into (−)-isoretronecanol, (−)-trachelanthamidine, and (−)-supinidine as outlined above (252). The optical purities of all these bases were in excess of 80%.

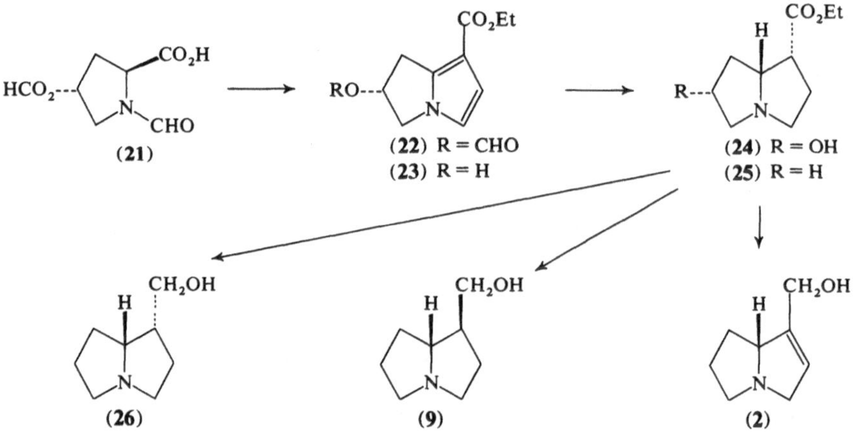

Scheme 6

A synthesis of (±)-supinidine (2) by TUFARIELLO and TETTE (305) also utilises 1,3-dipolar reactivity, in this case supplied by pyrroline-1-oxide. Addition of the nitrone (27) to methyl γ-hydroxycrotonate gave the bicyclic product (29) (Scheme 7). Hydrogenation of the mesylate derivative of (29) produced the pyrrolizidine ester (30), and dehydration and reduction steps yielded (±)-supinidine (2). TUFARIELLO and LEE (304) have extended this strategy into a synthesis of (±)-retronecine (18) by using a protected 3-ketonitrone (28).

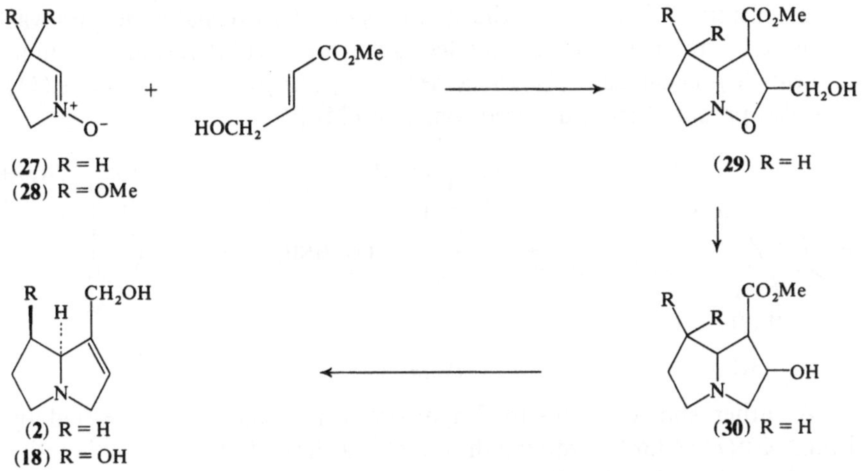

(27) R = H
(28) R = OMe

(29) R = H

(2) R = H
(18) R = OH

(30) R = H

Scheme 7

A number of necine bases have been synthesised by intramolecular opening of activated cyclopropanes by DANISHEFSKY *et al.* (*95*). Initially, (±)-isoretronecanol (**26**) was prepared as outlined in Scheme 8. Intramolecular homoconjugate addition of the amine (**31**) took place with complete inversion of stereochemistry to give the pyrrolizidinone (**32**). Removal of the hydrazide group from (**32**) and reduction of the product (**33**) yielded (±)-isoretronecanol (**26**). Analogous treatment of a stereoisomer of (**31**) afforded (±)-trachelanthamidine (**9**) (*96*).

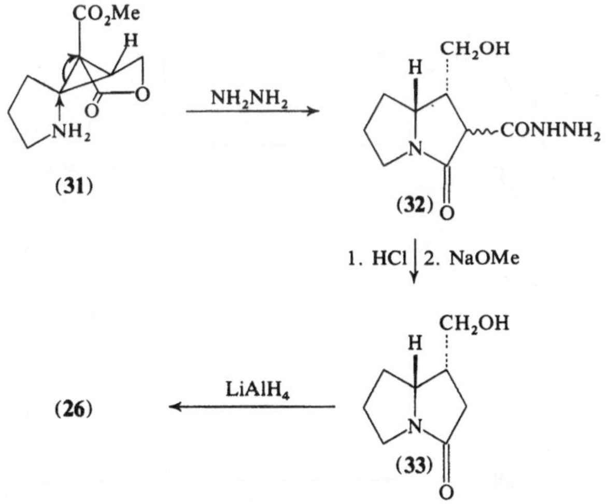

(31)

(32)

1. HCl │ 2. NaOMe

(26)

LiAlH₄

(33)

Scheme 8

Extension of this route to the synthesis of 7-hydroxylated necines was achieved by treatment of the activated cyclopropane (34) in similar fashion to yield ($\pm$)-hastanecine (35) (Scheme 9). ($\pm$)-Dihydroxyheliotridane (11) was also prepared from a stereoisomer of (34) (97).

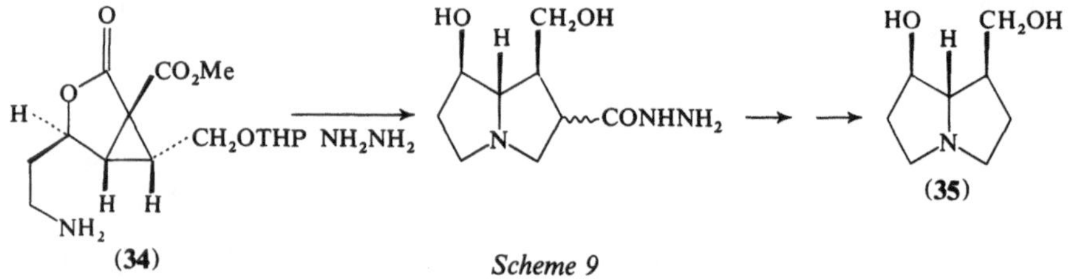

Scheme 9

A rather shorter route to 7-hydroxy-necines has been reported by BOHLMANN et al. (38), involving the direct insertion of oxygen functionality into the activated "allylic" position of the dihydropyrrolizine (36) using a peroxyester (Scheme 10). Hydrolysis and reduction steps then yielded the ($\pm$)-dihydropyrrolizinone (20).

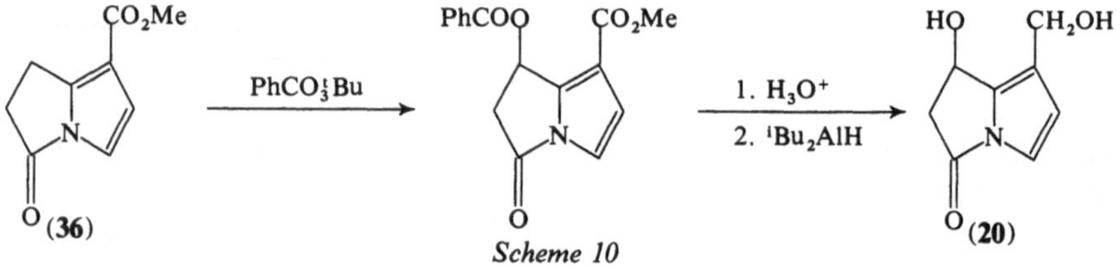

Scheme 10

An alternative route to the same necine (20) has recently been published by BOHLMANN et al. (171) using the Bestmann reaction and is outlined in Scheme 11.

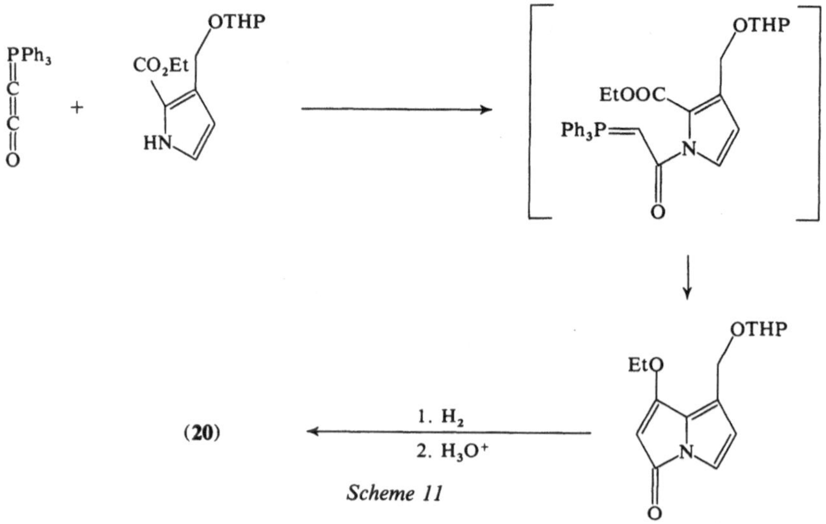

Scheme 11

III. The Necic Acids

In many cases, the constituent acid(s) of new pyrrolizidine alkaloids have not been isolated by hydrolysis or hydrogenolysis of the parent alkaloid and separately identified. Much reliance has been placed on identification of the whole alkaloid by spectroscopic methods. Accordingly only information pertaining to the separate acids will be dealt with here. Many of the acids are γ- or δ-hydroxyacids and consequently are often isolated as the corresponding γ- or δ-lactones.

The circular dichroism spectra of a number of necic acids have been recorded by CULVENOR *et al.* (*78*). The number of chromophores present in each necic acid prevents simple correlation of absolute configuration. However, comparisons between spectra of closely related acids have been helpful, particularly when one of the acids is of known absolute configuration.

1. C₆-Acids

(−)-2,3-Dihydroxy-3-methylpentanoic acid (**37**) is present in strigosine (Table 3 C). Stereospecific synthesis of (**37**) by CROUT and WHITEHOUSE (*76*) has established the (2*R*,3*R*) absolute configuration. The *erythro*-racemate (**37**) was obtained by *trans*-hydroxylation of (*E*)-3-methylpent-2-enoic acid with tungstic oxide and hydrogen peroxide. The (−)-isomer, obtained by resolution of the quinine salts of (±)-(**37**) was identical with the esterifying acid from strigosine. The absolute configuration of (−)-(**37**) was shown to be (2 *R*, 3 *R*) by degradation of the (−)-acid to the (2 *R*)-phthaloyl ester (**38**) (Scheme 12).

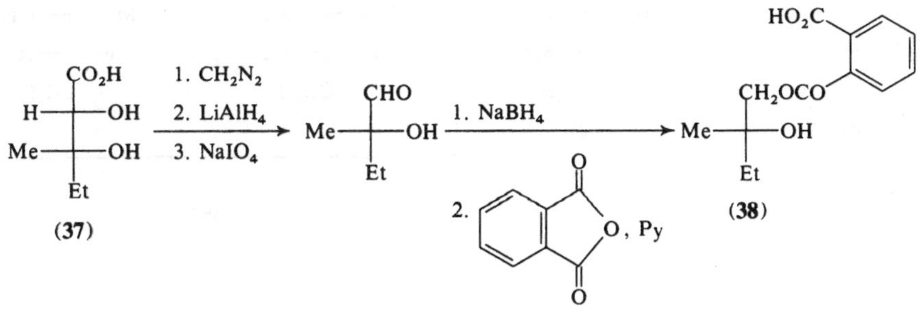

Scheme 12

2. C₈-Acids

a) (−)-Curassavic Acid

This is the first C_8-monocarboxylic acid to be discovered, and was obtained on alkaline hydrolysis of curassavine (Table 3 C) by MOHANRAJ *et al.* (*210*). Structure (**39**) for (−)-curassavic acid was deduced from the ^1H n.m.r. spectrum, and by oxidation of the methyl ester of (**39**) to the ester (**40**) (Scheme 13). The only stereochemical note is the assumption that the glycol group in (**39**) has the *erythro*-configuration, because the methyl ester of (−)-curassavic acid showed an electrophoretic mobility similar to that of the methyl ester of *erythro*-viridifloric acid*.

$$\text{MeCHOHCOHCHMeCH}_2\text{Me} \quad \xrightarrow[\text{2. NaIO}_4]{\text{1. CH}_2\text{N}_2} \quad \text{MeCH}_2\text{CHMeCOCO}_2\text{Me}$$

$$\underset{\text{CO}_2\text{H}}{\mid} \qquad \qquad \qquad \qquad \qquad (\textbf{40})$$

$$(\textbf{39})$$

Scheme 13

b) Crispatic Acid and Stereoisomers

The configurations of the optically inactive fulvinic (**41**) and crispatic acids (**42**) have been revised, as shown, to (*S*)-meso and (*R*)-meso respectively, by MATSUMOTO *et al.* (*199*) on the basis of asymmetric syntheses. The two other stereoisomeric forms of 3-hydroxy-2,3,4-trimethylglutaric acid, cromaduric and isocromaduric acids (**43**) are present in cromadurine and isocromadurine (Table 3 O ii). The absolute configuration of each enantiomer is not known.

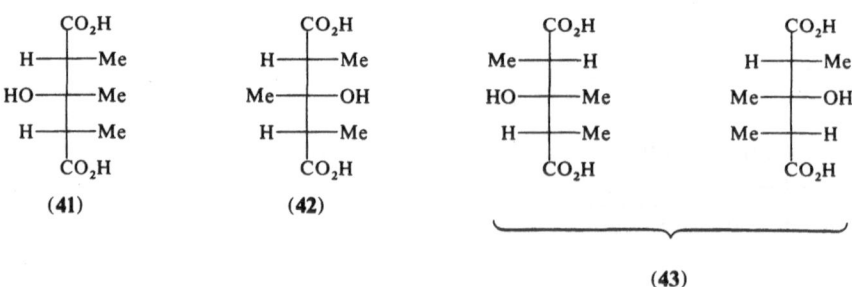

(**41**)   (**42**)   (**43**)

---

* In ref. *210a*, sign and magnitude of rotation of curassavic acid were given incorrectly. The values in ref. *210b* are correct.

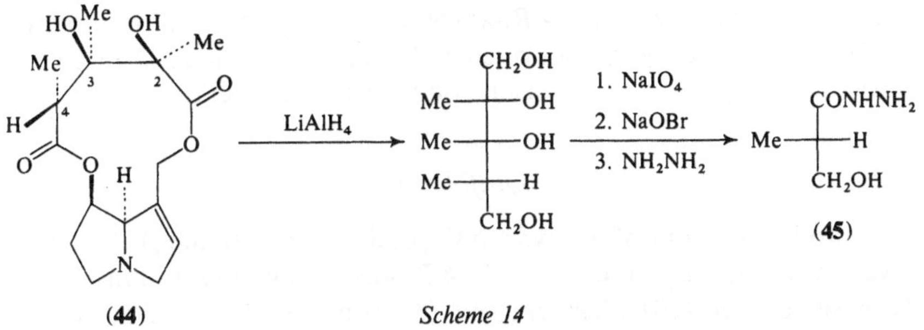

Scheme 14

c) Monocrotalic Acid

Monocrotalic acid is present in moncrotaline (44) and has three chiral centres. Degradation of monocrotaline to a derivative (45) of known absolute configuration (Scheme 14) was carried out by ROBINS and CROUT (247) to establish the (R)-configuration at C-4 of monocrotalic acid. The configuration at C-3 of monocrotalic acid (46) was deduced to be (R) since dehydration of the methyl ester of (46) proceeded extremely slowly, indicating a *cis*-arrangement of the hydrogen and hydroxy-groups at C-4 and C-3, respectively (Scheme 15). Finally ROBINS and CROUT (248) converted methyl anhydromonocrotalate (47) into a mixture (48) of the tritium-labeled *threo*-acids having (2R, 3R)- and (2R, 3S)-absolute configuration (Scheme 15). Aliquots of this radioactive mixture (48) were co-crystallised with the brucine salts of the authentic (2R, 3S)- and (2S, 3R)-acids. Only the former salt retained any radioactivity, thus establishing the (2R)-absolute configuration in monocrotalic acid (46).

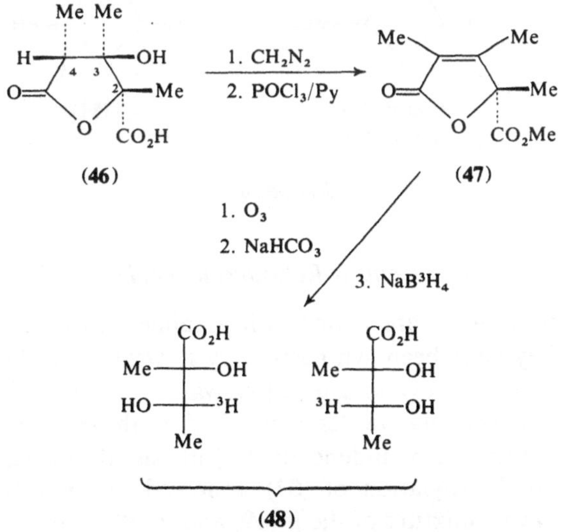

Scheme 15

Recently, the $(2R, 3R, 4R)$-absolute configuration of $(-)$-mono-crotalic acid (46) has been confirmed by Matsumoto et al. (202) as part of the synthesis of all eight stereoisomeric forms of this acid.

d) Latifolic Acid

Latifolic acid is one of the hydrolysis products of latifoline (Table 3 O i); it was shown to have the $(2S, 3S, 4R)$-absolute configuration (53) by Matsumoto et al. (201). The racemic (E)-isomer (49) was resolved with brucine, and the absolute configuration of one enantiomer was established by conversion of the methyl ester (50) into (S)-$(-)$-2-methylbutane-1,4-diol (51) (Scheme 16). The butyrolactones (52) and (54) formed by cis-hydroxylation of (50) were separated and identified by comparison with racemic material prepared earlier (200). Hydrolysis of (52) gave an acid which was identical with latifolic acid (53).

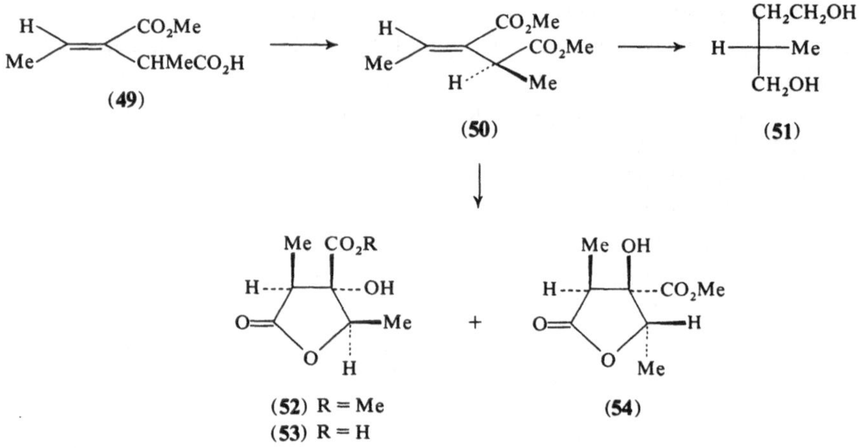

Scheme 16

e) α- and β-Retusanecic Acids

These epimeric acids are formed on alkaline hydrolysis of retusine (Table 3 K). They have been synthesised by Kiyooka and Hase (161) as shown in Scheme 17. The cyanolactone (56) was prepared by HCN addition to the ketone (55). The relative stereochemistry of the corresponding ethyl ester (57) was confirmed by independent synthesis from trans-opening of the epoxide (58). Methylation of (57), followed by hydrolysis and de-carboxylation gave a mixture of the β-(59) and α-(60) retusanecic acids, in which the β-form predominated.

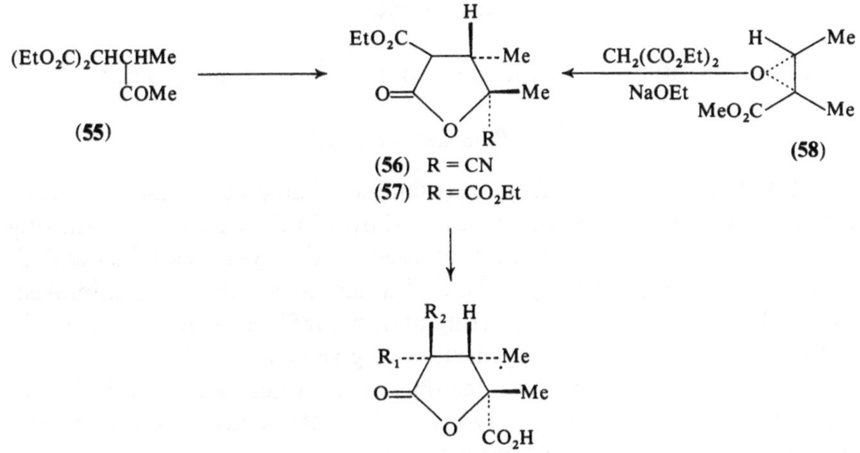

(55)

(56) R = CN
(57) R = CO_2Et

(58)

(59) R^1 = H, R^2 = Me
(60) R^1 = Me, R^2 = H

Scheme 17

3. C_9-Acids

Necic acids with skeletons containing nine carbon atoms are comparatively rare in pyrrolizidine alkaloids. Striatic (nilgiric) acid **(61)** was the first example discovered (*23*). It is present in nilgirine (Table 3 O iii). The geometrical isomer **(62)** and the *O*-acetyl derivative **(63)** of striatic acid have also been isolated from crotafoline (Table 3 P ii) (*72*) and crotastriatine (Table 3 O iii) (*31*), respectively. More recently, crotananic acid **(64)** [from crotananine (Table 3 O iii)] (*279*) and cronaburmic acid **(65)** [from cronaburmine (Table 3 O ii)] (*277*) have been identified. None of these acids has had its stereochemistry established.

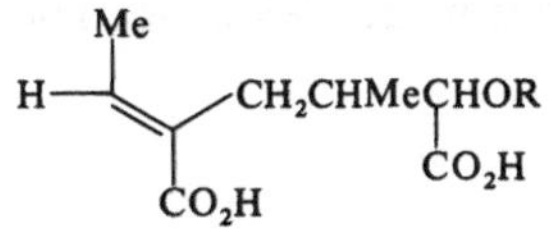

(61) R = H
(62) Geometrical isomer of (61)
(63) R = Ac

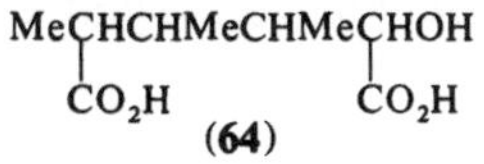

(64)

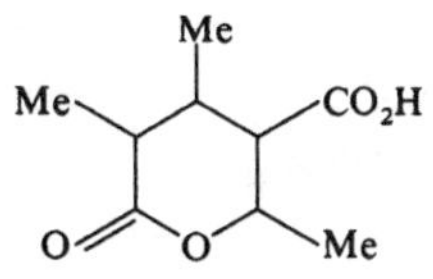

(65)

4. C_{10}-Acids

These constitute by far the most common type of necic acids.

a) Retusaminic Acid

This acid is derived from retusamine (Table 3 P i), and has been synthesised by KIYOOKA et al. (162) as shown in Scheme 18. The starting material (55) was the same as that used in the synthesis of α- and β-retusanecic acids (cf. Scheme 17), and a similar strategy was employed, except that ethylation of the starting material (55) preceded stereospecific cyanohydrin formation and cyclisation to give lactone (66). The absolute configuration of the resolved lactone acid, which was identical with retusaminic acid (67), was already known from the X-ray structure analysis of rétusamine carried out by WUNDERLICH (319).

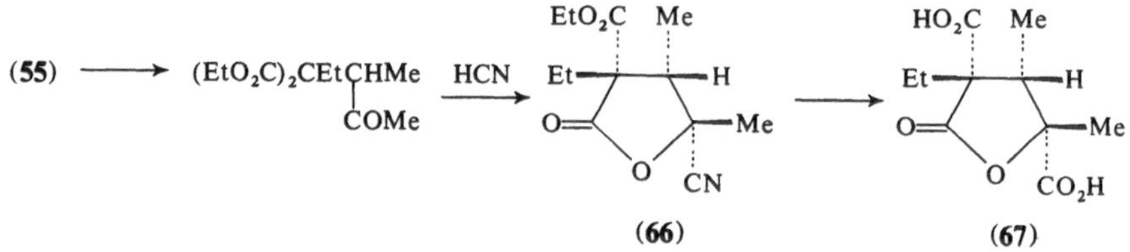

Scheme 18

b) Syneilesinolides A — C

Three different lactones are formed on alkaline hydrolysis of syneilesine (Table 3 P ii). Structures (68)—(70) for these syneilesinolides A – C were established by HIKICHI and FURUYA from spectroscopic data (135). Some stereochemical details were also deduced. Circular dichroism spectra of the three lactones, when compared with those of other necic acids, indicated that the stereochemistry at C-2 is R. In addition, the fact that the dilactone (70) forms readily demonstrates that the stereochemistry at C-4 is also R. Finally, from the ^{1}H n.m.r. coupling constant of 5.4 Hz between the protons at C-3 and C-4 of (70), the dihedral angle between these protons is estimated to be 36°, suggesting that the configuration at C-3 is R. The stereochemistry at C-5 has not been established.

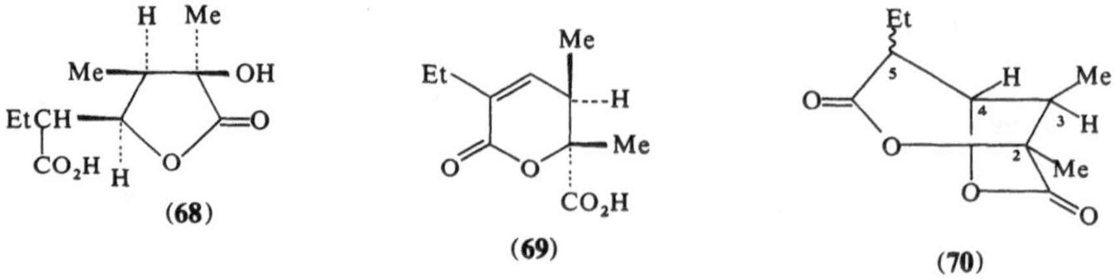

c) Ligularidenecic Acid

Ligularidenecic acid is a hydrolysis product of ligularidine (Table 3 P ii), obtained by HIKICHI et al. (133). Structure (71) for this δ-lactone was assigned when it was realised that the physical data were very similar to those reported earlier for a lactone synthesised by EDWARDS and MATSUMOTO (109), as part of their preparation of all the diastereomers of senecic and integerrinecic acids.

(71) $R^1 = Me$, $R^2 = CO_2H$
(72) $R^1 = CO_2H$, $R^2 = Me$

d) Crotaverric Acid

Crotaverric acid is also a δ-lactone, obtained from crotaverrine (Table 3 P ii) by SURI et al. (294). ^1H-N.m.r. spectroscopy showed that crotaverric acid is diastereomeric with the δ-lactone of integerrinecic acid (72). Since lead tetraacetate oxidation of both diacids gave the same ketoacid, it was deduced that crotaverric acid has the opposite stereochemistry to (72) at either C-2 or C-3.

e) Petasinecic Acid

Petasitenine (Table 3 P ii) yields petasinecic acid (73) on base hydrolysis. YAMADA et al. (321) oxidised petasinecic acid and jaconecic acid to give the same γ-lactone (74), thus establishing the stereochemistry of (73) at C-2, C-3, and C-5 (diastereomeric with jaconecic acid).

(73)   (74)

f) cis-Nemorensic Acid

The structure and absolute configuration of cis-nemorensic acid (75) obtained by alkaline hydrolysis of retroisosenine (Table 3 O iv) and doronenine (Table 3 O iv), has been determined by KIRFEL et al. (159), using X-ray crystallography.

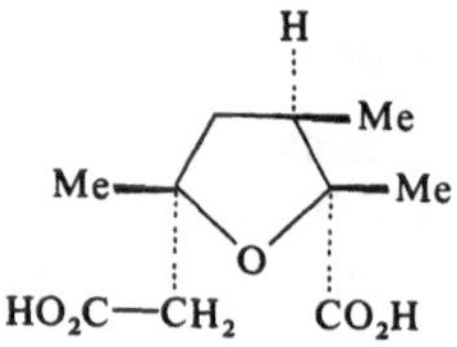

(75)

5. Orchid Acids

Cornucervine and the phalaenopsines (Table 3 C, D, and F) give (−)-2-isobutylmalic acid (76) and (−)-2-benzylmalic acid (77) on alkaline hydrolysis. Both acids have been assigned the R absolute configuration by Brandange et al. (53) by partial asymmetric synthesis of the enantiomers of both acids and comparison of their circular dichroism spectra with those of (S)-(+)-citromalic acid (78).

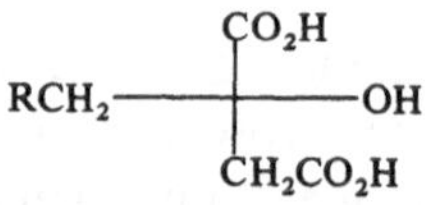

(76) R = CHMe$_2$
(77) R = Ph

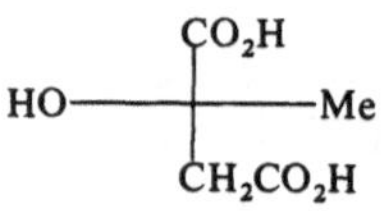

(78)

IV. The Pyrrolizidine Alkaloids

The structures of more than 200 known pyrrolizidine alkaloids are presented in Table 3, classified according to the necine they contain, with the fully saturated and less hydroxylated necines appearing first. For the macrocyclic diester alkaloids, further subdivision according to ring size is made. Attention should be drawn to several key points.

In their study of neutral, terpenoid metabolites from the Compositae, Bohlmann and his co-workers have discovered a series of dihydro-pyrrolizinone derivatives in *Senecio* species (Table 3 U). These are all macrocyclic diesters containing 12-membered rings, although it is interesting that macrocyclic diesters with both α and β stereochemistry at C-7 of the necine have been isolated. A related series of open chain esters, the senampelines (Table 3 T), have also been discovered. This work does indicate the need to carry out careful examination of both basic and neutral fractions from plant extracts when searching for pyrrolizidine derivatives. The extent of these new derivatives should also be ascertained by re-investigating species of other families known to contain pyrrolizidine alkaloids.

The most common ring size in pyrrolizidine alkaloids is 11 or 12 membered. However, a few examples of 13-membered macrocyclic diesters are known for platynecine (Table 3 M) and retronecine (Table 3 O iv), and the recent discovery of the first 14-membered derivatives of retronecine (Table 3 O v) is worthy of note.

Tissue cultures of *Crotalaria juncea* have been reported by KHANNA and MANOT (*156*) to produce the pyrrolizidine alkaloid junceine (Table 3 O ii).

1. Analytical Methods

A considerable number of improvements to existing procedures for the detection, isolation, and separation of pyrrolizidine alkaloid mixtures has been reported. The alkaloids often occur naturally in the form of their *N*-oxides, and these are usually reduced with zinc in acidic solution to liberate the free bases. An alternative method has been described by HUIZING and MALINGRÉ (*144*), which employs a redox polymer on an anion-exchange resin to give better yields of the free bases. Ion-exchange resins have been used for the extraction (*99*) and separation (*145*) of pyrrolozidine alkaloids.

Reproducibility of R_F values for pyrrolizidine alkaloids has been difficult to achieve on thin layer chromatography (t.l.c.) using silica gel plates even when these are impregnated with sodium hydroxide. To obviate this problem, the use of ion pair adsorption t.l.c. using chloride as the counter ion has been advocated by HUIZING and MALINGRÉ (*146*). Chloranil has been shown to be a sensitive reagent for pyrrolizidine alkaloids on thin layer chromatograms (*143*).

Through the work of SEGALL *et al.*, improved separations of mixtures of pyrrolizidine alkaloids have been obtained by high performance liquid chromatography (*234, 272, 274*). The *N*-oxides are also amenable to this treatment (*301*). The best separations were achieved on reverse-phase systems (*103, 273*).

Gas liquid chromatography — mass spectrometry of silylated derivatives is a very sensitive method for detection of pyrrolizidine alkaloids (*111*). The many alkaloids containing retronecine (**18**) can be readily determined either by ^{1}H n.m.r. spectroscopy utilising the downfield signal for the vinyl proton (*211*) or by gas liquid chromatography of fluorinated retronecine derivatives (*101*).

2. Spectroscopy

A major study of the ^{1}H n. m. r. spectroscopic behaviour of necine bases and the alkaloids has been carried out by CULVENOR *et al.* (*83*). In particular, from analysis of the coupling constants in the epimers retronecine (18) and heliotridine (79) it was deduced that retronecine and its ester derivatives have an *exo*-buckled conformation, whereas heliotridine and its esters are generally mixtures of *exo*- and *endo*-buckled conformers which are rapidly interconverting. The growing use of ^{13}C n. m. r. spectroscopy on pyrrolizidine alkaloids will greatly assist structure determinations, and may perhaps also provide useful information about the conformations of these alkaloids. ^{13}C N.m.r. assignments have been made for several pyrrolizidine alkaloids (*26, 104, 209*). Some confusion already exists, particularly regarding assignment for C-7 and C-8 of retronecine (18). Selective enrichment of these signals with ^{13}C (from biosynthetic experiments) has clarified this situation (*154*).

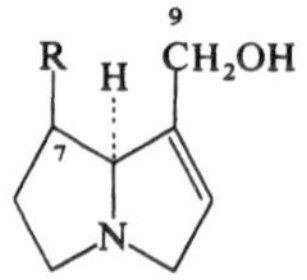

(18) R = β − OH
(79) R = α − OH

The mass spectra of pyrrolizidine alkaloids are very distinctive and often permit structural assignments to be made by comparison with mass spectra of alkaloids of known structure. The isomeric 1-hydroxy-methylpyrrolizidines give a base peak at *m/e* 83 (80) by typical fragmentation β to the nitrogen (*4, 213*). Characteristic ions at *m/e* 111 (81) and 80 (82) are observed for retronecine (18) and heliotridine (79), but measurement of the appearence and ionisation potentials has been used to distinguish between these epimers (*224*). Those alkaloids containing otonecine as base (Table 3 P) display characteristic fragment ions at *m/e* 168, 151, 150, 122, 110, and 94 (*2, 62*). The *N*-oxides of pyrrolizidine alkaloids usually show the presence of ions at M-16, M-17, and M-18 (*5*).

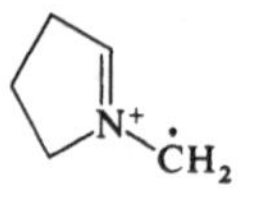

m/e 83

(80)

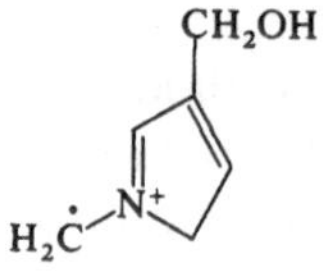

m/e 111

(81)

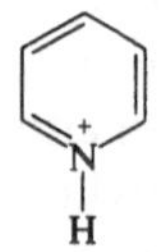

m/e 80

(82)

Further compilations of infrared spectra (*127*) and ultraviolet data (*126*) of pyrrolizidine alkaloids have been reported by GUPTA *et al.*

The circular dichroism spectra of a number of pyrrolizidine alkaloids have been recorded by CULVENOR *et al.* (*78*) and by HRBEK *et al.* (*141*).

Several generalisations were made. Saturated 1-substituted pyrrolizidines showed negative Cotton effects, while 1,7-disubstituted derivatives displayed positive effects. Unsaturated 1-substituted pyrrolizidines also exhibited positive Cotton effects, and those with a 1,7-disubstituted pattern, such as retronecine (18) or heliotridine (79) gave large positive Cotton effects.

X-ray crystallography has been used to establish the stereochemistry of about 20 pyrrolizidine alkaloids, and several interesting features regarding the conformation of these alkaloids have been revealed. X-ray studies on 12-membered macrocyclic diesters of retronecine (Table 3 O iii) such as jacobine (113, 225), swazine (177), retrorsine (64) and yamataimine (136) have shown that the ester carbonyl groups are nearly anti-parallel. A similar situation obtains in trichodesmine (300), which has an 11-membered macrocyclic diester. However, all of the remaining 11-membered macrocyclic diesters of retronecine which have been studied, including fulvine (295), axillarine (287), monocrotaline (285, 313), and incanine (299), have ester carbonyl groups which are syn-parallel and directed below the plane of the 11-membered ring. Macrocyclic diesters of otonecine (Table 3 P) which have been subjected to X-ray analysis include retusamine (319) (11-membered ring), clivorine (36), and senkirkine (35) (both 12-membered rings). The transannular N ... C = O distance in these three alkaloids is 1.64, 2.0, and 2.3 Å respectively. The last two values indicate very small interactions across the 12-membered rings, and the carbonyl bond lengths are close to the normal ketone value of 1.215 Å.

3. Synthesis

The first total syntheses of pyrrolizidine monoesters were accomplished by Russian workers (172). Trachelanthamine (83) and viridiflorine (84) were two of the monoesters prepared. The base portion, (−)-trachelanthamidine, was obtained by synthesis of a mixture of the two diastereomeric 1-hydroxymethylpyrrolizidines, followed by separation and resolution. (+)-Trachelanthic acid was prepared by cis-hydroxylation of trans-2-isopropylcrotonic acid, followed by resolution of the racemic product [resolution of the trans-hydroxylated product yielded (−)-viridifloric acid].

(83)
(84) Opposite stereochemistry at C*

Combination of (+)-trachelanthic acid and (−)-trachelanthamidine by a transesterification process gave trachelanthamine (83); viridiflorine (84) was formed from the same base and (−)-viridifloric acid in a similar manner.

Selective synthesis of diesters of the pyrrolizidine diols has proved more difficult. A number of semisynthetic pyrrolizidine diesters were prepared by CULVENOR et al. (81) by acylation of heliotridine (79) with various acyl chlorides. Selective acylation at C-9 of heliotridine and retronecine (18) is possible by conversion of the base into its 1-chloromethyl derivative, followed by nucleophilic substitution with the appropriate carboxylate anion.

Selective esterification at C-9 of retronecine (18) was also achieved by HOSKINS and CROUT (140) who used simple acids together with NN′-dicyclohexylcarbodiimide as coupling reagent. NN′-Carbonyldiimidazole was preferable when the acids to be used were αβ-unsaturated or sterically hindered at the α-position. Subsequent esterification at the C-7 position of the retronecine ester then yielded unsymmetrical diesters of retronecine.

The outstanding challenge in this area is the synthesis of the natural macrocyclic diester alkaloids. A first step in this direction was taken by ROBINS and SAKDARAT (251), when they managed to reconstruct an 11-membered macrocyclic diester (85) from (+)-retronecine and 3,3-dimethyl-glutaric anhydride. The mixture of 7- and 9-monoesters of retronecine initially produced was cyclised intramolecularly via the corresponding 2-pyridinethiol esters.

(85)

V. Pyrrolizidine Derivatives in the Lepidoptera

Male *Danaus* butterflies feed on plants containing pyrrolizidine alkaloids and convert the alkaloids into dihydropyrrolizines such as danaidone (86), danaidal (87), and hydroxydanaidal (88). These derivatives are used as pheromones (48). Danaid butterflies are also able to store pyrrolizidine alkaloids. This may act as a defence mechanism by rendering the butterflies less palatable to potential predators (*105, 106*). Moths of the families Ctenuchidae and Arctiidae also feed on plants containing pyrrolizidine alkaloids and store the alkaloids (*33, 125, 259*).

(86)

(87) R = H
(88) R = OH

A one-step synthesis of the pheromone danaidal (87) from *N*-formylproline and propargylic aldehyde has been reported by PIZZORNO and ALBONICO (*229*).

VI. Biosynthesis

1. Necine Bases

Retronecine (18) is the most common necine and all the biosynthetic studies reported have been carried out on this base. NOWACKI and BYERRUM (*216*) fed [2-^{14}C]-ornithine to *Crotalaria spectabilis* plants and obtained monocrotaline (44), which was shown to be radiocative only in the retronecine portion. BOTTOMLEY and GEISSMAN (*50*) extended this work by feeding [1,4-^{14}C]-putrescine and [2-^{14}C]- and [5-^{14}C]-ornithine to *Senecio douglasii* plants. The radioactive retronecine was degraded to yield C-9 as shown in Scheme 19. In all three experiments, 25% of the total necine activity was located at C-9. The conclusions from these results are that (i) at some stage in the biosynthetic pathway C-2 and C-5 of ornithine become equivalent, probably *via* putrescine, at least to form ring B of retronecine (18), and (ii) a later symmetrical "dimeric" intermediate may be involved. However, HUGHES *et al.* (*142*) reported different results when they fed [2-^{14}C]-ornithine to *Senecio isatideus* plants. The necine obtained was degraded by a series of Hofmann eliminations. As in the previous work, 26% of the necine activity was found at C-9, but there was no activity at C-5 and C-3. Instead, the rest of the activity was at C-(7 + 8). These workers concluded that two molecules of ornithine are incorporated into retronecine in a specific manner involving two different metabolic pools.

BALE and CROUT (*25a*) have carried out feeding experiments with arginine and ornithine labelled with different isotopes (^{3}H and ^{14}C). They found that ornithine is a slightly more efficient precursor than arginine for retronecine biosynthesis.

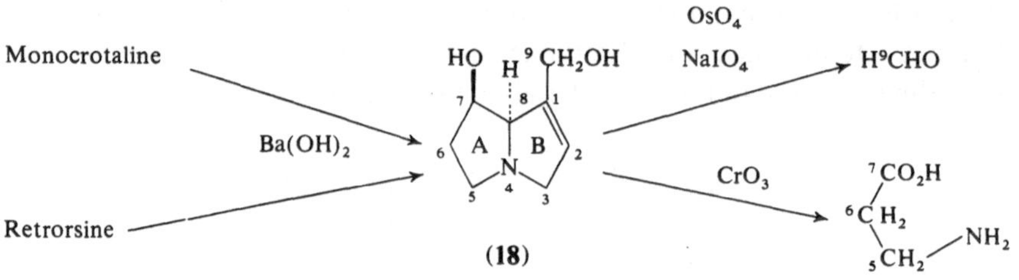

Scheme 19

Recently, ROBINS *et al.* began an investigation into the biosynthesis of retronecine. They recognised that further degradation of retronecine was necessary to establish whether the ornithine used to form ring A of retronecine also passed through a symmetrical intermediate. Thus, ROBINS and SWEENEY (*253*) isolated β-alanine from modified Kuhn-Roth oxidation of retronecine. This corresponds to C-$(5+6+7)$ of retronecine (Scheme 19). Feeding experiments were undertaken with *Senecio isatideus* plants which produce retrorsine. [5-^{14}C]-Ornithine, [1,4-^{14}C]-putrescine, *N*-(3-aminopropyl)-[1,4-^{14}C]-1,4-diaminobutane (spermidine), and *NN'*-bis-(3-aminopropyl)-[1,4-^{14}C]-diaminobutane (spermine), were incorporated well into retrorsine, and almost all of the activity was present in the derived retronecine. On further degradation, in all these experiments 22—25% of the necine activity was located at C-9, confirming previous results, and 22—24% of the total activity was in the β-alanine, supporting conclusion (ii) above.

The high total incorporation of ^{14}C radioactivity obtained in this work (up to 5.2%) suggested that ^{13}C-labelled precursors could be used to establish complete labelling patterns in retronecine with the aid of ^{13}C n. m. r. spectroscopy. Accordingly, KHAN and ROBINS (*154*) fed [1,4-^{13}C$_2$]-putrescine to *S. isatideus* plants. Enrichment factors of up to 1% ^{13}C were observed in the ^{13}C n. m. r. spectrum of retronecine hydrochloride for the signals corresponding to C-3, C-5, C-8, and C-9, and the enhancements of these four signals were nearly equal. Furthermore, the doubly-labelled

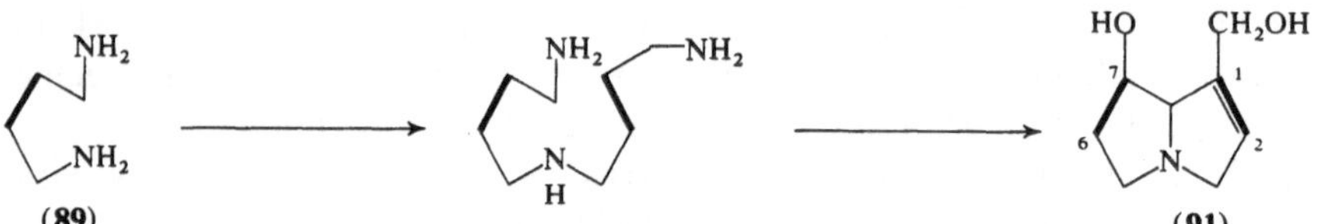

Scheme 20

precursor [2,3-^{13}C$_2$]-putrescine (**89**) was incorporated into retronecine (**91**) efficiently to produce a ^{13}C n.m.r. spectrum containing two pairs of doublets corresponding to C-1/C-2 and C-6/C-7 (Scheme 20). This suggests that in the biosynthetic pathway, two molecules of putrescine combine together to form a symmetrical intermediate, such as (**90**), which is then converted into retronecine.

Direct evidence for the involvement of a symmetrical "dimeric" intermediate in retronecine biosynthesis was obtained by KHAN and ROBINS (*155*). They fed [1-*amino*-^{15}N; 1-^{13}C]-putrescine to *S. isatideus* plants. The ^{13}C n.m.r. spectrum of the retronecine hydrochloride obtained showed enrichment factors of 0.4—0.5% for the signals at C-3, C-5, C-8, and C-9. In addition, the presence of ^{13}C-^{15}N species was evident from the presence of two doublets with enrichment factors of 0.2—0.25% corresponding to C-3/N-4 and C-5/N-4. These ^{13}C-^{15}N species were also observed in the ^{15}N n.m.r. spectrum. The enrichment of C-3 and C-5 of retronecine with approximately equal amounts of ^{13}C-^{14}N and ^{13}C-^{15}N species is compelling evidence for the participation of a symmetrical intermediate in retronecine biosynthesis.

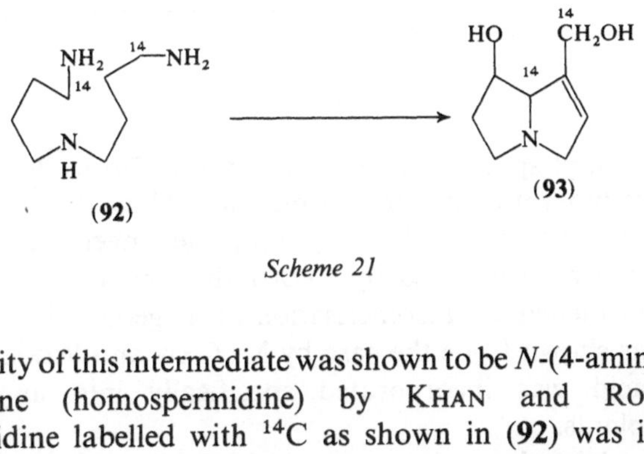

Scheme 21

The identity of this intermediate was shown to be *N*-(4-aminobutyl)-1,4-diaminobutane (homospermidine) by KHAN and ROBINS (*155*). Homospermidine labelled with ^{14}C as shown in (**92**) was incorporated efficiently into retronecine (**93**) (Scheme 21). On degradation (Scheme 19), 44% of the activity was found to be located at C-9, while the β-alanine portion was inactive. These results are consistent with the intact conversion of homospermidine into retronecine. Furthermore, the status of homospermidine as an intermediate in retronecine biosynthesis was tested by a trapping experiment. [5-^{14}C]-Ornithine was fed to *S. isatideus* plants and 24 hours later, the plants were harvested and extracted with trichloroacetic acid. Inactive homospermidine was added to this extract and the homospermidine was isolated as its phenylthiourea derivative. Recrystallisation to constant specific activity gave a homospermidine sample which retained 0.5% of the original activity fed. Therefore, homospermidine is confirmed as an intermediate in retronecine biosynthesis.

2. Necic Acids

Most of the necic acids are C_{10}-acids. Because they can apparently be divided into two isoprene units, they were originally believed to be derived from mevalonate, even though the oxidation patterns and mode of coupling of the C_5 units were quite different from those of terpenoid compounds. Indeed, acetate and mevalonate were found not to be specific precursors for the C_{10}-necic acid portions of pyrrolizidine alkaloids (74, 123). All of the acids investigated so far have been shown to be formed from common branched-chain amino acids.

Thus, CROUT (69) demonstrated that valine (94) was incorporated specifically into echimidinic acid (95) in *Cynoglossum officinale* plants. These plants produce heliosupine (Table 3 N) which has two esterifying acids, echimidinic acid and angelic acid. Cleavage of the diol (95) with sodium periodate gave acetone with 93% of the acid's activity, and from the acetone, iodoform was generated with the same specific activity. It is suggested that valine reacts with a C_2 "active acetaldehyde" to form echimidinic acid (Scheme 22).

$$
\begin{array}{c}
^{14}Me_2CH \\
| \\
CHNH_2 \\
| \\
CO_2H \\
\\
(94)
\end{array}
\quad + \quad
\begin{array}{c}
\text{'active} \\
\text{acetaldehyde'}
\end{array}
\quad \longrightarrow \quad
\begin{array}{c}
^{14}Me_2COH \\
| \\
HO{-}CHOHMe \\
| \\
CO_2H \\
\\
(95)
\end{array}
$$

Scheme 22

The same plant species was used by CROUT (70) to demonstrate that angelic acid is biosynthesized from isoleucine. Thus, [U-^{14}C]-L-isoleucine was specifically incorporated (98%) into the angelic acid portion of heliosupine. It was recognised by CROUT that the angelic acid could be formed from isoleucine *via* isomerisation of its geometrical isomer, tiglic acid. This was shown to be the case by McGAW and WOOLLEY (186). [1-^{14}C]-Tiglic acid was incorporated specifically into angelic acid in *C. officinale* plants.

Senecic acid has been the most widely studied of the C_{10}-acids and belongs to the main structural type found within this group of acids. Experiments by CROUT et al. (74) with a wide range of ^{14}C-labelled compounds demonstrates that isoleucine (96) (and its biological precursor threonine) is incorporated into the left-hand C_5 unit of senecic acid (97). Further experiments with [2-^{14}C]- and [6-^{14}C]-isoleucine were carried out by CROUT et al. (75), on *Senecio magnificus* plants which produce senecionine (Table 3 O iii). These precursors were specifically incorporated into the senecic acid component of senecionine, and degradations gave partial labelling patterns which were consistent with the biosynthesis of senecic acid from two isoleucine molecules with loss of both carboxyl carbons (Scheme 23). Furthermore, DAVIES and CROUT (98) showed that, of the four stereoisomers of isoleucine, only L-isoleucine is incorporated efficiently into senecic acid.

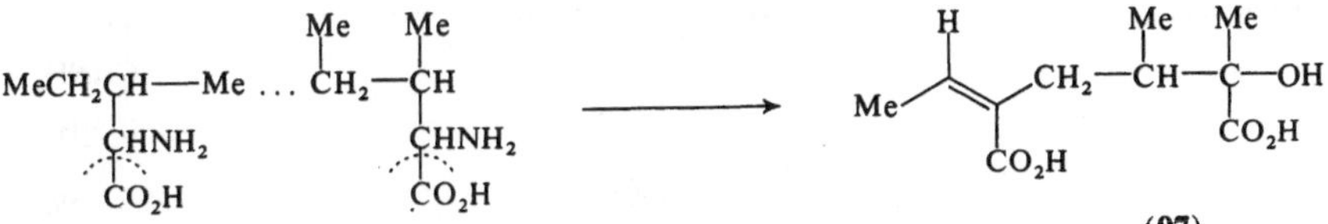

$$\underset{\textbf{(96)}}{\underset{\begin{array}{c}|\\ \text{CHNH}_2\\ |\\ \text{CO}_2\text{H}\end{array}}{\overset{\begin{array}{c}\text{Me}\\ |\end{array}}{\text{MeCH}_2\text{CH—Me}}} \quad \cdots \quad \underset{\textbf{(96)}}{\underset{\begin{array}{c}|\\ \text{CHNH}_2\\ |\\ \text{CO}_2\text{H}\end{array}}{\overset{\begin{array}{c}\text{Me} \quad \text{Me}\\ | \quad \quad |\end{array}}{\text{CH}_2\text{—CH}}}} \quad\longrightarrow\quad \underset{\textbf{(97)}}{\underset{\begin{array}{c}\text{CO}_2\text{H}\end{array}}{\overset{\begin{array}{c}\text{Me} \quad \text{Me}\\ | \quad \quad |\end{array}}{\text{Me}\diagup\diagdown\text{CH}_2\text{—CH—C—OH}}}}$$

Scheme 23

The mechanism by which these two isoleucine untis are coupled in senecic acid biosynthesis is not obvious. CROUT *et al.* (*25*) tested a number of C_5-intermediates of isoleucine metabolism as possible precursors for this coupling process. Eventually, they found that 2-amino-[3-^{3}H$_2$]-methylenepentanoic acid (**98**) was incorporated specifically into senecic acid (**97**) with about the same efficiency as for L-isoleucine. Unfortunately, the low radioactivity in the necic acid precluded further degradation to establish whether one or both halves of the senecic acid were labelled. There is also the possibility that the precursor (**98**) could be converted into isoleucine before incorporation. Thus, the status of (**98**) as an intermediate in senecic acid biosynthesis clearly requires further investigation. Nevertheless, CROUT (*73*) has postulated a mechanism for activation of (**98**) into a pyridoxal bound mesomeric nucleophile (**99**), which could then attack an unspecified electrophilic metabolite of isoleucine to form senecic acid (Scheme 24).

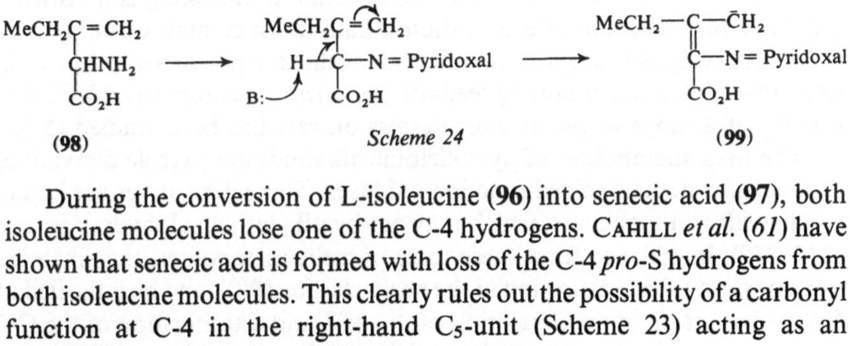

$$\underset{\textbf{(98)}}{\underset{\begin{array}{c}|\\ \text{CHNH}_2\\ |\\ \text{CO}_2\text{H}\end{array}}{\overset{}{\text{MeCH}_2\text{C}=\text{CH}_2}}} \longrightarrow \underset{}{\overset{}{\text{MeCH}_2\text{C}\overset{}{=}\text{CH}_2}} \longrightarrow \underset{\textbf{(99)}}{\overset{}{\text{MeCH}_2\text{—C—}\bar{\text{C}}\text{H}_2}}$$

Scheme 24

During the conversion of L-isoleucine (**96**) into senecic acid (**97**), both isoleucine molecules lose one of the C-4 hydrogens. CAHILL *et al.* (*61*) have shown that senecic acid is formed with loss of the C-4 *pro*-S hydrogens from both isoleucine molecules. This clearly rules out the possibility of a carbonyl function at C-4 in the right-hand C_5-unit (Scheme 23) acting as an electrophilic acceptor in senecic acic biosynthesis.

ROBINS *et al.* (*246*) demonstrated that L-threonine and L-isoleucine are specific precursors for monocrotalic acid (**46**), the C_8-necic acid component of monocrotaline (**44**), present in several *Crotalaria* species. Degradation of the monocrotalic acid produced gave results which were consistent with the incorporation of isoleucine into C-1, C-2, C-3, C-6, and C-7 of monocrotalic acid (Scheme 25). The remaining three carbon atoms may be derived from propionate.

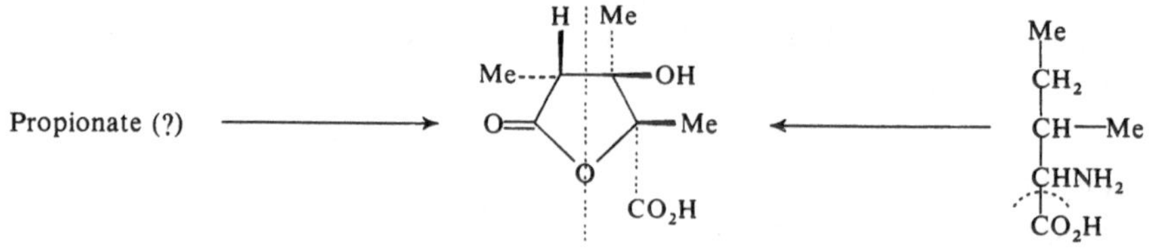

Scheme 25

VII. Pharmacology

Many pyrrolizidine alkaloids are toxic, acting specifically on the liver (*60, 147, 204*). Some are also carcinogenic (*16, 269*), and ingestion of these alkaloids by humans is now a recognised world-wide health problem. Consumption of wheat flour contaminated with seeds of *Heliotropium popovii* caused a severe outbreak of veno-occlusive disease in Afghanistan (*297*). Other cases have been reported in central India (*296*) and the U.S.A. (*284*). A number of plants containing pyrrolizidine alkaloids are still widely used as herbal remedies and in teas. Extracts of coltsfoot *(Tussilago farfara)* and comfrey *(Symphytum officinale)* are both carcinogenic (*138*), and the alkaloids present in these plants cause hepatic tumours in rats (*137*). Pyrrolizidine alkaloids have appeared in samples of honey (*100*) and cows' milk (*102*).

Loss of livestock due to pyrrolizidine alkaloid poisoning is a growing economic problem. The effects of these alkaloids on animals was discussed at a recent symposium (*150*). An outbreak of chicken poisoning in Australia was caused by food containing seeds of *Heliotropium europaeum* (*220*). The toxicity of a range of pyrrolizidine esters on rats has been studied (*81*).

The toxic metabolites of pyrrolizidine alkaloids are pyrrole derivatives produced by hepatic mixed function oxidases. These derivatives are highly reactive alkylating agents, which may act by alkylating sulphydryl groups (*242, 270*). In order to form these toxic pyrrolic metabolites, the alkaloids must possess 1,2-unsaturation in the necine [as in (**100**)], and be esterified at C-9. Substitution at the α-position of the acid and esterification of the C-7 hydroxy-group both increase the toxicity of the alkaloid.

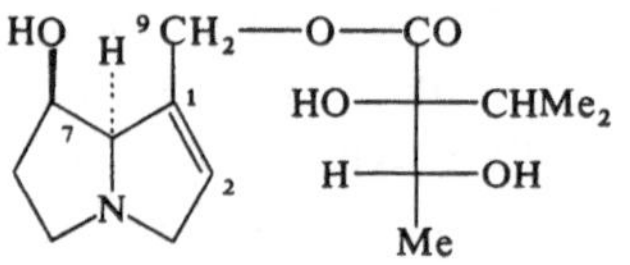

(**100**)

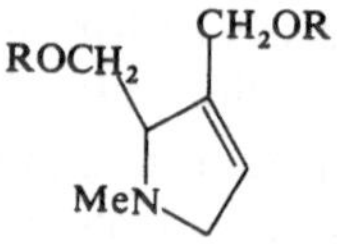

(**101**) R = H
(**102**) R = CONHEt

The *N*-oxide of indicine (**100**) does disply anti-tumor activity with less toxic effects than related pyrrolizidine alkaloids; it has recently been tested on humans (*176*). Some semi-synthetic quaternary pyrrolizidine derivatives block neuromuscular transmissions, and their synthesis and biological activities have recently been reviewed (*19, 261*).

MATTOCKS (*205*) has prepared several substituted 3-pyrrolines [*e.g.* synthanecine A (**101**)] which are monocyclic analogues of the pyrrolizidine alkaloids. Some of these derivatives [particularly the carbamate (**102**) of synthanecine A] can be dehydrogenated in animals to the corresponding pyrrole derivatives, which act as alkylating agents, and show similar toxic effects to those produced by pyrrolizidine alkaloids.

Table 1. *List of Plant Genera Containing Pyrrolizidine Alkaloids* (with Number of Species Investigated)

Family	Genera
Apocynaceae	*Alafia* (1), *Anodendron* (1), *Parsonsia* (4), *Urechtites* (1)
Boraginaceae	*Amsinckia* (3), *Anchusa* (2), *Asperugo* (1), *Caccinia* (1), *Cynoglossum* (9), *Echium* (5), *Ehretia* (1), *Heliotropium* (15), *Lappula* (2), *Lindelofia* (8), *Lithosperum* (1), *Macrotomia* (1), *Paracaryum* (1), *Paracynoglossum* (1), *Rindera* (4), *Solenanthus* (4), *Symphytum* (7), *Tournefortia* (2), *Trachelanthus* (2), *Trichodesma* (2), *Ulugbekia* (1)
Celastraceae	*Bhesa* (1)
Compositae	*Adenostyles* (2), *Brachyglottis* (1), *Cacalia* (4), *Doronicum* (1), *Emilia* (2), *Erechtites* (1), *Eupatorium* (4), *Farfugium* (1), *Kleinia* (1), *Ligularia* (4), *Petasites* (3), *Senecio* (118), *Syneilesis* (1), *Tussilago* (1)
Euphorbiaceae	*Phyllanthus* (1), *Securinega* (1)
Graminae	*Festuca* (1), *Lolium* (1), *Thelepogon* (1)
Leguminosae	*Adenocarpus* (4), *Crotalaria* (55), *Cytisus* (1)
Orchidaceae	*Chysis* (1), *Doritis* (1), *Hammarbya* (1), *Kingiella* (1), *Liparis* (7), *Malaxis* (2), *Phalaenopsis* (14), *Vanda* (4), *Vandopsis* (2)
Ranunculaceae	*Caltha* (2)
Rhizophoraceae	*Cassipourea* (2)
Santalaceae	*Thesium* (1)
Sapotaceae	*Mimusops* (1), *Planchonella* (2)
Scrophulariaceae	*Castilleja* (1)

Table 2. *Alkaloid Content of Plant Species Which Have Been Investigated*

	Alkaloid	References
Apocynaceae		
Alafia multiflora Stapf.	Alafine	(218)
Anodendron affine Druce	Anodendrine, alloanodendrine	(263)
Parsonsia eucalyptophylla (F. Muel.)	Lycopsamine, intermedine or indicine	(107)
P. heterophylla A. Cunn.	Parsonsine, heterophylline	(108)
P. spiralis Wall.	Parsonsine, heterophylline, spiraline spiranine, spiracine	(108)
P. straminea [(R. Br.) F. Muel.]	Lycopsamine, intermedine or indicine	(107)
Urechites karwinsky Muel.	Loroquine	(49)
Boraginaceae		
Amsinckia hispida (Ruiz et Pav.) I. M. Johnston	Intermedine, lycopsamine, echiumine	(89)
A. intermedia Fisch. et C. Mey	Intermedine, lycopsamine, echiumine, sincamidine	(89)
A. lycopsioides Lehm.	Intermedine, lycopsamine, echiumine	(89)
Anchusa arvensis (L.) Bieb.	Echinatine (or stereoisomer)	(222)
A. officinalis L.	Echinatine (or stereoisomer),	(222)
	Lycopsamine	(58)
Asperugo procumbens L.	Supinine, echinatine (or stereoisomer)	(222)
Caccinia glauca Savi	Retronecine-7,9-dibenzoate	(278)
Cynoglossum amabile Stapf. and Drummond	Amabiline, echinatine	(91)
C. australe R. Br.	Cynaustine, cynaustraline	(91)
C. creticum	Echinatine, heliosupine	(325)

Boraginaceae	Alkaloid	References
C. glochidiatum Wall. ex Lindl.	Amabiline	(289)
C. lanceolatum Forsk.	Cynaustraline, cynaustine	(289)
C. latifolium R. Br.	Latifoline, 7-angelylretronecine	(311)
C. officinale L.	Heliosupine, echinatine, cynoglossophine,	(311)
	Acetylheliosupine, 7-angelylheliotridine	(221)
C. pictum Ait.	Pictumine	(195)
C. viridiflorum Pallas ex Lehm.	Viridiflorine	(310)
Echium diffusum	Retronecine and heliotridine diesters	(18)
E. italicum L.	Echimidine	(60)
— var. biebersteinii	Retronecine and heliotridine diesters	(18)
E. lycopsis L. (E. plantagineum L.)	Echiumine, echimidine	(311)
E. vulgare L.	Heliosupine	(189)
	Asperumine	(149)
	Echinatine, heliosupine, acetylheliosupine (or stereoisomers of these)	(222)
Ehretia aspera Willd.	Ehretinine	(292)
Heliotropium amplexicaule Vahl.	Indicine	(60)
H. arbainense	Heliotrine, lasiocarpine, europine	(325)
H. arborescens L. (H. peruvianum L.)	Lasiocarpine (?)	(198)
H. arguzioides Kar. et Kir.	Heliotrine	(207)
	Trichodesmine	(8)
H. curassavicum L.	Curassavine, coromandalin, heliovicine	(210)
	Heliotrine, lasiocarpine, 7-angelylheliotridine	(236)
H. dasycarpum Ledeb.	Heliotrine	(8)

Table 2 (*continued*)

Boraginaceae	Alkaloid	References
Heliotropium eichwaldii Steud.	Heliotrine,	(60)
	Lasiocarpine, 7-angelylheliotrine	(293)
H. europaeum L.	Heliotrine, supinine, lasiocarpine, europine, heleurine	(311)
	Dipyrrolizidine	(92)
	Acetyllasiocarpine	(85)
H. indicum L.	Indicine	(311)
	Acetylindicine, indicinine	(203, 206)
	Heliotrine, lasiocarpine, echinatine, supinine, heleurine	(139)
H. lasiocarpum Fisch. et C. Mey	Heliotrine, lasiocarpine	(310)
H. maris-mortui	Europine	(324)
	Lasiocarpine	(325)
H. olgae Bunge	Heliotrine	(160)
	Incanine	(276)
H. ramosissimum	Heliotrine	(129)
H. rotundifolium	Europine	(324)
H. supinum L.	Supinine, heliosupine, echinatine, 7-angelylheliotridine esters	(311)
Lappula glochidiata	Echinatine	(291)
L. intermedia	Lasiocarpine	(197)
Lindelofia anchusioides	Lindelofine	(310)
L. augustifolia (Schrenk) Brand	Amabiline, echinatine	(289)
L. macrostyla (Bunge) M. Pop.	Lindelofine, lindelofamine	(311)
L. olgae (Regel et Smirnov) Brand	Viridiflorine	(10)
L. pterocarpa (Rupr.) M. Pop.	Viridiflorine	(13)
L. spectabilis Lehm.	Echinatine, monocrotaline, 7-acetylechinatine	(293)
L. stylosa (Kar. et Kir.) Brand	Viridiflorine, lindelofine, echinatine	(9, 160)

Boraginaceae	Alkaloid	References
Lindelofia tshimganica	Carategine, echinatine, viridiflorine	(11)
Lithosperum officinale L.	Retronecine or heliotridine ester	(222)
Macrotomia echioides Boiss	Macrotomine	(310)
Paracaryum himalayense (Klotzsch) C. B. Clarke	Viridiflorine	(13)
Paracynoglossum imeretinum (Kusnez.) Popov	Heliosupine, echinatine	(194)
Rindera austroechinata M. Pop.	Echinatine	(11)
R. baldschuanica Kusnezov	Rinderine, echinatine, trachelanthamine, turkestanine	(11)
R. echinata Regel	Echinatine	(311)
R. oblongifolia M. Pop.	Carategine, echinatine, turkestanine	(11)
Solenanthus circinatus Ledeb.	Carategine	(13)
S. coronatus Regel	Echinatine	(9)
S. karateginus Lipsky	Carategine, echinatine	(13)
S. turkestanicus (Regel et Smirnov) Kusnezov	Rinderine, turkestanine	(160)
Symphytum asperum Lepechin	Asperumine, heliosupine, echinatine	(193)
	Acetyl derivatives of echinatine and heliosupine (or stereosiomers)	(222)
S. caucasicum	Asperumine, lasiocarpine, echinatine, echimidine	(196)
S. officinale L.	Echimidine, symphytine	(114)
	Lasiocarpine, heliosupine, echinatine, viridiflorine	(192)
	Acetyl derivatives of echimidine or heliosupine (or stereosiomers)	(222)
— var. *ochroleucum* DC.	Echinatine, acetyl derivative (or stereosiomers)	(222)

Table 2 (continued)

	Alkaloid	References
Boraginaceae		
Symphytum orientale	Anadoline	(306)
	Symphytine, echimidine	(307)
S. tuberosum	Echimidine, anadoline	(308)
Symphytum x uplandicum Nyman	Heliotridine or retronecine esters	(222)
	Echimidine, symphytine, lycopsamine, intermedine, 7-acetyllycopsamine, 7-acetylindicine, symlandine, uplandicine	(79)
Tournefortia sarmentosa Lam.	Supinine	(311)
T. sibirica L.	Turneforcine	(310)
Trachelanthus hissaricus Lipsky	Trachelanthamine, viridiflorine	(12)
T. korolkovii Lipsky	Trachelanthamine	(310)
Trichodesma incanum Bunge	Trichodesmine	(311)
T. zeylanicum (Burm f.) R. Br.	Supinine	(217)
Ulugbekia tschimganica (B. Fedtsh.) Zak.	Uluganine	(157)
Celastraceae		
Bhesa archboldiana (Merr. et Perry) Ding Hou	9-angelylretronecine	(84)
Compositae		
Adenostyles alliariae	Platyphylline, seneciphylline	(320)
A. rhombifolia (Willd.) Pim.	Platyphylline, seneciphylline, sarracine	(228)
Brachyglottis repanda Forst. et Forst. f.	Senecionine, senkirkine	(60)
Cacalia hastata L. subsp. *orientalis* Kitamura	Hastacine	(310)
	Integerrimine	(132)

References, pp. 187—203

Compositae	Alkaloid	References
Cacalia robusta Tolmatch	Hastacine	(175)
C. yatabei Maxim.	Yamataimine	(136)
Doronicum macrophyllum	Doronine	(15)
Emilia flammea Cass.	Emiline	(173)
E. sonchifolia DC.	Senecionine	(60)
Erechtites heiracifolia (L.) Raf. ex DC.	Senecionine, seneciphylline	(310)
Eupatorium cannabinum L.	Echinatine, supinine	(223)
E. maculatum L.	Echinatine, trachelanthamidine	(303)
E. serotinum Michx	Supinine, rinderine	(183)
E. stoechadosmum Hance	Lindelofine, supinine	(116)
Farfugium japonicum Kit.	Senkirkine	(118)
Kleinia kleinioides (Sch. Bip.) M. R. F. Taylor	Isosenaetnine, dehydroisosenaetnine	(39)
Ligularia brachyphylla Hand.-Mazz.	Clivorine, ligularine, ligudentine	(167)
L. clivorum Maxim.	Clivorine	(170)
L. dentata (A. Gray) Hara	Clivorine, ligularine, ligudentine, Ligularidine	(167) (133)
L. elegans Cass. [*L. macrophylla* (Ledeb.) DC.]	Clivorine, ligularine	(167)
Petasites hybridus (L.) P. Gaertn. et al.	Senecionine	(60)
P. japonicus Maxim.	Petasitenine, neopetasitenine, Fukinotoxin (= petasitenine) Senkirkine, petasinine, petasinoside Platyphylline, senkirkine, senecionine	(321) (117) (322) (311)
P. laevigatus (Willd.) Reichenb. (*Nardosmia laevigata*)		

Table 2 (continued)

Compositae	Alkaloid	References
Senecio adnatus DC.	Platyphylline	(310)
S. aegyptius L.	Senecionine	(122)
	Erucifoline	(169a)
S. aetnensis Jan.	Senaetnine	(40)
S. alpinus (L.) Scop.	Integerrimine, jacozine	(164)
S. amphilobus (C. Koch)	Macrophylline	(94)
S. ampullaceus Hook.	Senecionine, seneciphylline, retrorsine	(310)
S. angulatus L. f.	Angularine, rosmarinine	(311)
S. antieuphorbium (Sch. Bip.)	Senkirkine, integerrimine	(258)
	Senaetnine, isosenaetnine	(46)
S. aquaticus Hill	Seneciphylline	(311)
S. aucheri DC.	Pterophorine, senaetnine	(42)
	(plus geometrical isomers)	
S. aureus L.	Senecionine	(310)
S. auricola Bourg.	Neosenkirkine	(219)
S. barbellatus DC.	Retrorsine, seneciphylline	(310)
(*S. retrorsus* and *S. latifolius*)		
S. barbertonicus Klatt.	Dehydroisosenaetnine isomer	(45)
S. bipinnatisectus Belcher	Retrorsine	(314)
(*Erechtites atkinsoniae*)		
S. brachypodus DC.	Rosmarinine	(310)
S. brasiliensis DC.	Jacobine, senecionine, seneciphylline	(310)
S. brasiliensis Less.	Integerrimine, retrorsine,	(311)
(*S. ambrosioides*)	Seneciphylline, senecionine	
S. bupleuroides DC.	Retrorsine	(311)
S. campestris (Retz.) DC.	Campestrine, senecionine	(311)
var. *maritimus*		
S. carthamoides Greene	Senecionine, seneciphylline	(311)

Compositae	Alkaloid	References
Senecio chrysanthemoides DC.	Seneciphylline	(311)
S. cineraria DC.	Jacobine, seneciphylline, senecionine	(311)
	Otosenine	(128)
	Retrorsine	(163)
S. cissampelinus (DC.) Sch. Bip.	Senampelines A – D	(43)
S. colaminus Cuatr.	Pterophorine	(41)
S. crispatis DC.	7-Angelylheliotridine	(311)
(*S. rivularis*)		
S. cruentus DC.	Alkaloids $C_{18}H_{25}O_5N$ and $C_{18}H_{25}O_6N$	(63)
S. desfontanei Druce	Seneciphylline	(122)
(*S. coronopifolius*)		
S. discolor DC.	Retrorsine, senecionine	(311)
S. doria L. (*S. paucifolius*)	Seneciphylline, alkaloid $C_7H_{13}O_2N$	(65)
S. doronicum L.	Doronenine, bulgarsenine	(255)
S. douglasii DC. (*S. longilobus, S. riddellii, S. spartioides*)	Seneciphylline, senecionine, retrorsine, riddelline	(311)
S. durieui Gay	Integerrimine	(219)
S. eremophilus Richards	Senecionine, seneciphylline, retrorsine, riddelline	(310)
S. erraticus (Bertol.) subsp. barbaraeifolius Krock	Senecionine, seneciphylline, otosenine, integerrimine, erucifoline	(310)
S. erucifolius L.	Senecionine, seneciphylline	(311)
	Erucifoline	(174)
	Retrorsine	(112)
S. floridanus Sch. Bip. (*Brachyglottis floridiana*)	Floricaline, floridanine, florosenine, otosenine	(62)
S. fluviatilis Wallr.	Otosenine, seneciphylline, florosenine	(169)
S. francheti C. Winkl.	Franchetine, sarracine	(12)
S. fremontii Torr. et A. Gray	Senecionine, seneciphylline	(311)
S. fuchsii C. C. Gmel.	Fuchsisenecionine	(254)

Table 2 *(continued)*

Compositae	Alkaloid	References
Senecio glabellus (Turcz.) DC.	Senecionine	*(310)*
S. glaberrimus DC.	Retrorsine	*(310)*
S. graminifolius Jacq.	Retrorsine, graminifoline	*(310)*
S. grisebachii Baker	Retrorsine	*(212)*
S. halimifolius L.	Rosmarinine	*(312)*
S. hygrophyllus R. A. Dyer et C. A. Smith	Platyphylline, rosmarinine, hygrophylline	*(311)*
S. ilicifolius Thunb.	Senecionine, seneciphylline, retrorsine	*(310)*
S. inaequidens DC.	Pterophorine, inaequidenine	*(40)*
S. incanus (L.) subsp. *carniolicus* (Willd.) Br.-Bl.	Integerrimine, jacozine, seneciphylline	*(164)*
S. integerrimus Nutt.	Integerrimine, senecionine	*(310)*
S. isatideus DC.	Retrorsine	*(310)*
S. jacobaea L.	Seneciphylline, senecionine, jacobine, jacoline, jaconine, jacozine	*(311)*
S. kaempferi DC.	Sarracine	*(310)*
S. kirkii Hook. f. ex T. Kirk	Senkirkine	*(311)*
	O-Acetylsenkirkine	*(57)*
S. kleinia Sch. Bip.	Integerrimine	*(311)*
	Senkirkine	*(257)*
S. kubensis (Grossheim)	Seneciphylline	*(152)*
S. lautus Forst. f. ex Willd.	Senecionine	*(60)*
S. longiflorus Sch. Bip.	Senecionine, seneciphylline	*(311)*
	Senaetnine	*(42)*
S. longilobus Benth.	Seneciphylline, retrorsine, riddelline	*(310)*
S. macrophyllus Bieb.	Macrophylline	*(311)*
S. magnificus F. Muell.	Senecionine, integerrimine	*(311)*
S. mikanoides (Walp.) Otto.	Sarracine	*(311)*
	Senampelines C – G	*(42)*
S. morrisonensis Hayata	Integerrimine	*(77)*

Compositae	Alkaloid	References
Senecio nebrodensis L. var. *sicula*	Senecionine, integerrimine	(230)
S. nemorensis L.		
— var. *bulgaricus* (Vel.) Stoj. et Stef.	Nemorensine, bulgarsenine, retroisosenine	(214)
— subsp. *fuchsii* Gmelin	Senecionine	(316)
— — var. *nova* (Zlatnik)	Nemorensine	(165)
— subsp. *Jaquinianus* (Rchb.) Durard	Nemorensine	(165)
— var. *subdecurrens* Griseb.	Nemorensine	(165)
	Retroisosenine, bulgarsenine	(168)
S. othonnae Bieb.	Otosenine, onetine, seneciphylline	(311)
	Doronine, floridanine	(153)
S. othonniformis Fourcade	Bisline, isoline, retrorsine	(67)
S. palmatus Pall.	Seneciphylline	(311)
(*S. cannabifolius*)		
S. paludosus L.	Seneciphylline, jacobine	(311)
S. palustris (L.) Hook.	Alkaloid $C_{18}H_{27}O_5N$	(310)
S. pampeanus Cabrera	Senecionine	(311)
S. paniculatus Berg. (*S. grandifolius*)	Platyphylline, senecionine	(60)
S. paucicalyculatus Klatt.	Paucicaline, $C_{18}H_{22}NO_8$, retrorsine	(311)
S. paucitigulatus R. A. Dyer et C. A. Smith	Rosmarinine	(310)
S. pellucidus (*S. ruderalis*)	Retrorsine	(60)
S. petasitis DC.	Senecionine	(122)
S. platyphylloides Somm. et Lev	Platyphylline, seneciphylline	(60)
S. praealtus Bertol.	Seneciphylline	(311)
(*S. borysthenicus*)		
S. propinquus Schischk.	Seneciphylline	(151)
S. procerus L. var. *procerus* Stoj. Stef. et Kit.	Procerine	(148)
S. pseudo-arnica Less.	Senecionine	(310)

Table 2 *(continued)*

Compositae	Alkaloid	References
Senecio pterophorus DC.	Senecionine, seneciphylline, retrorsine	*(310)*
S. pubigerus L.	Pterophorine	*(43)*
S. pulviniformis Hieron.	Pterophorine	*(44)*
S. quadridentatus Labill.	Pterophorine (and epimer)	*(41)*
(Erechtites quadridentata)	Senecionine, seneciphylline, retrorsine	*(60)*
S. racemosus DC.	Seneciphylline	*(60)*
S. renardii Winkl.	Seneciphylline, senkirkine, otosenine	*(310)*
S. rhombifolius (Willd.) Sch. Bip.	Sarracine, platyphylline,	*(311)*
(S. platyphyllus)	seneciphylline, neoplatyphylline	
S. riddellii Torr. et Gray	Retrorsine, riddelline	*(310)*
var. *parksii* Cory		
S. rosmarinifolius L. f.	Rosmarinine	*(310)*
S. rudbeckiafolius Meyer et Walp.	Pterophorine	*(41)*
S. ruwenzoriensis S. Moore	Ruwenine, ruzorine	*(310)*
S. sarracenicus L.	Sarracine	*(310)*
S. scandens	Senecionine, seneciphylline	*(32)*
S. sceleratus Schweickerdt	Retrorsine, sceleratine,	*(310)*
	sceleratinic ester of retronecine	
S. schweizovii Korsh.	Macrophylline	*(60)*
S. spathulatus A. Rich.	Integerrimine, senecionine	*(314)*
S. squalidus L.	Senecionine, integerrimine	*(310)*
S. stenocephalus Maxim.	Seneciphylline	*(310)*
S. subalpinus Koch	Senecionine, seneciphylline	*(60)*
	Integerrimine, jacozine	*(164)*
S. swaziensis Compton	Swazine, retrorsine	*(124)*
S. sylvaticus L.	Sarracine	*(315)*
S. taiwensis Hayata	Rosmarinine	*(184)*
S. tomentosus	Senecionine, otosenine	*(311)*

	Alkaloid	References
Compositae		
Senecio tournefortia Lapeyr.	Platyphylline	(309)
S. triangularis Hook.	Senecionine	(60)
S. venosus Harvey	Retrorsine	(310)
S. vernalis Waldst. et Kit.	Senecivernine, senecionine, senkirkine, retrorsine	(256)
S. viminalis Bremek.	Senecionine, retrorsine	(311)
S. viscosus L.	Senecionine, integerrimine	(311)
S. vulgaris L.	Senecionine, seneciphylline, retrorsine	(310)
S. warszewiczii A. Br. et Bouché	Pterophorine	(42)
Syneilesis palmata Maxim.	Syneilesine	(134)
	Acetylsyneilesine, senecionine	(135)
Tussilago farfara L.	Senecionine	(60)
	Senkirkine	(82)
Euphorbiaceae		
Phyllanthus piruri L.	(+)-Norsecurinine	(260)
Securinega virosa	(−)-Norsecurinine, and dihydro derivative	(262)
Graminae		
Festuca arundinacea Schreb.	Loline (festucine)	(323)
Lolium cuneatum Nevski	Loline, lolinine, norloline,	(311)
	N-Formylnorloline	(30)
	N-Methylloline, N-acetylnorloline, N-formylloline	(28)
	Lolidine	(29)
Thelepogon elegans Roth.	Thelepogine, thelepogidine	(311)

Table 2 (continued)

Leguminosae	Alkaloid	References
Adenocarpus argyrophyllus	Decorticasine	(17)
A. decorticanus Boiss.	Decorticasine	(241, 281)
A. grandiflorus Boiss.	Decorticasine	(208)
A. hispanicus DC.	Decorticasine	(240)
Crotalaria aegyptiaca Benth.	Monocrotaline,	(187)
	7β-hydroxy-1-methylenepyrrolizidine	
C. agatiflora Schweinf.	Crosemperine	(325)
	Madurensine, 7-acetylmadurensine,	(93)
	anacrotine, 6-acetylanacrotine,	
	and four other diesters of crotanecine	
C. albida Heyne ex Roth.	Croalbidine	(266)
(*C. montana*)		
C. anagyroides H. B. et K.	1-Methylenepyrrolizidine	(311)
	Senecionine, anacrotine	(22)
C. aridicola Domin.	Methyl ethers of supinidine and	(311)
	retronecine	
C. axillaris Ait.	Axillarine, axillaridine	(71)
C. barbata R. Graham	Crobarbatine	(232)
C. brevifolia	Usaramine, integerrimine	(268)
Crotalaria burhia Buch-Ham.	Crotalarine (croburhine), monocrotaline	(14, 237)
C. candicans W. et A. 184	Crocandine, isocrocandine	(280)
C. crassipes Hook.	Retusamine	(88)
C. crispata F. Muell. ex Benth.	Monocrotaline, fulvine, crispatine	(311)
C. cylindrocarpa DC.	7β-Hydroxy-1-methylenepyrrolizidine	(227)
C. damarensis Engl.	1-Methylenepyrrolizidine	(311)
C. dura Wood et Evans	Dicrotaline	(310)
C. fulva Roxb.	Fulvine	(311)
C. globifera E. Mey.	Dicrotaline	(310)
C. goreensis Guill. et Perr.	7β-Hydroxy-1-methylenepyrrolizidines	(311)

Leguminosae	Alkaloid	References
Crotalaria grahamiana R. Wight et Walk. Arn.	Monocrotaline, grahamine, monocrotalinine	(21, 235)
C. *grandistipulata* Harms.	1-Methylenepyrrolizidine	(227)
C. *grantiana* Harv.	Grantianine, grantaline,	(60)
	1-hydroxymethyl-1,2-epoxypyrrolizidine	
C. *incana* L.	Integerrimine	(310)
	Usaramine, anacrotine	(60)
C. *intermedia* Kotschy	Integerrimine, usaramine	(23)
C. *juncea* L.	Senecionine, seneciphylline,	(311)
	riddelline, trichodesmine, junceine	
C. *laburnifolia* L.	Anacrotine, madurensine	(265)
— subsp. *eldomae* (Bak. f.) Polhill	Anacrotine, madurensine, senkirkine,	(72)
	hydrosenkirkine, crotafoline	
C. *lachnophora* A. Rich.	1-Methylenepyrrolizidine	(227)
C. *leiloba* Bartl.	Monocrotaline	(233)
C. *leschenaulti* DC.	Monocrotaline, crispatine	(23)
C. *madurensis* R. Wight	Fulvine, madurensine	(22)
	Crispatine	(130)
	Cromadurine	(239)
	Isocromadurine	(238)
C. *maypurensis* H. B. et K.	7β-Hydroxy-1-methylenepyrrolizidines	(60)
C. *medicaginea* Lam.	Methyl ethers of supinidine and retronecine	(264, 275)
C. *mitchelli* Benth.	Monocrotaline, retusamine	(88)
C. *mucronata* Desv.	Usaramine, integerrimine	(34)
(C. *striata*)	Crotastriatine (=nilgirine acetate)	(121)
C. *mysorensis* Roth.	Monocrotaline	(60)
C. *nana* Burm.	Crotananine	(279)
	Cronaburmine	(277)
C. *natalitia* Meissner	1-Methylenepyrrolizidine	(227)
C. *novae-hollandae* DC.	Monocrotaline, retusamine	(88)

Table 2 (continued)

Leguminosae	Alkaloid	References
Crotalaria paniculata Willd.	Fulvine	(288)
C. podocarpa DC.	7β-Hydroxy-1-methylenepyrrolizidine	(227)
C. quinquefolia L.	Monocrotaline	(60)
C. retusa	Monocrotaline, retusine, retusamine, retronecine	(311)
C. rhodesiae	1-Methylenepyrrolizidine	(60)
C. rubiginosa Willd. (*C. wightiana*)	Junceine, trichodesmine	(24)
C. sagittalis L.	Monocrotaline	(317)
C. semperflorens Vent.	Crosemperine	(20)
C. spartioides DC.	Retrorsine	(59)
C. spectabilis Roth. (*C. sericea*)	Monocrotaline, spectabiline, retusine	(311)
C. stipularia Desv.	Monocrotaline	(233)
C. stolzii (Baker f.) Milne-Redh. ex Polhill	1-Methylenepyrrolizidine	(227)
C. tetragona Roxb.	Integerrimine, trichodesmine	(233)
C. trifoliastrum Willd.	Methyl ethers of supinidine and retronecine	(311)
C. usaramoensis E. G. Baker	Integerrimine, usaramine, senecionine, retrorsine	(90)
C. verrucosa L.	1-Methylenepyrrolizidine	(60)
	Crotaverrine, O-acetylcrotaverrine	(294)
C. walkeri Arn.	O-Acetylsenkirkine, isosenkirkine,	(23)
	Crotaverrine, O-acetylcrotaverrine	(290)
Cytisus laburnum	Laburnine, laburnamine	(311)

Orchidaceae	Alkaloid	References
Chysis bractescens Lindl.	Chysine	*(185)*
Doritis pulcherrima	Phalaenopsine La or T	*(56)*
(Phalaenopsis esmeralda)		
Hammarbya paludosa (L.) O. K.	Paludosine, hammarbine	*(180, 181)*
Kingiella taenialis (Lindl.) Rolfe	Phalaenopsine La	*(52)*
Liparis auriculata	Malaxine	*(215)*
L. bicallosa Schltr.	Malaxine	*(178, 215, 298)*
L. hachijoensis Nakai	Malaxine	*(215)*
L. keitaoensis Hay.	Keitaonine, keitine	*(181)*
L. kumokiri F. Maekwa	Kumokirine	*(215)*
L. loeselii (L.) L. C. Rich	Auriculine	*(180)*
L. nervosa Lindl.	Nervosine	*(215)*
Malaxis congesta comb. nov. (Rchb. f.)	Malaxine	*(178, 298)*
M. grandifolia Schltr.	Grandifoline	*(182)*
Phalaenopsis amabilis Bl.	Phalaenopsine T	*(54)*
P. amboinensis	Phalaenopsine La	*(56)*
P. aphrodite	Phalaenopsine T	*(56)*
P. cornu-cervi Rchb. f.	Cornucervine	*(55)*
P. equestris Rchb. f.	Phalaenopsine Is, Phalaenopsine T	*(56)*
P. fimbriata	Phalaenopsine T	*(56)*
P. heiroglyfica	Phalaenopsine La or T	*(56)*
P. lueddemanniana	Phalaenopsine La or T	*(56)*
P. mannii Rchb. f.	Phalaenopsine La	*(54)*
P. sanderiana Rchb. f.	Phalaenopsine T, Phalaenopsine La	*(56)*
P. schilleriana	Phalaenopsine La	*(56)*
P. stuartiana Rchb. f.	Phalaenopsine T, Phalaenopsine La	*(56)*
P. sumatrana	Phalaenopsine La	*(56)*
P. violacea	Phalaenopsine La or T	*(56)*
Vanda cristata Lindl.	Acetyllaburnine	*(179)*

Table 2 (continued)

	Alkaloid	References
Orchidaceae		
Vanda helvola Bl.	Acetyllaburnine, acetyllindelofidine	(51)
V. hindsii Lindl.	Acetyllaburnine	(51)
V. luzonica Loher.	Laburnine or enantiomer	(51)
Vandonopsis gigantea Pfitz.	Acetyllaburnine, acetyllindelofidine	(51)
V. lissochiloides Pfitz.	Acetyllaburnine, acetyllindelofidine	(51)
Ranunculaceae		
Caltha biflora DC.	Senecionine	(282)
C. leptosepala DC.	Senecionine	(282)
Rhizophoraceae		
Cassipourea gummiflua Tulasne	Cassipourine	(318)
— var. verticellata Lewis	Cassipourine	(66)
Santalaceae		
Thesium minkwitzianum B. Fedtsch.	Thesine, thesinine, thesinicine, (+)-isoretronecanol	(311)
Sapotaceae		
Mimusops elengi L.	1-Hydroxymethylpyrrolizidine tiglate	(60)
Planchonella anteridifera Lam.	Planchonelline, three esters of laburnine	(131)
P. thyrsoidea White ex Walker	Planchonelline, two esters of laburnine	(131)
Scrophulariaceae		
Castilleja rhexifolia Rydb.	Senecionine	(283)

Table 3. *The Structures of the Pyrrolizidine Alkaloids*

A. Simple pyrrolizidine bases and loline group	Alkaloid	References
	1-Methylenepyrrolizidine	(311)
	R = α–H	(311)
	R = β–H	(311)
	R = H	(87)
	R = Me	(86)
	Chysine	(185)
	Loroquine	(49)

Table 3 (continued)

A. Loline group	Alkaloid	References
	$R^1 = Me$, $R^2 = H$: Loline (Festucine)	(27, 311)
	$R^1 = R^2 = H$: Norloline	(311)
	$R^1 = Me$, $R^2 = Ac$: Lolinine	(311)
	$R^1 = H$, $R^2 = COEt$: Decorticasine	(281)
	$R^1 = H$, $R^2 = CHO$: N-Formylnorloline	(30)
	$R^1 = R^2 = Me$: N-Methylloline	(28)
	$R^1 = H$, $R^2 = Ac$: N-Acetylnorloline	(28)
	$R^1 = Me$, $R^2 = CHO$: N-Formylloline	(28)
	(?) Lolidine	(29)

B. Retronecanol group	Alkaloid	References
	Ehretinine	(292)

C. (−)-Trachelanthamidine group	Alkaloid	References
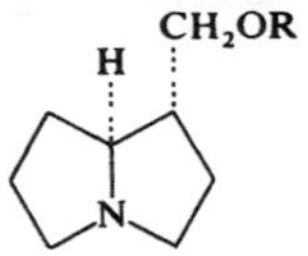 CH₂OR	R = H: Trachelanthamidine	(311)
	R 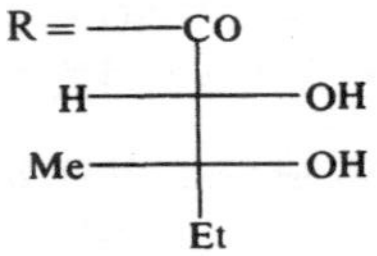 = (Strigosyl): Strigosine	(76, 311)
	R = COC(OH)(CMe₂OH)CHOHMe	(310)
	Macrotomine (Macrotomy):	
	R 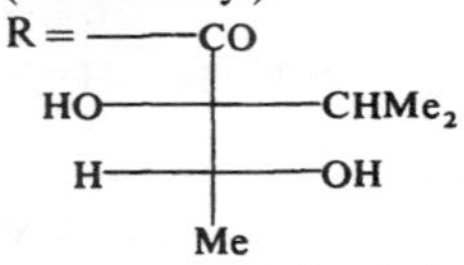 = [(+)-Trachelanthyl]: Trachelanthamine	(310)
	R = (−)-Trachelanthyl: Heliovicine	(210)
	R 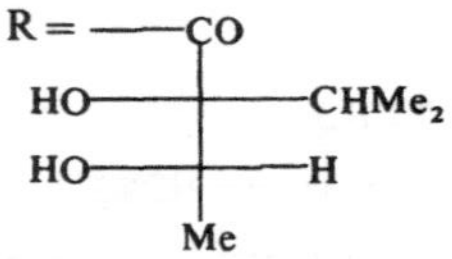 = [(−)-Viridifloryl]: Viridiflorine	(310)
	R = (+)-Viridifloryl: Coromandalin	(210)
	R = COC(OH)(CHOHMe)CHMeEt Curassavine	(210)
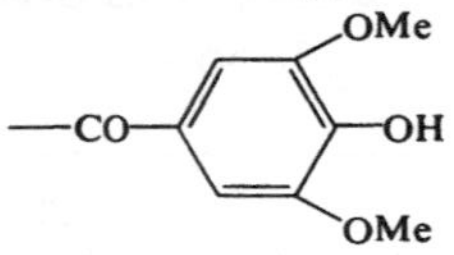	[(+)-Homoviridifloryl]: Alafine	(218)

(Macrotomine structure)

```
      Et
       |
H —— CO
Me —— OH
       |
      OH
```

```
      Me
       |
HO —— CHMe₂
H  —— CO
       |
      OH
```

```
      Me
       |
HO —— CHMe₂
HO —— CO
       |
      H
```

```
      OMe
   /        OH
R = —CO
   \        OMe
```

11*

Table 3 (*continued*)

C. (−)-Trachelanthamidine group	Alkaloid	References
	R = MeO₂C ... —COCH₂, OH, CH₂CHMe₂ Cornucervine	(53, 55)
	R = PhCH₂ ... CO, OH, CH₂CO₂Me Phalaenopsine T	(53, 54)

D. (+)-Laburrine group	Alkaloid	References
	R = H: Laburnine	(51, 310)
	R = Ac	(179)
	R = (Z)-COCH=CHMe (Angelyl)	(131)
	R = (E)-COCH=CHMe: Planchonelline	(131)
	R = COPh	(131)
	R as for Phalaenopsine T: Phalaenopsine La	(53, 54)
	R¹ = H, R² = Glu: Malaxine	(178, 215)
	R¹ = OMe, R² = H: Keitine	(181)
	R¹ = OMe, R² = Glu: Keitoaine	(181)
	R¹ = CH₂CH=CMe₂, R² = Glu: Auriculine	(180, 215)
	R¹ = CH₂CH=CMe₂, R² = Glu-Ara: Grandifoline	(182)
	Glu = β-D-glucopyranosyl	
	Glu-Ara = 2-O-β-D-glucopyranosyl-L-α-arabinosyl	

E. (+)-Isoretronecanol group	Alkaloid	References

R = H: Isoretronecanol (lindelofidine) — (311)
R = Ac — (51)
R = (−)-viridifloryl: Cynaustraline — (91)
R = (+)-trachelanthyl: Lindelofine — (310)
R = β-angelyl- (or tiglyl)-trachelanthyl: Lindelofamine — (310)
R as for auriculine: Paludosine — (180)
R as for keitoaine: Hammarbine — (181)

R = (E)-COCH=CH— ⬡ —OH: Thesinine — (311)

Thesine — (311)

F. (−)-Isoretronecanol group	Alkaloid	References

R as for Phalaenopsine T: Phalaenopsine Is — (53, 56)

Table 3 (continued)

Alkaloid	References

G. (−)-Supinidine group

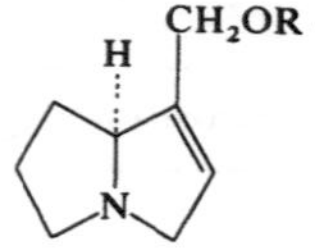

R = Me
R = viridifloryl: Amabiline
R = trachelanthyl: Supinine
R = COC(OH)CHMe₂CH(OMe)Me (heliotryl): Heleurine

Alkaloid	References
	(264, 275)
	(91)
	(310)
	(311)

H. (+)-Supinidine group

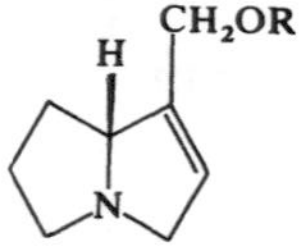

R = (−)-viridifloryl: Cynaustine

Alkaloid	References
	(91)

I. Macronecine group

Macrophylline

Alkaloid	References
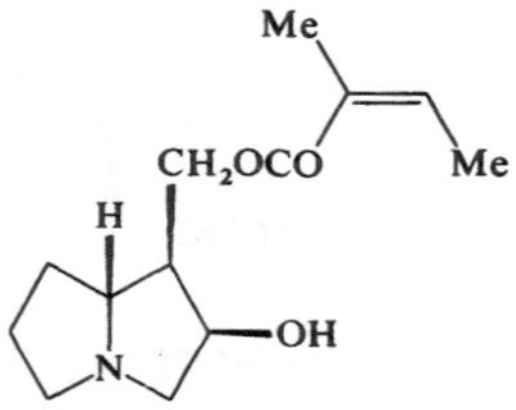	(3, 311)

Alkaloid	References

J. Petasinecine group

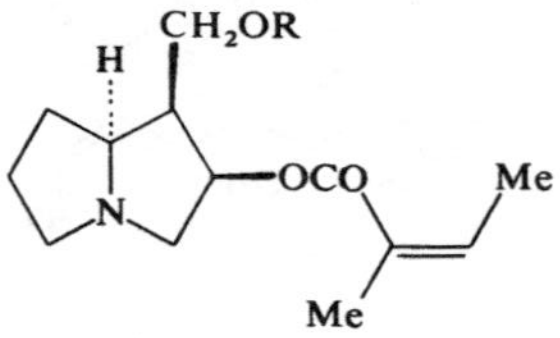

R = H: Petasinine (322)

R = CO– : Petasinoside (322)

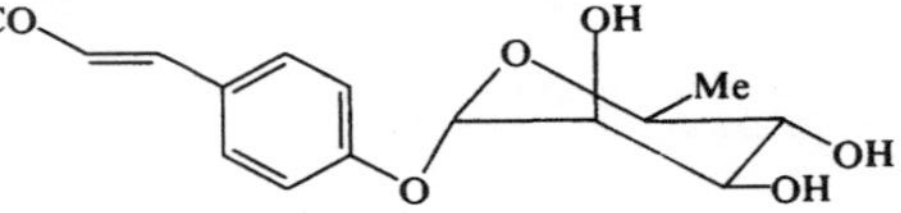

Alkaloid	References

K. Turneforcidine group

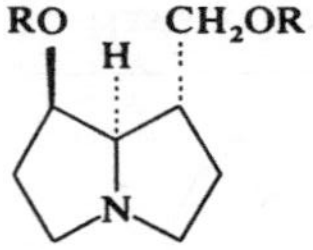

R = H or (Z)-COCMe=CHME: Turneforcine (3, 310)

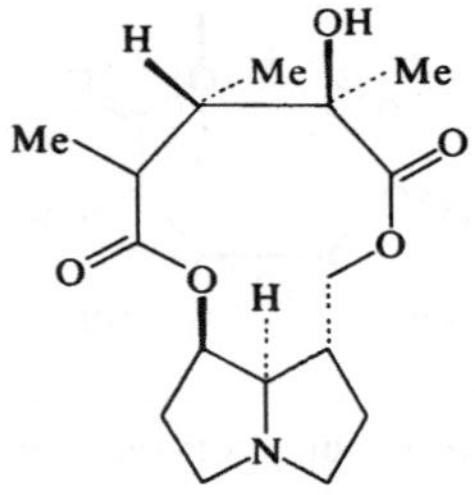

Retusine (3, 161, 311)

Table 3 (continued)

K. Turneforcidine group	Alkaloid	References
	Crocandine	(280)
	Isocrocandine	(280)

L. Hastanecine group	Alkaloid	References
	R = trachelanthyl	(3, 60)
	Hastacine	(3, 310)

M. Platynecine group	Alkaloid	References
	Sarracine	(310)
	R = H: Platyphylline R = H, geometrical isomer: Neoplatyphylline R = OH: Hygrophylline	(310) (310) (311)
	Nemorensine	(165, 168)
	Bulgarsenine	(214, 286)

Table 3 (continued)

N. Heliotridine group	Alkaloid	References
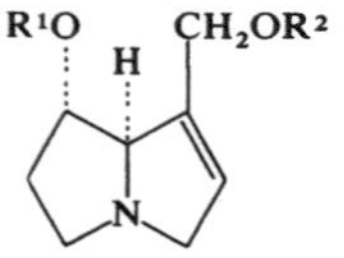	R^1 = H, R^2 = Me	(264, 275)
	R^1 = (Z)-COCMe=CHMe, R^2 = H: 7-Angelylheliotridine	(311)
	R^1 = R^2 = Angelyl: Asperumine	(190, 191)
	R^1 = H, R^2 = (−)-viridifloryl: Echinatine	(311)
	R^1 = H, R^2 = trachelanthyl: Rinderine	(10, 183)
	R^1 = H, R^2 = heliotryl: Heliotrine	(310)
	R^1 = R^2 = COC(OH)(CMe$_2$OH)CH(OMe)Me (lasiocarpyl): Europine	(311)
	R^1 = Ac, R^2 = (−)-viridifloryl: 7-Acetylechinatine	(293)
	R^1 = Angelyl, R^2 = (−)-viridifloryl: 7-Angelylheliotridine viridiflorate	(311)
	R^1 = Angelyl, R^2 = trachelanthyl: 7-Angelylheliotridine trachelanthate	(311)
	R^1 = Angelyl, R^2 = heliotryl: 7-Angelylheliotrine	(293)
	R^1 = Angelyl, R^2 = COC(OH)(CMe$_2$OH)CH(OH)Me (echimidinyl): Heliosupine	(311)
	R^1 = Angelyl, R^2 = β-acetylechimidinyl: Acetylheliosupine	(221)
	R^1 = Angelyl, R^2 = lasiocarpyl: Lasiocarpine	(310)
	R^1 = Angelyl, R^2 = β-acetyllasiocarpyl: Acetyllasiocarpine	(85)

R^1O, CH$_2$OR^2

O. Retronecine group	Alkaloid	References
(i) Acyclic Derivatives		
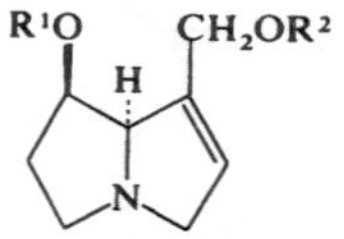	R^1 = R^2 = H: Retronecine	(311)
	R^1 = H, R^2 = Me	(311)
	R^1 = Ac, R^2 = Me	(87)
	R^1 = Angelyl, R^2 = H: 7-Angelylretronecine	(311)
	R^1 = H, R^2 = Angelyl: 9-Angelylretronecine	(84)
	R^1 = R^2 = COPh: Retronecine-7,9-dibenzoate	(278)
	R^1 = H, R^2 = (−)-viridifloryl: Lycopsamine	(89)
	R^1 = H, R^2 = (+)-trachelanthyl: Intermedine	(89)
	R^1 = H, R^2 = (−)-trachelanthyl: Indicine	(203, 206)
	R^1 = H, R^2 = β-acetyl-(−)-trachelanthyl: Acetylindicine	(203)
	R^1 = H, R^2 = β-tiglyl-trachelanthyl: Anadoline	(80, 306)

R^1O, CH$_2$OR^2

O. Retronecine group	Alkaloid	References
	R¹ = Ac, R² = (−)-viridifloryl: 7-Acetyllycopsamine	(79)
	R¹ = Ac, R² = (+)-trachelanthyl: 7-Acetylintermedine	(79)
	R¹ = Ac, R² = echimidinyl: Uplandicine	(79)
	R¹ = Angelyl, R² = (−)-viridifloryl: Symlandine	(79)
	R¹ = Angelyl, R² = (−)-trachelanthyl: Echiumine	(89)
	R¹ = Angelyl, R² = echimidinyl: Echimidine	(311)
	R¹ = Angelyl, R² = [structure] : Latifoline	(201, 311)
	R¹ = Tiglyl, R² = (−)-viridifloryl: Symphytine	(114, 115)

(ii) 11-Membered Cyclic Diesters

	R¹ = R³ = H, R² = OH: Dicrotaline	(310)
	R¹ = Me, R² = H, R³ = CHOHMe: Cronaburmine	(277)

	R¹ = R² = R³ = H: Crobarbatine	(232)
	R¹ = H, R² = Me, R³ = OAc: Spectabiline	(311)
	R¹ = Me, R² = Et, R³ = OH: Crotalarine (Croburhine)	(14)

Table 3 (continued)

(ii) Retronecine group	Alkaloid	References
	R¹ = OH, R² = Me, R³ = H: Fulvine	(201, 311)
	R¹ = Me, R² = OH, R³ = H: Crispatine	(201, 311)
	(or reverse mode of connection of acid)	
	Diastereomers of crispatine: Cromadurine	(239)
	Isocromadurine	(238)
	R¹ = OH, R² = Me, R³ = OH: Monocrotaline	(247, 248, 310)
	R¹, R³ = —O—CHMe—O— R² = Me: Monocrotalinine (possibly an artefact)	(235)
	R¹ = OCOCHMeEt, R² = Me, R³ = OH: Grahamine	(21)
	R¹ = OH, R² = R³ = Me: Trichodesmine	(247, 248, 300, 311)
	R¹ = H, R² = Me, R³ = Me: Incanine	(299, 311)
	R¹ = OH, R² = H, R³ = CHOHMe: Axillarine	(71, 287)
	R¹ = H, R² = Et: Axillaridine	(71)
	R¹ = Me, R² = CH₂OH: Junceine	(311)

O (ii) Retronecine group	Alkaloid	References

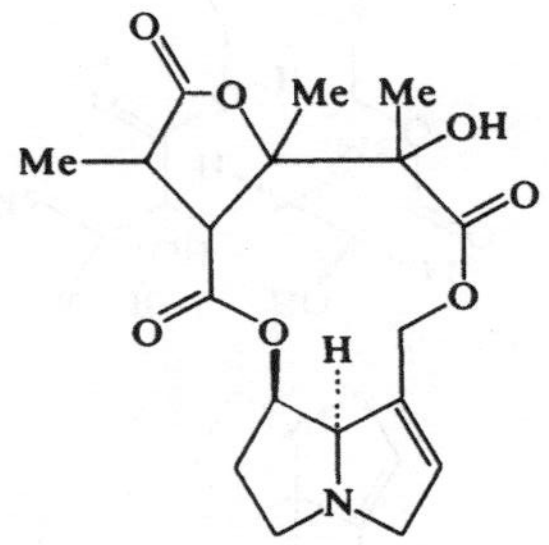

(?) Grantianine

(310)

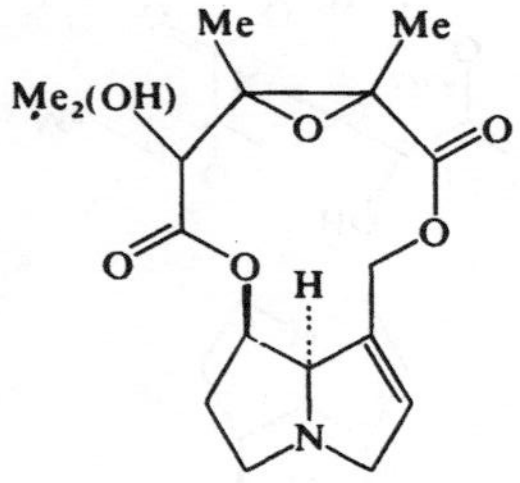

(?) Grantaline

(60)

(iii) 12-Membered Cyclic Diesters

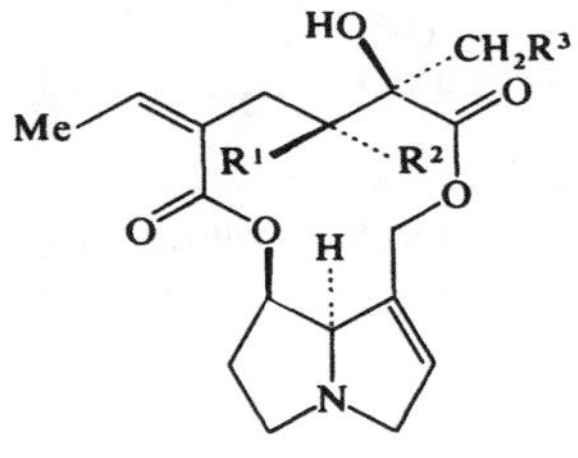

$R^1 = R^3 = H$, $R^2 = Me$: Senecionine

$R^1 = H$, $R^2 = Me$, $R^3 = OH$: Retrorsine

R^1, $R^2 = CH_2$, $R^3 = H$: Seneciphylline

R^1, $R^2 = CH_2$, $R^3 = OH$: Riddelline

(310)
(310)
(310)
(310)

Table 3 *(continued)*

O	(iii) Retronecine group	Alkaloid	References
		$R^1 = R^3 = H$, $R^2 = Me$: Integerrimine	*(310)*
		$R^1 = H$, $R^2 = Me$, $R^3 = OH$: Usaramine (Mucronatine)	*(90, 268)*
		R^1, $R^2 = CH_2$, $R^3 = H$: Spartiodine	*(310)*
		(stereochemistry at C-12 unknown)	
		$R^1 = H$, $R^2 = Me$: Jacobine	*(255, 311)*
		R^1, $R^2 = CH_2$: Jacozine	*(311)*
		$R^1 = OH$, $R^2 = H$: Jacoline	*(311)*
		$R^1 = Cl$, $R^2 = OH$: Jaconine	*(311)*

O (iii) Retronecine group	Alkaloid	References
	Erucifoline	(271)
	R¹ = OH, R² = Et, R³ = H: Bisline R¹ = OH, R² = Et, R³ = Ac: Isoline R¹ = Et, R² = R³ = H: Yamataimine	(67) (67, 68) (136)
	Swazine	(124, 177)

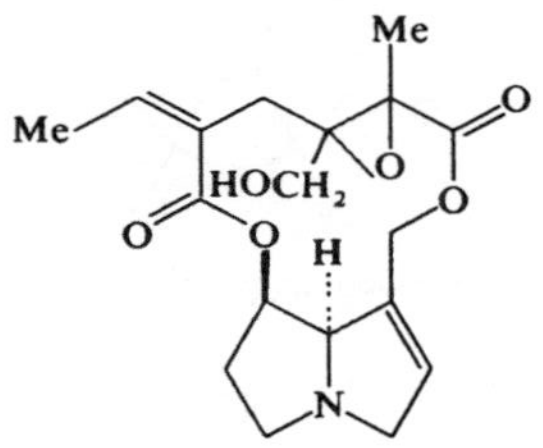

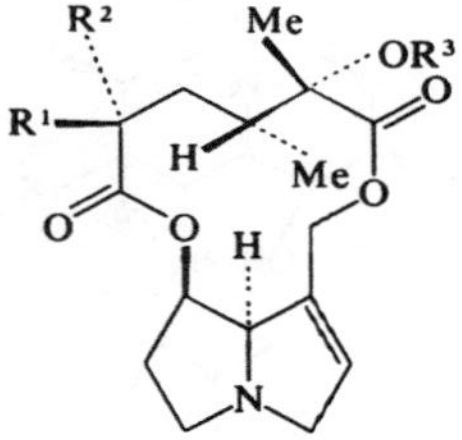

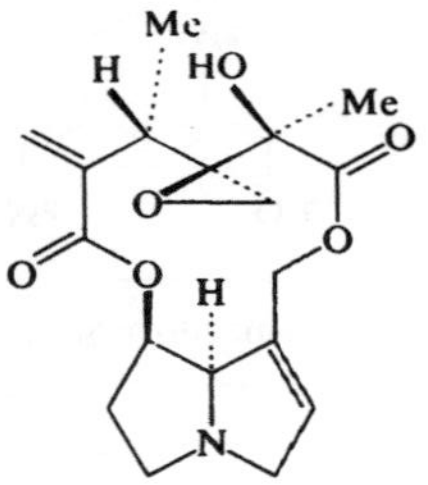

Structures:

Erucifoline: retronecine macrocyclic diester with Me, epoxide (O), HOCH₂, and =CHMe (ethylidene) substituents, N-bridged bicyclic system.

R¹ = OH, R² = Et, R³ = H: Bisline
R¹ = OH, R² = Et, R³ = Ac: Isoline
R¹ = Et, R² = R³ = H: Yamataimine

Swazine: macrocyclic diester with Me, HO, Me, and exocyclic methylene substituents.

Table 3 *(continued)*

O. Retronecine group	Alkaloid	References
	R = H: Nilgirine R = Ac: Crotastriatine	(23) (31)
	R¹ = Me, R² = R³ = H: Crotananine R¹ = Me, R² = OH, R³ = CH₂OH: Sceleratine R¹ = Me, R² = OH, R³ = CH₂Cl R¹, R² = CH₂, R³ = Me: Senecivernine	(279) (310) (310) (256)

(iv) 13-Membered Cyclic Diesters

	Retroisosenine	(214)

	Alkaloid	References
O. (*iv*) Retronecine group		
	Doronenine	(158, 255)
(*v*) 14-Membered Cyclic Diesters		
	$R^1 = R^2 = R^3 = H$: Parsonsine $R^1 = Me$, $R^2 = R^3 = H$: Heterophylline $R^1 = R^2 = H$, $R^3 = OH$: Spiraline $R^1 = Me$, $R^2 = H$, $R^3 = OH$: Spiranine $R^1 = Me$, $R^2 = R^3 = OH$: Spiracine	(108, 110, 120) (108) (108) (108) (108)

	Alkaloid	References
P. Otonecine group		
(*i*) 11-Membered Cylic Diesters		
	Retusamine	(311, 319)

Table 3 (continued)

P. (i) Otonecine group	Alkaloid	References
	Emiline	(173, 302)
	Crosemperine	(20)

P. Otonecine group	Alkaloid	References

(ii) 12-Membered Cyclic Diesters

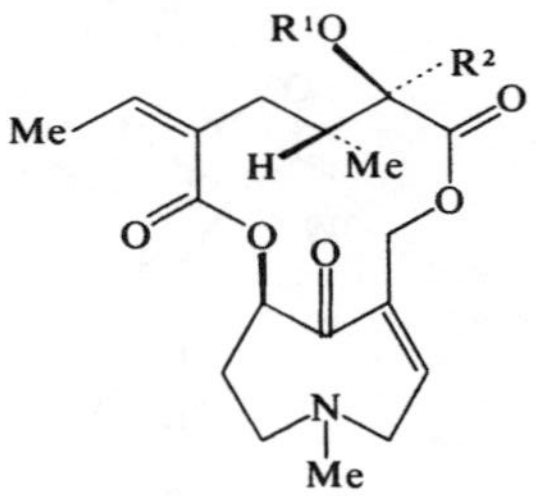

$R^1 = H$, $R^2 = Me$: Senkirkine
$R^1 = Ac$, $R^2 = Me$: *O*-Acetylsenkirkine
$R^1 = H$, $R^2 = CH_2OH$: Hydroxysenkirkine

(35, 311)
(57)
(72)

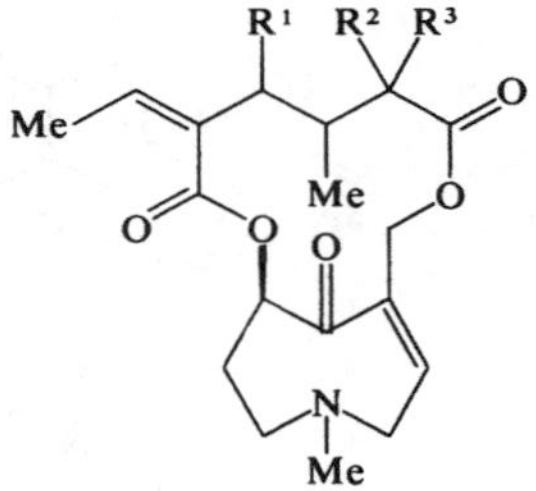

$R^1 = R^2 = OAc$, $R^3 = Me$: Ligularine
$R^1 = R^2 = H$, $R^3 = OH$: Crotafoline

(167)
(72)

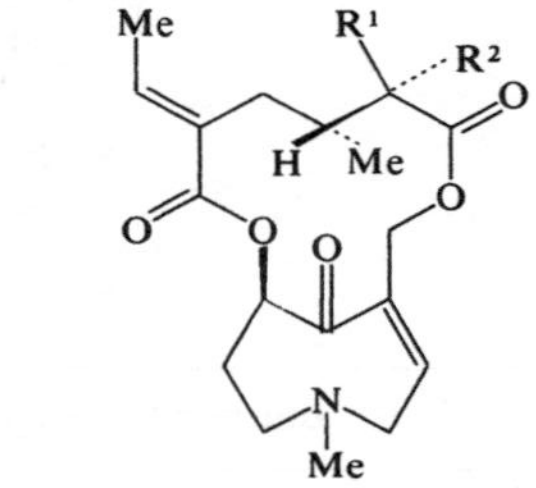

$R^1 = Me$, $R^2 = OAc$: Ligularidine
$R^1 = OH$, $R^2 = Me$: Neosenkirkine

(133)
(219)

12*

Table 3 (continued)

P (ii) Otonecine group	Alkaloid	References
	R = H: Crotaverrine R = Ac: O-Acetylcrotaverrine	(294) (294)
	R = H, α-epoxide: Otosenine R = Ac, α-epoxide: Florosenine R = H, β-epoxide: { Petasitenine { Fukinotoxin R = Ac, β-epoxide: Neopetasitenine	(62, 226, 311) (62) (321) (117) (321)
	R¹ = OH, R² = H: Onetine R¹ = Cl, R² = Ac: Doronine R¹ = OH, R² = Ac: Floridanine R¹ = OAc, R² = Ac: Floricaline	(311) (15) (62) (62)

P (ii) Otonecine group	Alkaloid	References
	R = H: Syneilesine R = Ac: Acetylsyneilesine	(134, 135) (135)
	Clivorine	(37, 166, 170)

Q. Croalbinecine group	Alkaloid	References
	Croalbidine	(266, 267)

Table 3 *(continued)*

R. Rosmarinecine group	Alkaloid	References
	$R^1 = H$, $R^2 = Me$: Rosmarinine $R^1 = R^2 = CH_2$: Angularine	(310) (311)

S. Crotanecine group	Alkaloid	References
(i) 12-Membered Cyclic Diesters 	$R^1 = R^3 = H$, $R^2 = Me$: Anacrotine (Crotalaburnine) $R^1 = Ac$, $R^2 = Me$, $R^3 = H$: 6-Acetylanacrotine $R^1 = Ac$, $R^2 = H$, $R^3 = Me$: 6-Acetyl-*trans*-anacrotine $R^1 = $ angelyl, $R^2 = H$, $R^3 = Me$: 6-Angelyl-*trans*-anacrotine	(22, 93) (93) (93) (93)
(ii) 13-Membered Cyclic Diesters 	$R^1 = R^2 = H$: Madurensine $R^1 = Ac$, $R^2 = H$: 7-Acetylmadurensine $R^1 = Ac$, $R^2 = H$, geometrical isomer: 7-Acetyl-*cis*-madurensine $R^1 = H$, $R^2 = OH$: Crotaflorine	(22, 93) (93) (93). (93)

T. Senampeline group	Alkaloid	References

R = COCH=CMe$_2$ (senecioyl): Senampeline A ⎫ Not separated (42, 43)
R = tiglyl: Senampeline B ⎭

R^1 = tiglyl, R^2 = senecioyl: Senampeline C ⎫ not separated (42, 43)
R^1 = R^2 = tiglygl: Senampeline D ⎭
R^1 = angelyl, R^2 = tiglyl: Senampeline E ⎫ (41)
R^1 = angelyl, R^2 = senecioyl: Senampeline F ⎬ not separated
R^1 = R^2 = angelyl: Senampeline G ⎭

U. Dihydropyrrolizinone group	Alkaloid	References

R^1 = H, R^2 = Me: Senaetnine (40, 45)
R^1 = H, R^2 = Me, geometrical isomer (42)
R^1, R^2 = CH$_2$: Dehydrosenaetnine (45)

Table 3 (continued)

U. Dihydropyrrolizinone group	Alkaloid	References
	$R^1 = H$, $R^2 = Me$: Isosenaetnine R^1, $R^2 = CH_2$: Dehydroisosenaetnine	(39, 45) (39, 45)
	Pterophorine C-7 Epimer	(40, 43) (41)
	(?) Inaequidenine	(40)

V. Miscellaneous group	Alkaloid	References

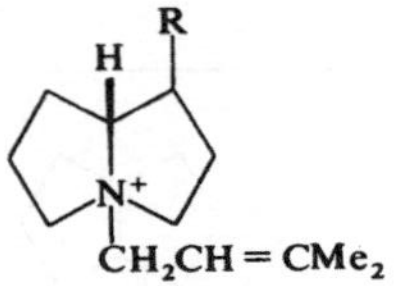

$R = \beta\text{-CO}_2^-$: Anodendrine — (263)

$R = \alpha\text{-CO}_2^-$: Alloanodendrine — (263)

$CH_2CH = CMe_2$

$CH_2OCOC(OH)(CHMe_2)CHMeOMe$ — (92)

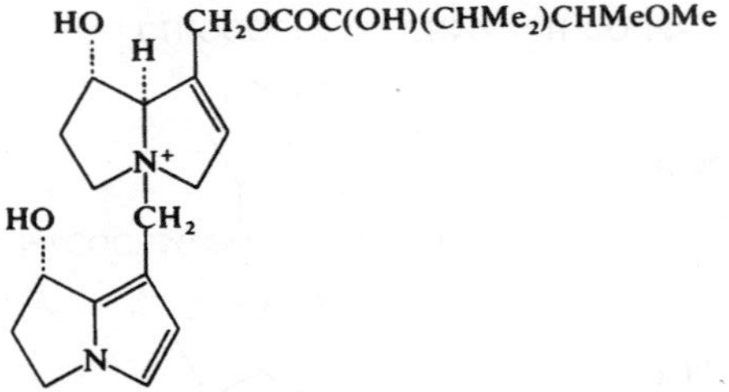

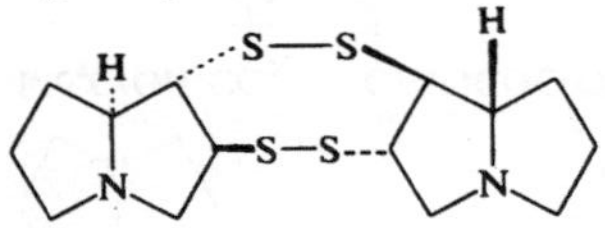

Cassipourine
(Absolute configuration unknown) — (66, 119, 318)

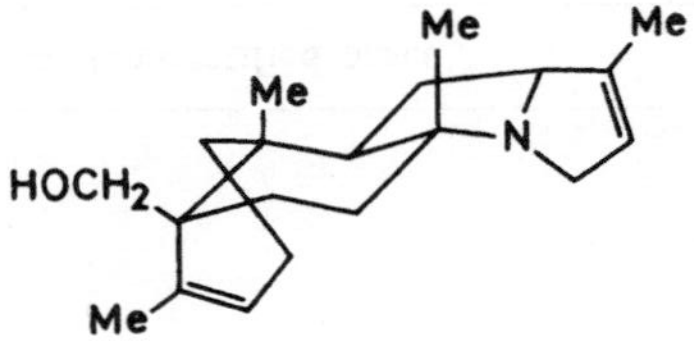

Thelepogine — (311)

Table 3 *(continued)*

W. Unidentified bases	Alkaloid	References
$CH_2OCOCH=CMe_2$ HO (pyrrolizidine structure)	Fuchsisenecionine	*(254)*
$Me_2C(OH)CO_2$ $CH_2OCOC(OH)(CHMe_2)CHOHMe$ (pyrrolizidine structure)	Uluganine	*(157)*
$MeCOCHMeCO_2$ CH_2OH OH (pyrrolizidine structure)	Procerine	*(148)*
$CH_2OCOC(OAc)(CHMe_2)CH_2CO_2Me$ (pyrrolizidine structure)		*(108)*

References

1. AASEN, A. J., and C. C. J. CULVENOR: The Saturated Pyrrolizidinediols. A Partial Synthesis of Turneforcidine. Aust. J. Chem. **22**, 2657 (1969).
2. ⁀— — The Saturated Pyrrolizidinediols. II. The Total Synthesis and Stereochemistry of ꞯMacronecine. J. Org. Chem. **34**, 4143 (1969).
3. AASEN, A. J., C. C. J. CULVENOR, and L. W. SMITH: The Saturated Pyrrolizidinediols. I. Spectral Studies and the Conversion of an Ester of Dihydroxyheliotridane into the (+)-Enantiomer of Hastanecine. J. Org. Chem. **34**, 4137 (1969).
4. ABDULLAEV, U. A., YA. U. RASHKES, KH. SHAKHIDOYATOV, and S. YU. YUNUSOV: Mass Spectra of Pyrrolizidine Alkaloids of the Heliotridane Series. Khim. Prir. Soedin. 1972, 634 [Chem. Abstr. **78**, 84612 (1973)]; RASKHES, YA. V., U. A. ABDULLAEV, and S. YU. YUNUSOV: Mass Spectra of Pyrrolizidine Alkaloids. Khim. Prir. Soedin. 1978, 153 [Chem. Abstr. **89**, 163803 (1978)].
5. ABDULLAEV, U. A., YA. V. RASHKES, and S. YU. YUNUSOV: Mass Spectra of N-Oxides of Pyrrolizidine Alkaloids. Khim. Prir. Soedin. 1974, 620 [Chem. Abstr. **82**, 73270 (1975)].
6. — — — Fragmentation of a Macrocyclic Ring of Alkaloids with an Otonecine Nucleus. Khim. Prir. Soedin. 1976, 66 [Chem. Abstr. **85**, 177726 (1976)].
7. ADAMS, R., S. MIYANO, and M. D. NAIR: Synthesis of Substituted Pyrrolidines and Pyrrolizidines. J. Am. Chem. Soc. **83**, 3323 (1961).
8. AKRAMOV, S. T., F. KIYAMITDINOVA, and S. YU. YUNUSOV: Alkaloids of *Heliotropium dasycarpum* and *H. arguzioides*. Dokl. Akad. Nauk Uz. SSR **4**, 30 (1961) [Chem. Abstr. **60**, 16209 (1964)].
9. — — — Alkaloids from *Lindelofia stylosa*. Dokl. Akad. Nauk Uz. SSR 1961, 35 [Chem. Abstr. **61**, 4700 (1964)].
10. — — — Alkaloids of *Solenanthus turkestanicus, Lindelofia olgae*, and *Trachelanthus korolkovii*. Dokl. Akad. Nauk Uz. SSR **19**, 29 (1962) [Chem. Abstr. **61**, 11005 (1964)].
11. — — — Study of *Rindera* and *Lindelofia*. Dokl. Akad. Nauk Uz. SSR **22**, 35 (1965) [Chem. Abstr. **63**, 16770 (1965)].
12. — — — The Alkaloids from *Senecio francheti, Trachelanthus hissoricus*, and *T. korolkovii*. Khim. Prir. Soedin. 1967, 351 [Chem. Abstr. **68**, 47001 (1968)].
13. AKRAMOV, S. T., A. S. SAMATOV, and S. YU. YUNUSOV: Alkaloids of *Solenanthus circinatus, Paracaryum himalayense*, and *Lindelofia pterocarpa*. Dokl. Akad. Nauk. Uz. SSR **21**, 28 (1964).
14. ALI, M. A., and G. A. ADIL: Isolation and Structure of Crotalarine, a New Alkaloid from *Crotolaria burhia*. Pakistan J. Sci. Ind. Res. **16**, 227 (1973).
15. ALIEVA, SH. A., U. A. ABDULLAEV, M. V. TELEZHENETSKAYA, and S. YU. YUNUSOV: Alkaloids of *Doronicum macrophyllum*. Khim. Prir. Soedin. 1976, 194 [Chem. Abstr. **85**, 108841 (1976)].
16. ALLAN, J. R., I. C. HSU, and L. A. CARSTENS: Dehydroretronecine-induced Rhabdomyosarcomas in Rats. Cancer Res. **35**, 997 (1975).
17. ALONSO DE LAMA, J. M., and I. RIBAS: Alkaloids of the Papilionaceae. XXIII. Alkaloids of *Adenocarpus argyrophyllus* from the Cañaveral Mountains of Cáceres, Spain. Anales Real Espan. Fis. Quim. (Madrid) **49 B**, 711 (1953) [Chem. Abstr. **49**, 4681 (1955).
18. AMAL, H., and O. ATES: *Echium italicum* var. *biebersteinii*. Instanbul Univ. Eczacilik Fak. Mecm. **7**, 85 (1971) [Chem. Abstr. **77**, 72582 (1972)].
19. ATAL, C. K.: Semisynthetic Derivatives of Pyrrolizidine Alkaloids of Pharmacodynamic Importance. A Review. Lloydia **41**, 312 (1978).
20. ATAL, C. K., C. C. J. CULVENOR, R. S. SAWHNEY, and L. W. SMITH: Crosemperine, a New Otonecine Ester from *Crotalaria semperflorens* Vent. Aust. J. Chem. **20**, 805 (1967).
21. — — — — The Alkaloids of *Crotalaria grahamiana*. Grahamine, the 3′-[(−)-2-methylbutyryl]ester of Monocrotaline. Aust. J. Chem. **22**, 1773 (1969).

22. ATAL, C. K., K. K. KAPUR, C. C. J. CULVENOR, and L. W. SMITH: A New Pyrrolizidine Aminoalcohol in Alkaloids from *Crotalaria* Species. Tetrahedron Lett. 1966, 537.

23. ATAL, C. K., and R. S. SAWHNEY: The Pyrrolizidine Alkaloids from Indian *Crotalarias*. Indian J. Pharm. **35**, 1 (1973).

24. ATAL, C. K., R. K. SHARMA, C. C. J. CULVENOR, and L. W. SMITH: Alkaloids of *Crotalaria rubiginosa* Willd. Trichodesmine and Junceine. Aust. J. Chem. **19**, 2189 (1966).

25. BALE, N. M., R. CAHILL, N. M. DAVIES, M. B. MITCHELL, E. H. SMITH, and D. WHITEHOUSE: Biosynthesis of the Necic Acids of the Pyrrolizidine Alkaloids. Further Investigations of the Formation of Senecic and Isatinecic Acids in *Senecio* Species. J. Chem. Soc., Perkin Trans. I 1978, 101.

25a. BALE, N. M., and D. H. G. CROUT: Determination of the Relative Rates of Incorporation of Arginine and Ornithine into Retronecine During Pyrrolizidine Alkaloid Biosynthesis. Phytochemistry **14**, 2617 (1975).

26. BARREIRO, E. J., A. DE LIMA PEREIRA, L. NELSON, L. F. GOMES, and A. J. R. DA SILVA: Carbon-13 Nuclear Magnetic Resonance of Pyrrolizidine Alkaloids: a Reassignment. J. Chem. Res. (S) 1980, 330.

27. BATES, R. B., and S. R. MOREHEAD: Absolute Configurations of Pyrrolizidine Alkaloids of the Loline Group. Tetrahedron Lett. 1972, 1629.

28. BATIROV, E. KH., S. A. KHAMIDKHODZHAEV, V. M. MALIKOV, and S. YU. YUNUSOV: Study of Alkaloids from *Lolium cunateum*. Khim. Prir. Soedin. 1976, 60 [Chem. Abstr. **85**, 59556 (1976)].

29. BATIROV, E. KH., V. MALIKOV, and S. YU. YUNUSOV: Lolidine, a New Chlorine Containing Alkaloid from *Lolium cuneatum* Seeds. Khim. Prir. Soedin. 1976, 63 [Chem. Abstr. **85**, 74883 (1976)].

30. — — — Alkaloids from *Lolium cuneatum* Seeds. Khim. Prir. Soedin. 1976, 120 [Chem. Abstr. **86**, 13792 (1977)].

31. BATRA, V., R. N. GANDHI, and T. RAJAGOPALAN: Structure of Crotastriatine. Indian J. Chem. **13**, 989 (1975).

32. BATRA, V., and T. R. RAJAGOPALAN: Alkaloidal Constituents of *Senecio scandens*. Curr. Sci. **46**, 141 (1977).

33. BENN, M., J. DE GRAVE, C. GNANASUNDERAM, and R. HUTCHINS: Host-plant Pyrrolizidine Alkaloids in *Nyctemera annulata* Boisduval: Their Persistence Through the Life Cycle and Transfer to a Parasite. Experientia **35**, 731 (1979).

34. BHACCA, N. S., and R. K. SHARMA: Mucronatinine, a New Alkaloid from *Crotalaria mucronata* Desv. I. Tetrahedron **24**, 6319 (1968).

35. BIRNBAUM, G. I.: The Nature of Intramolecular N . . . C = O Interactions. Crystal Structure of the *Senecio* Alkaloid Senkirkine. J. Am. Chem. Soc. **96**, 6165 (1974).

36. BIRNBAUM, K. B.: Structure and Absolute Configuration of the Alkaloid Clivorine. Acta Crystallogr. Sect. B **28**, 2825 (1972).

37. BIRNBAUM, K. B., A. KLÁSEK, P. SEDMERA, G. SNATZKE, L. F. JOHNSON, and F. ŠANTAVÝ: Revised Structure of the Alkaloid Clivorine. Tetrahedron Lett. 1971, 3421.

38. BOHLMANN, F., W. KLOSE, and K. NIKISCH: Synthese des Dehydroheliotridins und des 3-Oxo-dehydroheliotridins. Tetrahedron Lett. 1979, 3699.

39. BOHLMANN, F., and K.-H. KNOLL: Zwei neue Acylpyrrole aus *Kleinia kleinioides*. Phytochemistry **17**, 599 (1978).

40. BOHLMANN, F., K.-H. KNOLL, C. ZDERO, P. K. MAHANTA, M. GRENZ, A. SUWITA, D. EHLERS, N. L. VAN, W.-R. ABRAHAM, and A. A. NATU: Terpen-Derivate aus *Senecio*-Arten. Phytochemistry **16**, 965 (1977).

41. BOHLMANN, F., and C. ZDERO: Neue C_{10}-Säureamide, Furanoeremophilane und andere Inhaltsstoffe aus Bolivianischen *Senecio*-Arten. Phytochemistry **18**, 125 (1979).

42. BOHLMANN, F., C. ZDERO, D. BERGER, A. SUWITA, P. MAHANTA, and C. JEFFREY: Neue

Furanoeremophilane und weitere Inhaltsstoffe aus Südafrikanischen *Senecio*-Arten. Phytochemistry **18**, 79 (1979).

43. BOHLMANN, F., C. ZDERO, and M. GRENZ: Natürlich vorkommende Terpen-Derivate, 78. Weitere Inhaltsstoffe aus Südafrikanischen *Senecio*-Arten. Chem. Ber. **110**, 474 (1977).

44. BOHLMANN, F., C. ZDERO, and A. A. NATU: Weitere Bisabolen-Derivate und andere Inhaltsstoffe aus Südafrikanischen *Senecio*-Arten. Phytochemistry **17**, 1757 (1978).

45. BOHLMANN, F., C. ZDERO, and G. SNATZKE: Zur Stereochemie der Acylpyrrole aus *Senecio*-Arten. Chem. Ber. **111**, 3009 (1978).

46. BOHLMANN, F., and J. ZIESCHE: Neue Germacren-Derivate aus *Senecio*-Arten. Phytochemistry **18**, 1489 (1979).

47. BOPPRÉ, M.: Chemical Communication, Plant Relationships, and Mimicry in the Evolution of Danaid Butterflies. Entomol. Exp. Appl. **24**, 264 (1978).

48. BOPPRÉ, M., R. L. PETTY, D. SCHNEIDER, and J. MEINWALD: Behaviorally Mediated Contacts Between Scent Organs: Another Prerequisite for Pheromone Production in *Danaus chrysippus* Males (Lepidoptera). J. Comp. Physiol., Sect. A **126**, 97 (1978).

49. BORGES DEL CASTILLO, J., A. G. ESPANA DE AGUIRRE, J. L. BRETÓN, A. G. GONZÁLEZ, and J. TRUJILLO: Loroquin, A New Necine Isolated from *Urechtites karwinsky* Mueller (1-Hydroxy-methylene-7-keto-dihydropyrrolizine). Tetrahedron Lett. 1970, 1219.

50. BOTTOMLEY, W., and T. A. GEISSMAN: Pyrrolizidine Alkaloids. The Biosynthesis of Retronecine. Phytochemistry **3**, 357 (1964).

51. BRANDÄNGE, S., and I. GRANELLI: Studies on Orchidaceae Alkaloids. XXXVI. Alkaloids from some *Vanda* and *Vandopsis* Species. Acta Chem. Scand. **27**, 1096 (1973).

52. BRANDÄNGE, S., I. GRANELLI, and B. LÜNING: Orchidaceae Alkaloids. XVIII. Isolation of Phalaenopsin La from *Kingiella taenialis* (Lindl.) Rolfe. Acta Chem. Scand. **24**, 354 (1970).

53. BRANDÄNGE, S., S. JOSEPHSON, and S. VALLEN: Orchidaceae Alkaloids. XXXVIII. Asymmetric Synthesis of 2-isobutylmalic acid and 2-(cyclohexylmethyl)malic acid. Acta Chem. Scand. **27**, 3668 (1973).

54. BRANDÄNGE, S., and B. LÜNING: Studies on Orchidaceae Alkaloids. XII. Pyrrolizidine Alkaloids from *Phalaenopsis amabilis* Bl. and *Ph. mannii* Rchb. f. Acta Chem. Scand. **23**, 1151 (1969).

55. BRANDÄNGE, S., B. LÜNING, C. MOBERG, and E. SJÖSTRAND: Orchidaceae Alkaloids. XXIV. Pyrrolizidine Alkaloid from *Phalaenopsis cornucervi*. Acta Chem. Scand. **25**, 349 (1971).

56. — — — — Orchidaceae Alkaloids. XXX. Fourteen *Phalaenopsis* Species. New Pyrrolizidine Alkaloid from *Phalaenopsis equestris*. Acta Chem. Scand. **26**, 2558 (1972).

57. BRIGGS, L. H., R. C. CAMBIE, B. J. CANDY, G. M. O'DONOVAN, R. H. RUSSELL, and R. N. SEELYE: Alkaloids of New Zealand *Senecio* Species. Part II. Senkirkine. J. Chem. Soc. 1965, 2492.

58. BROCH-DUE, A. I., and A. A. AASEN: Alkaloids of *Anchusa officinalis* L. Identification of the Pyrrolizidine Alkaloid Lycopsamine. Acta Chem. Scand. Ser. B. **34**, 75 (1980).

59. BRUEMMERHOFF, S. W. D., and H. L. DE WAAL: Retrorsine from *Crotalaria spartioides*. J. S. African Chem. Inst. **14**, 101 (1961).

60. BULL, L. B., C. C. J. CULVENOR, and A. T. DICK: The Pyrrolizidine Alkaloids. North-Holland Publ., Amsterdam, 1968.

61. CAHILL, R., D. H. G. CROUT, M. B. MITCHELL, and U. S. MÜLLER: Isoleucine Biosynthesis and Metabolism: Stereochemistry of the Formation of L-isoleucine and of its conversion into Senecic and Isatinecic Acids in *Senecio* Species. J. Chem. Soc., Chem. Commun. 1980, 419.

62. CAVA, M. P., K. V. RAO, J. A. WEISBACH, R. F. RAFFAUF, and B. DOUGLAS: The Alkaloids of *Cacalia floridana*. J. Org. Chem. **33**, 3570 (1968).

63. CHU, Y.-L., and J.-H. CHU: Pyrrolizidine Alkaloids. II. The Alkaloids of *Senecio cruentus*. Yao Hsueh Hsueh Pao 11, 168 (1964) [Chem. Abstr. 61, 1904d (1964)].

64. COLEMAN, P. C., E. D. COUCOURAKIS, and J. A. PRETORIUS: Crystal Structure of Retrorsine. S. African J. Chem. 33, 116 (1980); STOECKLI-EVANS, H.: Retrorsine Hydrobromide Ethanol Solvate: A Pyrrolizidine Alkaloid. Acta Crystallogr., Sect. B 35, 2798 (1979).

65. CONSTANTINESCU, E., and D. ALBULESCU: The Isolation of an Alcamine from *Senecio doria*. Farmacia 9, 139 (1961) [Chem. Abstr. 56, 14396 (1962)].

66. COOKS, R. G., F. L. WARREN, and D. H. WILLIAMS: Rhizophoraceae Alkaloids. Part III. Cassipourine. J. Chem. Soc. (C) 1967, 286.

67. COUCOURAKIS, E. D., and C. G. GORDON-GRAY: The *Senecio* Alkaloids. Suggested Structures for Isoline and Bisline, Two New Alkaloids from *Senecio othonniformis* Fourcade. J. Chem. Soc. (C) 1970, 2312.

68. COUCOURAKIS, E. D., C. G. GORDON-GRAY, and C. G. WHITELEY: The *Senecio* Alkaloids. The Structure and Absolute Configuration of Isoline. J. Chem. Soc., Perkin Trans. I 1972, 2339.

69. CROUT, D. H. G.: Pyrrolizidine Alkaloids. The Biosynthesis of Echimidinic Acid. J. Chem. Soc. (C) 1966, 1968.

70. — Pyrrolizidine Alkaloids. Biosynthesis of the Angelate Component of Heliosupine. J. Chem. Soc. (C) 1967, 1233.

71. — The Structure of Axillarine, a Novel Pyrrolizidine Alkaloid from *Crotalaria axillaris* Ait. Chem. Commun. 1968, 429; — Structures of Axillarine and Axillaridine, Novel Pyrrolizidine Alkaloids from *Crotalaria axillaris* Ait. J. Chem. Soc. (C) 1969, 1379.

72. — Pyrrolizidine and Seco-pyrrolizidine Alkaloids of *Crotalaria laburnifolia* L. Subspecies *eldomae*. J. Chem. Soc., Perkin Trans. I 1972, 1602.

73. — The Pyrrolizidine Alkaloids, their Physiological Activity and Biosynthesis in Higher Plants. Chimia 30, 270 (1976).

74. CROUT, D. H. G., M. H. BENN, H. IMASEKI, and T. A. GEISSMAN: Pyrrolizidine Alkaloids. The Biosynthesis of Seneciphyllic Acid. Phytochemistry 5, 1 (1966).

75. CROUT, D. H. G., N. M. DAVIES, E. H. SMITH, and D. WHITEHOUSE: Pyrrolizidine Alkaloids. The Biosynthesis of Senecic Acid. J. Chem. Soc., Perkin Trans. I 1972, 671; — — — — Biosynthesis of the C_{10} Necic Acids of the Pyrrolizidine Alkaloids. Chem. Commun. 1970, 635.

76. CROUT, D. H. G., and D. WHITEHOUSE: Absolute Configuration of 2,3-Dihydroxy-3-methylpentanoic Acid; an Intermediate in the Biosynthesis of Isoleucine, and its Identity with the Esterifying Acid of the Pyrrolizidine Alkaloid Strigosine. J. Chem. Soc., Perkin Trans. I 1977, 544.

77. CULVENOR, C. C. J.: Pyrrolizidine Alkaloids — Occurrence and Systematic Importance in Angiosperms. Bot. Notiser 131, 473 (1978).

78. CULVENOR, C. C. J., D. H. G. CROUT, W. KLYNE, W. P. MOSE, J. D. RENWICK, and P. M. SCOPES: Circular Dichroism of Pyrrolizidine Alkaloids and Related Compounds. J. Chem. Soc. (C) 1971, 3653.

79. CULVENOR, C. C. J., J. A. EDGAR, J. L. FRAHN, and L. W. SMITH: The Alkaloids of *Symphytum* × *uplandicum* (Russian Comfrey). Aust. J. Chem. 33, 1105 (1980).

80. CULVENOR, C. C. J., J. A. EDGAR, J. L. FRAHN, L. W. SMITH, A. ULUBELEN, and S. DOGANCA: The Structure of Anadoline. Aust. J. Chem. 28, 173 (1975).

81. CULVENOR, C. C. J., J. A. EDGAR, M. V. JAGO, A. OUTTERIDGE, J. E. PETERSON, and L. W. SMITH: Hepato- and Pneumotoxicity of Pyrrolizidine Alkaloids and Derivatives in Relation to Molecular Structure. Chem.-Biol. Interactions 12, 299 (1976).

82. CULVENOR, C. C. J., J. A. EDGAR, L. W. SMITH, and I. HIRONO: The Occurrence of Senkirkine in *Tussilago farfara*. Aust. J. Chem. 29, 229 (1976).

83. CULVENOR, C. C. J., M. L. HEFFERNAN, and W. G. WOODS: Nuclear Magnetic Resonance

Spectra of Pyrrolizidine Alkaloids. I. The Spectra of Retronecine and Heliotridine. Aust. J. Chem. **18**, 1605 (1965); CULVENOR, C. C. J., and W. G. WOODS: Nuclear Magnetic Resonance Spectra of Pyrrolizidine Alkaloids. II. The Pyrrolizidine Nucleus in Ester and Non-ester Alkaloids and their Derivatives. Aust. J. Chem. **18**, 1625 (1965).

84. CULVENOR, C. C. J., S. R. JOHNS, J. A. LAMBERTON, and L. W. SMITH: The Isolation of Calycanthine and Pyrrolizidine Alkaloids from *Bhesa archboldiana* (Celastraceae): An Unusual Co-occurrence of Alkaloidal Types. Aust. J. Chem. **23**, 1279 (1970).

85. CULVENOR, C. C. J., S. R. JOHNS, and L. W. SMITH: Acetyllasiocarpine, an Alkaloid from *Heliotropium europaeum*. Aust. J. Chem. **28**, 2319 (1975).

86. CULVENOR, C. C. J., J. D. MORRISON, A. J. C. NICHOLSON, and L. W. SMITH: Alkaloids of *Crotalaria trifoliastrum* Willd. II. 1-Methoxymethyl-1,2-epoxypyrrolizidine. Aust. J. Chem. **16**, 131 (1963).

87. CULVENOR, C. C. J., G. M. O'DONOVAN, and L. W. SMITH: Alkaloids of *Crotalaria trifoliastrum* Willd. and *C. aridicola* Domin. III. Additional Pyrrolizidine Derivatives. Aust. J. Chem. **20**, 757 (1967).

88. — — — The Identity of the Amino Alcohol of Retusamine with Otonecine. Aust. J. Chem. **20**, 801 (1967).

89. CULVENOR, C. C. J., and L. W. SMITH: The Alkaloids of *Amsinckia* Species: *A. intermedia* Fisch. et Mey., *A. hispida* (Ruiz et Pav.) Johnst. and *A. lycopsioides* Lehm. Aust. J. Chem. **19**, 1955 (1966).

90. — — Usaramine, a New Pyrrolizidine Alkaloid from *Crotalaria usaramoensis* E. G. Baker. Aust. J. Chem. **19**, 2127 (1966).

91. — — The Alkaloids of *Cynoglossum australe* R. Br. and *C. amabile* Stapf. and Drummond. Aust. J. Chem. **20**, 2499 (1967).

92. — — A Quaternary N-Dihydropyrrolizinomethyl Derivative of Heliotrine from *Heliotropium europaeum*. Tetrahedron Lett. 1969, 3603.

93. — — Crotanecine Ester Alkaloids of *Crotalaria agatiflora*. Anales de Quim. **68**, 883 (1972).

94. DANILOVA, A. V., and L. M. UTKIN: Structure of the Alkaloid Macrophylline. Zh. Obshch. Khim. **30**, 345 (1960) [Chem. Abstr. **54**, 22698 (1960)].

95. DANISHEFSKY, S.: Electrophilic Cyclopropanes in Organic Synthesis. Acc. Chem. Res. **12**, 66 (1979).

96. DANISHEFSKY, S., R. MCKEE, and R. K. SINGH: Kinetically Controlled Total Syntheses of *dl*-Trachelanthamidine and *dl*-Isoretronecanol. J. Am. Chem. Soc. **99**, 4783 (1977).

97. — — — Stereospecific Total Synthesis of *dl*-Hastanecine and *dl*-Dihydroxyheliotridane. J. Am. Chem. Soc. **99**, 7711 (1977).

98. DAVIES, N. M., and D. H. G. CROUT: Pyrrolizidine Alkaloid Biosynthesis. Relative Rates of Incorporation of the Isomers of Isoleucine into the Necic Acid Component of Senecionine. J. Chem. Soc., Perkin Trans. I 1974, 2079.

99. DEAGEN, J. T., and M. L. DEINZER: Improvements in the Extraction of Pyrrolizidine Alkaloids. Lloydia **40**, 395 (1977).

100. DEINZER, M. L., P. A. THOMSON, D. M. BURGETT, and D. L. ISAACSON: Pyrrolizidine Alkaloids: Their Occurrence in Honey from Tansy Ragwort (*Senecio jacobaea* L.). Science **195**, 497 (1977).

101. DEINZER, M. L., P. THOMSON, D. GRIFFIN, and E. DICKINSON: A Sensitive Analytical Method for Pyrrolizidine Alkaloids. The Mass Spectra of Retronecine Derivatives. Biomed. Mass Spectrom. **5**, 175 (1978).

102. DICKINSON, J. O., M. P. COOKE, R. R. KING, and P. A. MOHAMED: Milk Transfer of Pyrrolizidine Alkaloids in Cattle. J. Am. Vet. Med. Assoc. **169**, 1192 (1976).

103. DIMENNA, G. P., T. P. KRICK, and H. J. SEGALL: Rapid High-performance Liquid Chromatography Isolation of Monoesters, Diesters and Macrocyclic Diester

Pyrrolizidine Alkaloids from *Senecio jacobaea* and *Amsinckia intermedia*. J. Chromatogr. **192**, 474 (1980).

104. DREWES, S. E., I. ANTONOWITZ, P. T. KAYE, and P. C. COLEMAN: ^{13}C Nuclear Magnetic Resonance Spectra of the *Senecio* Alkaloids, Retrorsine, Swazine, Isoline, and Hygrophylline. J. Chem. Soc., Perkin Trans. I 1981, 287.

105. EDGAR, J. A., M. BOPPRÉ, and D. SCHNEIDER: Pyrrolizidine Alkaloid Storage in African and Australian Danaid Butterflies. Experientia **35**, 1447 (1979).

106. EDGAR, J. A., P. A. COCKRUM, and J. L. FRAHN: Pyrrolizidine Alkaloids in *Danaus plexippus* L. and *D. chrysippus* L. Experientia **32**, 1532 (1976).

107. EDGAR, J. A., and C. C. J. CULVENOR: Pyrrolizidine Alkaloids in *Parsonsia* Species (Family Apocynaceae) which Attract Danaid Butterflies. Experientia **31**, 393 (1975).

108. EDGAR, J. A., N. J. EGGERS, A. J. JONES, and G. B. RUSSELL: Unusual Macrocyclic Pyrrolizidine Alkaloids from *Parsonsia heterophylla* A. Cunn and *Parsonsia spiralis* Wall. (Apocynaceae). Tetrahedron Lett. 1980, 2657.

109. EDWARDS, J. D., and T. MATSUMOTO: *Senecio* Alkaloids. Synthesis of Decanecic Acids. J. Org. Chem. **32**, 1837 (1967).

110. EGGERS, N. J., and G. J. GAINSFORD: Parsonsine ($C_{22}H_{33}NO_8$): A Pyrrolizidine Alkaloid from *Parsonsia heterophylla* A. Cunn. Cryst. Struct. Commun. **8**, 597 (1979).

111. EVANS, J. V., A. PENG, and C. J. NIELSEN: The Gas Chromatographic Mass Spectrometric Analysis of the New Antitumor Drug Indicine N-oxide Utilizing a Novel Reaction Accompanying Trimethylsilylation. Biomed. Mass Spectrom. **6**, 38 (1979).

112. FERRY, S., and J. L. BRAZIER: Sur les Alcaloides de Quelques *Senecio* Indigènes II. Identification et Structures. Ann. Pharm. Fr. **34**, 133 (1976).

113. FRIDRICHSONS, J., A. M. MATHIESON, and D. J. SUTOR: Crystal Structure of Jacobine Bromohydrin. Acta Crystallogr., Sect. B **16**, 1075 (1963).

114. FURUYA, T., and K. ARAKI: Constituents of Crude Drugs. I. Alkaloids of *Symphytum officinale*. Chem. Pharm. Bull. (Japan) **16**, 2512 (1968).

115. FURUYA, T., and M. HIKICHI: Alkaloids and Triterpenoids of *Symphytum officinale*. Phytochemistry **10**, 2217 (1971).

116. — — Lindelofine and Supinine: Pyrrolizidine Alkaloids from *Eupatorium stoechadosmum*. Phytochemistry **12**, 225 (1973).

117. FURUYA, T., M. HIKICHI, and Y. IITAKA: Fukinotoxin, a New Pyrrolizidine Alkaloid from *Petasites japonicus*. Chem. Pharm. Bull. (Japan) **24**, 1120 (1976).

118. FURUYA, T., K. MURAKAMI, and M. HIKICHI: Constituents of Crude Drugs. III. Senkirkine, a Pyrrolizidine Alkaloid from *Farfugium japonicum*. Phytochemistry **10**, 3306 (1971).

119. GAFNER, G., and L. J. ADMIRAAL: The Crystal and Molecular Structure of the Alkaloid Cassipourine. Acta Crystallogr. Sect. B **25**, 2114 (1969).

120. GAINSFORD, G. J.: The Second Form of the Alkaloid Parsonsine, $C_{22}H_{33}NO_8$. Cryst. Struct. Commun. **9**, 173 (1980).

121. GANDHI, R. N., T. R. RAJAGOPALAN, and T. R. SESHADRI: A New Alkaloid from the Seeds of *Crotalaria striata*. Curr. Sci. **37**, 285 (1968).

122. GHARBO, S. A., and A. M. HABIB: Phytochemical Investigation of Egyptian *Senecio*. II. Alkaloids of *Senecio aegyptius*, *S. desfontainei*, *S. vulgaris*, *S. petasitis*, and *S. mikanoides*. Lloydia **32**, 503 (1969).

123. GORDON-GRAY, C. G., and F. D. SCHLOSSER: The Senecio Alkaloids. The Biosynthesis of the Necic Acids. J. South African Chem. Inst. **23**, 13 (1970).

124. GORDON-GRAY, C. G., R. B. WELLS, N. HALLAK, M. B. HURSTHOUSE, S. NEIDLE, and T. P. TOUBE: Swazine, a New Alkaloid from *Senecio swaziensis* Compton: A Chemical and X-ray Crystal Study of a Novel Spiro Dilactone. Tetrahedron Lett. 1972, 707; GORDON-GRAY, C. G., and R. B. WELLS: Structure and Absolute Configuration of Swazine. J. Chem. Soc., Perkin Trans. I 1974, 1556.

125. Goss, G. J.: The Interaction Between Moths and Plants Containing Pyrrolizidine Alkaloids. Environ. Entomol. **8**, 487 (1979).
126. Gupta, V. P., S. K. Handoo, and R. S. Sawhney: Experimental Analysis of Ultraviolet Spectra of Some Pyrrolizidine Esters. Indian J. Pure Appl. Phys. **13**, 776 (1975); Simanek, V., A. Klásek, and F. Šantavý: Ultraviolet Absorption Spectroscopy of Some Pyrrolizidine Alkaloids. Collect. Czech. Chem. Commun. **34**, 1832 (1969).
127. Gupta, V. P., S. K. Handoo, and R. S. Sawhney: Experimental Analysis of the Infrared Spectrum of Some Pyrrolizidine Esters. Curr. Sci. **44**, 451 (1975).
128. Habib, A. A. M.: Senecionine, seneciphylline, jacobine, and otosenine from *Senecio cineraria*. Planta Med. **26**, 279 (1974).
129. — Phytochemical Investigation of *Heliotropium ramosissimum*. Bull. Fac. Sci., Riyadh Univ. **7**, 67 (1975) [Chem. Abstr. **85**, 119597 (1976)].
130. Habib, A. A. M., M. R. I. Saleh, and M. Farag: Isolation of Crispatine and Fulvine from *Crotalaria madurensis*. Lloydia **34**, 455 (1971).
131. Hart, N. K., and J. A. Lamberton: Pyrrolizidine Alkaloids from *Planchonella* Species (Family Sapotaceae). I. The Alkaloids of *Planchonella thyrsoidea* C. T. White and *P. anteridifera* (White and Francis) H. J. Lam. Aust. J. Chem. **19**, 1259 (1966).
132. Hayashi, K., A. Natorigawa, and H. Mitsuhashi: Integerrimine from *Cacalia hastata* L. subsp. *orientalis* Kitamura. Chem. Pharm. Bull. (Japan) **20**, 201 (1972).
133. Hikichi, M., Y. Asada, and T. Furuya: Ligularidine, a New Pyrrolizidine Alkaloid from *Ligularia dentata*. Tetrahedron Lett. **1979**, 1233.
134. Hikichi, M., and T. Furuya: Syneilesine, a New Pyrrolizidine Alkaloid from *Syneilesis palmata*. Tetrahedron Lett. **1974**, 3657.
135. — — Studies on Constituents of Crude Drugs. VII. Syneilesine and Acetylsyneilesine from *Syneilesis palmata*. Chem. Pharm. Bull. (Japan) **24**, 3178 (1976).
136. Hikichi, M., T. Furuya, and Y. Iitaka: Yamataimine, a New Pyrrolizidine Alkaloid from *Cacalia yatabei*. Tetrahedron Lett. **1978**, 767.
137. Hirono, I., M. Haga, M. Fujii, S. Matsuura, N. Matsubara, M. Nakayama, T. Furuya, M. Hikichi, H. Takanashi, E. Uchida, S. Hosaka, and I. Ueno: Induction of Hepatic Tumors in Rats by Senkirkine and Symphytine. J. Natl. Cancer Inst. **63**, 469 (1979).
138. Hirono, I., H. Mori, and M. Haga: Carcinogenic Activity of *Symphytum officinale*. J. Natl. Cancer Inst. **61**, 865 (1978).
139. Hoque, M. S., A. Ghani, and H. Rashid: Alkaloids of *Heliotropium indicum* Linn. Grown in Bangladesh. Bangladesh Pharm. J. **5**, 13 (1976).
140. Hoskins, W. M., and D. H. G. Crout: Pyrrolizidine Alkaloid Analogues. Preparation of Semisynthetic Esters of Retronecine. J. Chem. Soc., Perkin Trans. I **1977**, 538.
141. Hrbek, J., L. Hruban, A. Klásek, N. K. Kochetkov, A. M. Likhosherstov, F. Šantavý, and G. Snatzke: Pyrrolizidine Alkaloids. XVIII. Chiroptical Properties of some Pyrrolizidine Alkaloids. Collect. Czech. Chem. Commun. **37**, 3918 (1972).
142. Hughes, C. A., R. Letcher, and F. L. Warren: The *Senecio* Alkaloids. Part XVI. The Biosynthesis of the "Necine" Bases from Carbon-14 Precursors. J. Chem. Soc. **1964**, 4974.
143. Huizing, H. J., F. de Boer, and T. M. Malingré: Chloranil, a Sensitive Detection Reagent for Pyrrolizidine Alkaloids on Thin-Layer Chromatograms. J. Chromatogr. **195**, 407 (1980); Molyneux, R. J., and J. N. Roitman: Specific Detection of Pyrrolizidine Alkaloids on Thin-Layer Chromatograms. J. Chromatogr. **195**, 412 (1980).
144. Huizing, H. J., and T. M. Malingré: Reduction of Pyrrolizidine *N*-oxides by the Use of a Redox Polymer. J. Chromatogr. **173**, 187 (1979).
145. — — Purification and Separation of Pyrrolizidine Alkaloids from Boraginaceae on a Polystyrene-divinylbenzene Resin. J. Chromatogr. **176**, 274 (1979).

146. Huizing, H. J., and T. M. Malingré: Ion-pair Adsorption Chromatography of Pyrrolizidine Alkaloids. J. Chromatogr. **205**, 218 (1981).

147. Huxtable, R. J.: New Aspects of the Toxicology and Pharmacology of Pyrrolizidine Alkaloids. Gen. Pharmacol. **10**, 159 (1979).

148. Jovčeva, R. J., A. Boeva, H. Potěšilová, A. Klásek, and F. Šantavý: Alkaloids from *Senecio procerus* L., var. *procerus* Stoj. Stef. et Kit. Collect. Czech. Chem. Commun. **43**, 2312 (1978).

149. Karimov, A., M. V. Telezhenetskaya, K. L. Lutfullin, and S. Yu. Yunusov: Alkaloids of *Echium vulgare* and *Berberis oblonga*. Khim. Prir. Soedin. 1975, 433 [Chem. Abstr. **84**, 14662 (1976)].

150. Keeler, R. F., K. R. van Kampen, and L. F. James (eds.): Effects of Poisonous Plants on Livestock. New York: Academic Press. 1978.

151. Khalilov, D. S.: Phytochemical Study of Different Species of *Senecio* from Azerbaidzhan Flora. Izvest. Akad. Nauk Azerb. SSR, Ser. Biol. Nauk 1971, 122 [Chem. Abstr. **77**, 123781 (1972)].

152. Khalilov, D. S., and M. V. Telezhenetskaya: Alkaloids of *Senecio*. Khim. Prir. Soedin. 1973, 128 [Chem. Abstr. **78**, 15643 (1973)].

153. Khalilov, D. S., M. V. Telezhenetskaya, and S. Yu. Yunusov: Doronine from *Senecio othonnae*. Khim. Prir. Soedin. 1977, 866 [Chem. Abstr. **88**, 101593 (1978)].

154. Khan, H. A., and D. J. Robins: Pyrrolizidine Alkaloid Biosynthesis; Incorporation of ^{13}C-Labelled Putrescines into Retronecine. J. Chem. Soc., Chem. Commun. 1981, 146.

155. — — Pyrrolizidine Alkaloids. Evidence for N-(4-Aminobutyl)-1,4-diaminobutane (Homospermidine) as an Intermediate in Retronecine Biosynthesis. J. Chem. Soc., Chem. Commun. 1981, 554.

156. Khanna, P., and S. K. Manot: Production of Pyrrolizidine Alkaloid from *in vitro* Tissue Culture of *Crotalaria juncea* Linn. Indian J. Exp. Biol. **15**, 807 (1977).

157. Khasanova, M. A., U. A. Abdullaev, M. V. Telezhenetskaya, and S. Yu. Yunusov: Structure of Uluganine. Khim. Prir. Soedin. 1974, 809 [Chem. Abstr. **82**, 140349 (1975)].

158. Kirfel, A., G. Will, H. Wiedenfeld, and E. Röder: (1aR,6bR,10R,11R)-9,15-Dioxo-10-hydroxy-10,11,13-trimethyl-1a,2,3,6b-tetrahydro-5H-Pyrrolizino-[1a,6b,6a — b,c]-1,8-dioxa-15-*cis*-tridecene, $C_{18}H_{25}NO_5$. Cryst. Struct. Commun. **9**, 353 (1980).

159. — — — — (2S,4R,5R)-2-Carboxymethyl-5-Carboxy-2,4,5-trimethyl-2,3,4,5-Tetrahydrofuran, $C_{10}H_{16}O_5$. Cryst. Struct. Commun. **9**, 363 (1980).

160. Kiyamitdinova, F., S. T. Akramov, and S. Yu. Yunusov: Alkaloids from the Family of Boraginaceae. Khim. Prir. Soedin. 1967, 411 [Chem. Abstr. **68**, 75730 (1968)].

161. Kiyooka, S., and T. Hase: Pyrrolizidine Alkaloids: The Synthesis and Stereochemistry of α- and β-Retusanecic Acids. Bull. Chem. Soc. (Japan) **46**, 3609 (1973).

162. Kiyooka, S., T. Hase, and J. D. Edwards: Pyrrolizidine Alkaloids. The Total Synthesis of Retusaminic Acid. Chem. Lett. 1973, 963.

163. Klásek, A., V. A. Mnatsakanyan, and F. Šantavý: Alkaloids from *Senecio cineraria* D. C. Collect. Czech. Chem. Commun. **40**, 2524 (1975).

164. Klásek, A., T. Reichstein, and F. Šantavý: Die Pyrrolizidin-Alkaloide aus *Senecio alpinus* (L.) Scop., *S. subalpinus* Koch und *S. incanus* L. subsp. *carniolicus* (Willd.) Br.-Bl. Helv. Chim. Acta **51**, 1088 (1968).

165. Klásek, A., P. Sedmera, A. Boeva, and F. Šantavý: Pyrrolizidine Alkaloids. XX. Nemorensine, an Alkaloid from *Senecio nemorensis* L. Collect Czech. Chem. Commun. **38**, 2504 (1973).

166. Klásek, A., P. Sedmera, and F. Šantavý: Pyrrolizidine Alkaloids. XV. Structure of the Alkaloid Clivorine. Collect. Czech. Chem. Commun. **35**, 956 (1970).

167. — — — Pyrrolizidine Alkaloids. XVI. Alkaloids from Some Plants of the Genus *Ligularia*. Collect. Czech. Chem. Commun. **36**, 2205 (1971).

168. Klásek, A., P. Sedmera, J. Vokoun, A. Boeva, S. Dvorackova, and F. Šantavý:

Oxynemorensine, an Alkaloid from *Senecio nemorensis* L., var. *subdecurrens* Griseb. Collect. Czech. Chem. Commun. **45**, 548 (1980).

169. KLÁSEK, A., B. ŠULA, and F. ŠANTAVÝ: Pyrrolizidine Alkaloids. XXI. Alkaloids from *Senecio fluviatilis* Wallr. Collect. Czech. Chem. Commun. **38**, 2658 (1973).

169a. KLÁSEK, A., V. SVÁROVSKY, S. S. AHMED, and F. ŠANTAVÝ: Isolation of Pyrrolizidine Alkaloids from *Senecio aegyptius* L. and *S. desfontainei* Druce (Syn. *S. coronopifolius* Desf.). Coll. Czech. Chem. Commun. **33**, 1738 (1968).

170. KLÁSEK, A., P. VRUBLOVSKÝ, and F. ŠANTAVÝ: Isolation of Pyrrolizidine Alkaloids from the Plants *Senecio rivularis* D. C. and *Ligularia clivorum* Maxim. Collect. Czech. Chem. Commun. **32**, 2512 (1967).

171. KLOSE, W., K. NIKISCH, and F. BOHLMANN: Synthesis of 5,7a-Didehydroheliotridin-3-one, the Parent Compound of a New Class of Pyrrolizidine Alkaloids. Chem. Ber. **113**, 2694 (1980).

172. KOCHETKOV, N. K., A. M. LIKHOSHERSTOV, and V. N. KULAKOV: The Total Synthesis of Some Pyrrolizidine Alkaloids and Their Absolute Configuration. Tetrahedron **25**, 2313 (1969).

173. KOHLMUENZER, S., H. TOMCZYK, and A. SAINT-FERMIN: Alkaloid Emiline, a New Otonecine Ester from *Emilia flammea*. Diss. Pharm. Pharmacol. **23**, 419 (1971) [Chem. Abstr. **76**, 96972 (1972)].

174. KOMPIŠ, I., and F. ŠANTAVÝ: Alkaloide des raukenblättrigen Kreuzkrauts (*Senecio erucifolius* L.). Collect. Czech. Chem. Commun. **27**, 1413 (1962).

175. KONOVALOV, V. S., and G. P. MEN'SHIKOV: Alkaloids of *Cacalia hastata*. Zh. Obshch. Khim. **15**, 328 (1945) [Chem. Abstr. **40**, 3760 (1946)].

176. KOVACH, J. S., M. M. AMES, G. POWIS, C. G. MOERTEL, R. G. HAHN, and E. T. CREAGAN: Toxicity and Pharmacokinetics of a Pyrrolizidine Alkaloid, Indicine *N*-oxide, in Humans. Cancer Res. **39**, 4540 (1979).

177. LAING, M., and P. SOMMERVILLE: The Crystal and Molecular Structure of the Methiodide of Swazine, an Alkaloid from *Senecio swaziensis* Compton. Tetrahedron Lett. 1972, 5183.

178. LEANDER, K., and B. LÜNING: Studies on Orchidaceae Alkaloids. VII. Structure of a Glucosidic Alkaloid from *Malaxis congesta* comb. nov. (Rchb. f.). Tetrahedron Lett. 1967, 3477.

179. LINDSTRÖM, B., and B. LÜNING: Studies on Orchidaceae Alkaloids. XIII. A New Alkaloid, Laburnine Acetate, from *Vanda cristata* Lindl. Acta Chem. Scand. **23**, 3352 (1969).

180. — — Orchidaceae Alkaloids. XXIII. Alkaloids of *Liparis loeselii* and *Hammarbya paludosa*. Acta Chem. Scand. **25**, 895 (1971).

181. — — Orchidaceae Alkaloids. XXXV. Alkaloids from *Hammarbya paludosa* and *Liparis keitaoensis*. Acta Chem. Scand. **26**, 2963 (1972).

182. LINDSTRÖM, B., B. LÜNING, and K. SÜRALA-HANSEN: Orchidaceae Alkaloids. XXVI. New Glycosidic Alkaloid from *Malaxis grandifolia*. Acta Chem. Scand. **25**, 1900 (1971).

183. LOCOCK, R. A., J. L. BEAL, and R. W. DOSKOTCH: Alkaloid Constituents of *Eupatorium serotinum*. Lloydia **29**, 201 (1966).

184. LU, S. T., C. N. LIN, T. S. WU, and D. C. SHIEH: Alkaloids of *Senecio taiwanensis* and *S. morrisonensis*. J. Chinese Chem. Soc. (Taipei) **19**, 127 (1972) [Chem. Abstr. **77**, 161936 (1972)].

185. LÜNING, B., and H. TRÄNKER: Studies on Orchidaceae Alkaloids. X. A Pyrrolizidine Alkaloid from *Chysis bractescens* Lindl. Acta Chem. Scand. **22**, 2324 (1968).

186. McGAW, B. A., and J. G. WOOLLEY: The Biosynthesis of Angelic Acid in *Cynoglossum officinale*. Phytochemistry **18**, 1647 (1979).

187. MAHRAN, G., G. WASSEL, B. EL-MENSHAWI, G. EL-HOSSARY, and A. SAEED:

Pyrrolizidine Alkaloids of *Crotalaria aegyptiaca* and *Crotalaria madurensis*. Acta Pharm. Suec. **16**, 333 (1979).

188. MAN'KO, I. V.: Alkaloids of *Cynoglossum officinale*. Ukr. Khim. Zh. **25**, 627 (1959) [Chem. Abstr. **54**, 12494 (1960)].

189. — Alkaloids of *Cynoglossum officinale* and *Echium vulgare* and Standard Drug Preparation from *C. officinale*. Farm. Zh. (Kiev) **19**, 22 (1964) [Chem. Abstr. **64**, 4125 (1966)].

190. MAN'KO, I. V., M. P. ·KOROTKOVA, and N. M. SHEVTSOVA: Alkaloids of Some *Symphytum* Species. Rast. Resur. **5**, 508 (1969) [Chem. Abstr. **72**, 87175 (1970)].

191. MAN'KO, I. V., and B. K. KOTOVSKII: Asperumine, a New Alkaloid from the Plant *Symphytum asperum*, and its Structure. J. Gen. Chem. U.S.S.R. **40**, 2506 (1970).

192. MAN'KO, I. V., B. K. KOTOVSKII, and Y. G. DENISOV: Level of Alkaloids in *Symphytum officinale* Dependent on the Phase of Plant Development. Rast Resur. **6**, 409 (1970) [Chem. Abstr. **74**, 61608 (1971)].

193. — — — Accumulation of Alkaloids in *Symphytum asperum* Dependent on the Phase of Plant Development. Rast. Resur. **6**, 582 (1970) [Chem. Abstr. **74**, 84023 (1971)].

194. MAN'KO, I. V., and L. G. MARCHENKO: Alkaloids from *Paracynoglossum imeretinum*. Khim. Prir. Soedin. 537 (1971) [Chem. Abstr. **75**, 126598 (1971)].

195. — — Pictumine, a New Alkaloid from *Cynoglossum pictum*. Khim. Prir. Soedin. 1972, 655 [Chem. Abstr. **78**, 84611 (1973)].

196. MAN'KO, I. V., Z. V. MEL'KUMOVA, and U. F. MALYSHEVA: Accumulation of Alkaloids in Different Organs of *Symphytum caucasicum*. Rast. Resur. **8**, 538 (1972) [Chem. Abstr. **78**, 82085 (1973)].

197. MAN'KO, I. V., and P. N. VASIL'KOV: *Lappula intermedia* Alkaloids. I. Trudy Leningrad, Khim. Farm. Inst. **26**, 166 (1968) [Chem. Abstr. **73**, 73849 (1970)].

198. MARQUEZ, V. C.: Chromatographic Separation of the Alkaloids of *Bulnesia retamo, Heliotropium arborescens*, and *Cestrum auriculatum*. Biol. Soc. Quim. Peru **27**, 161 (1961) [Chem. Abstr. **61**, 15032 (1964)].

199. MATSUMOTO, T., K. FUKUI, and J. D. EDWARDS: Pyrrolizidine Alkaloids. Synthesis and Configuration of Fulvinic and Crispatic Acids. Chem. Lett. 1973, 283.

200. MATSUMOTO, T., T. OKABE, and K. FUKUI: Senecio Alkaloids. The Synthesis and Configuration of (±)-Latifolic Acid. Chem. Lett. 1972, 29.

201. — — — Pyrrolizidine Alkaloids. The Absolute Configuration of Latifolic Acid and its Stereoisomers. Chem. Lett. 1973, 773.

202. MATSUMOTO, T., M. TAKAHASHI, and Y. KASHIHARA: Pyrrolizidine Alkaloids. The Synthesis and Absolute Configuration of All Stereoisomers of Monocrotalic Acid. Bull. Chem. Soc. Japan **52**, 3329 (1979).

203. MATTOCKS, A. R.: Minor Alkaloids of *Heliotropium indicum* L. J. Chem. Soc. (C) 1967, 329.

204. — Toxicity and Metabolism of Senecio Alkaloids. "Phytochemical Ecology" ed. J. B. HARBORNE, p. 179. London and New York: Academic Press. 1972.

205. — Pyrrolizidine Alkaloid Analogues. Part 2. Further Hydroxymethyl-1-methyl-3-pyrrolines (Synthanecines), and the Preparation and Esterification of Some Hydroxymethylpyrroles. J. Chem. Soc., Perkin Trans. I 1978, 896; — The Distribution of Radioactivity in Rats given [^{3}H]2,3-Bisacetoxymethyl-1-methylpyrrole, a Pneumotoxic Analogue of Labile Pyrrolizidine Alkaloid Metabolites. Toxicol. Lett. **3**, 79 (1979).

206. MATTOCKS, A. R., R. SCHOENTAL, H. C. CROWLEY, and C. C. J. CULVENOR: Indicine: the Major Alkaloid of *Heliotropium indicum* L. J. Chem. Soc. 1961, 5400.

207. MEDVEDEVA, R. G., and Z. M. ZOLOTAVINA: Dynamics of Alkaloids in *Heliotropium arguzioides*. Trudy Inst. Bot., Akad. Nauk Kaz. SSR **29**, 181 (1971) [Chem. Abstr. **76**, 56583 (1972)].

208. MÉNDEZ, M. R., and I. RIBAS: Papilionaceous Alkaloids. XXX. Alkaloids of *Adenocarpus grandiflorus*. Anales Real. Soc. Espan. Fis. Quim. (Madrid) **54B**, 161 (1958) [Chem. Abstr. **52**, 17313e (1958)].

209. MODY, N. V., R. S. SAWHNEY, and S. W. PELLETIER: Carbon-13 Nuclear Magnetic Resonance Spectral Assignments for Pyrrolizidine Alkaloids. J. Natural Products **42**, 417 (1979).

210. a) MOHANRAJ, S., P. S. SUBRAMANIAN, C. C. J. CULVENOR, J. A. EDGAR, J. L. FRAHN, L. W. SMITH, and P. A. COCKRUM: Curassavine, an Alkaloid from *Heliotropium curassavicum* Linn. with a C_8 Necic Acid Skeleton. J. Chem. Soc., Chem. Commun. 1978, 423; b) SUBRAMANIAN, P. S., S. MOHANRAJ, P. A. COCKRUM, C. C. J. CULVENOR, J. A. EDGAR, J. L. FRAHN, and L. W. SMITH: The Alkaloids of *Heliotropium curassavicum*. Aust. J. Chem. **33**, 1357 (1980).

211. MOLYNEUX, R. J., A. E. JOHNSON, J. N. ROITMAN, and M. E. BENSON: Chemistry of Toxic Range Plants. Determination of Pyrrolizidine Alkaloid Content and Composition in *Senecio* species by Nuclear Magnetic Resonance Spectroscopy. J. Agric. Food Chem. **27**, 494 (1979).

212. MONTIDOME, M., and P. C. FERREIRA: Alkaloids from the Genus *Senecio*. Rev. Fac. Farm. Bioquim. Univ. Sao Paulo **4**, 175 (1966) [Chem. Abstr. **67**, 79670 (1967)].

213. NEUNER-JEHLE, N., H. NESVADBA, and G. SPITELLER: Anwendung der Massenspektrometrie zur Strukturaufklärung von Alkaloiden, 6. Mitt. Monatsh. Chem. **96**, 321 (1965).

214. NGHIA, N. T., P. SEDMERA, A. KLÁSEK, A. BOEVA, L. DRJANOVSKA, L. DOLEJŠ, and F. ŠANTAVÝ: Bulgarsenine and Retroisosenine, Alkaloids from *Senecio nemorensis* L., var. *bulgaricus* (Vel.) Stoj. et Stef. Collect. Czech. Chem. Commun. **41**, 2952 (1976).

215. NISHIKAWA, K., M. MIYAMURA, and Y. HIRATA: Chemotaxonomical Alkaloid Studies. Structures of *Liparis* Alkaloids. Tetrahedron **25**, 2723 (1969).

216. NOWACKI, E., and R. U. BYERRUM: A Study on the Biosynthesis of the *Crotalaria* Alkaloids. Life Sci. **1**, 157 (1962).

217. O'KELLY, J., and K. SARGEANT: Supinine from the Seeds of *Trichodesma zeylanicum*, R. Br. J. Chem. Soc. 1961, 484.

218. PAIS, M., F. X. JARREAU, P. FOUCHÉ, and R. GOUTAREL: Adulteration of *Strophanthus gratus* Seeds. Alafine, a New Alkaloid Isolated from the Seeds of *Alafia* species and *Alafia multiflora* (Apocynaceae). Ann. Pharm. Fr. **29**, 57 (1971).

219. PANIZO, F. M., and B. RODRIGUEZ GONZALES: Chemical Study of Peninsular *Senecios*. III. Alkaloids of *S. durieui* and *S. auricola*. Anales de Quim. **70**, 1043 (1974).

220. PASS, D. A., G. G. HOGG, R. G. RUSSELL, J. A. EDGAR, I. M. TENCE, and L. RIKARD-BELL: Poisoning of Chickens and Ducks by Pyrrolizidine Alkaloids of *Heliotropium europaeum*. Aust. Vet. J. **55**, 284 (1979).

221. PEDERSEN, E.: Minor Pyrrolizidine Alkaloids from *Cynoglossum officinale*. Dansk. Tidsskr. Farm. **44**, 287 (1970) [Chem. Abstr. **74**, 72780 (1971)].

222. — Pyrrolizidine Alkaloids in Danish Species of the Family Boraginaceae. Arch. Pharm. Chem. Sci. Ed. **3**, 55 (1975).

223. — Echinatine and Supinine: Pyrrolizidine Alkaloids from *Eupatorium cannabinum*. Phytochemistry **14**, 2086 (1975).

224. PEDERSEN, E., and E. LARSEN: Mass Spectrometry of Some Pyrrolizidine Alkaloids. Org. Mass Spectrom. **4**, 249 (1970).

225. PÉREZ-SALAZAR, A.: Crystal Structure of the Alkaloid Jacobine by X-Ray Diffraction. Anales de Quim. **74**, 196 (1978); PÉREZ-SALAZAR, A., F. H. CANO, and S. GARCÍA-BLANCO: The Alkaloid Jacobine ($C_{18}H_{25}NO_6$): A Refinement. Cryst. Struct. Commun. 1978, 105.

226. PÉREZ-SALAZAR, A., F. H. CANO, J. FAYOS, S. MARTÍNEZ-CARRERA, and S. GARCÍA-

BLANCO: The Alkaloid Otosenine. Evidence of a Weak $-N: \cdots C - O$ Intra-Annular Bond. Acta Crystallogr., Sect. B **33**, 3525 (1977).

227. PILBEAM, D. J., R. M. POLHILL, and E. A. BELL: Free Amino Acids and Alkaloids of South American, Asian and Australian *Crotalaria* species. Bot. J. Linn. Soc. **79**, 259 (1979).

228. PIMENOV, M. G., L. D. YAKHONTOVA, D. PAKALNE, and L. A. SAPUNOVA: Intraspecific Chemical Differentiation of *Adenostyles rhombifolia* in Northern Caucasus. Rastit. Resur. **11**, 72 (1975) [Chem. Abstr. **82**, 152218 (1975)].

229. PIZZORNO, M. T., and S. M. ALBONICO: Cycloaddition to Propargylic Aldehyde: a Single Step Synthesis of a Pheromone. Chem. Ind. (London) 1978, 349.

230. PLESCIA, S., G. DAIDONE, and V. SPRIO: Compositae. *Senecio nebrodensis* L. var *sicula*. Phytochemistry **15**, 2026 (1976).

231. PORTER, L. A., and T. A. GEISSMAN: Angularine, a New Pyrrolizidine Alkaloid from *Senecio angulatus* L. J. Org. Chem. **27**, 4132 (1962).

232. PURI, S. C., R. S. SAWHNEY, and C. K. ATAL: Genus Crotalaria. XIII. Crobarbatine, a New Pyrrolizidine Alkaloid of *C. barbata*. Experientia **29**, 390 (1973).

233. — — Genus *Crotalaria*. XVI. Pyrrolizidine Alkaloids of *C. leiloba* Bartl., *C. stipularia* Desv. and *C. tetragona* Roxb. J. Indian Chem. Soc. **51**, 628 (1974).

234. QUALLS, C. W., and H. J. SEGALL: Rapid Isolation and Identification of Pyrrolizidine Alkaloids *(Senecio vulgaris)* by Use of High-Performance Liquid Chromatography. J. Chromatogr. **150**, 202 (1978).

235. RAJAGOPALAN, T. R., and V. BATRA: Monocrotalinine, a New Alkaloid from *Crotalaria grahamiana*. Indian J. Chem., Sect. B **15**, 455 (1977).

236. — — Alkaloidal Constituents of *Heliotropium curassavicum*. Indian J. Chem., Sect. B **15**, 494 (1978).

237. RAO, P. G., R. S. SAWHNEY, and C. K. ATAL: Genus *Crotalaria*: Part XXIII — Croburhine, a Novel Pyrrolizidine Alkaloid from *Crotalaria burhia* Buch.-Ham. Indian J. Chem. **13**, 835 (1975).

238. — — Genus *Crotalaria* XXI. Isocromadurine, a Novel Pyrrolizidine Alkaloid of *C. madurensis* R. Wight. Experientia **31**, 878 (1975).

239. — — Genus *Crotalaria*: Part XX — Cromadurine, a New Pyrrolizidine Alkaloid from *Crotalaria madurensis* R. Wight. Indian J. Chem. **13**, 870 (1975).

240. RIBAS, I., and J. M. ALONSO DE LAMA: Alkaloids of the Papilionaceae. XII. Alkaloids of *Adenocarpus hispanicus* from the Guadarrama Mountains, Madrid. Farmacognosia (Madrid) **13**, 367 (1953).

241. RIBAS, I., and J. J. BARREIRO: Alkaloids of the Papilionaceae. XVIII. Alkaloids of *Adenocarpus decorticans*. Anales Asoc. Quim. Arg. **41**, 27 (1953) [Chem. Abstr. **48**, 3987i (1954)].

242. ROBERTSON, K. A., J. L. SEYMOUR, M.-T. HSIA, and J. R. ALLEN: Covalent Interaction of Dehydroretronecine, a Carcinogenic Metabolite of the Pyrrolizidine Alkaloid, Monocrotaline, with Cysteine and Glutathione. Cancer Res. **37**, 3141 (1977).

243. ROBINS, D. J.: Advances in Pyrrolizidine Chemistry. Adv. Heterocyclic Chem. **24**, 247 (1969).

244. — Senecioneae — Chemical Review *in* The Biology and Chemistry of the Compositae (HEYWOOD, V. H., J. B. HARBORNE, and B. L. TURNER, eds.), p. 831. London and New York: Academic Press. 1977.

245. — Pyrrolizidine Alkaloids *in* The Alkaloids (GRUNDON, M. F., ed.). The Chemical Society, London, 1978—1981, Vols. 8—11; see also earlier volumes: SAXTON, J. E.: 1971—1975, Vols. 1—5; CROUT, D. H. G.: 1976—1977, Vols. 6—7.

246. ROBINS, D. J., N. M. BALE, and D. H. G. CROUT: Pyrrolizidine Alkaloids. Biosynthesis of Monocrotalic Acid, the Necic Acid Component of Monocrotaline. J. Chem. Soc., Perkin Trans. I 1974, 2082.

247. ROBINS, D. J., and D. H. G. CROUT: Pyrrolizidine Alkaloids. The Stereochemistry of Monocrotalic and Trichodesmic Acids. J. Chem. Soc. (C) 1969, 1386.

248. — — Pyrrolizidine Alkaloids. The Absolute Configuration at C-2 in Monocrotalic Acid. J. Chem. Soc. (C) 1970, 1334.

249. ROBINS, D. J., and S. SAKDARAT: Synthesis of the 8β-Pyrrolizidine Bases (+)-Isoretronecanol, (+)-Laburnine, and (+)-Supinidine. J. Chem. Soc., Chem. Commun. 1979, 1181.

250. — — Synthesis of the Pyrrolizidine Base, (±)-Supinidine. J. Chem. Soc., Perkin Trans. I 1979, 1734.

251. — — Pyrrolizidine Alkaloids. Synthesis of 13,13-Dimethyl-1,2-didehydrocrotalanine. J. Chem. Soc., Chem. Commun. 1980, 282.

252. — — Synthesis of Optically Active Pyrrolizidine Bases. J. Chem. Soc., Perkin Trans. I 1981, 909.

253. ROBINS, D. J., and J. R. SWEENEY: Pyrrolizidine Alkaloids. Evidence for the Involvement of Spermidine and Spermine in the Biosynthesis of Retronecine. J. Chem. Soc., Chem. Commun. 1979, 120.

254. RÖDER, E., and H. WIEDENFELD: Isolierung und Strukturaufklärung des Alkaloids Fuchsisenecionin aus Senecio fuchsii. Phytochemistry 16, 1462 (1977).

255. RÖDER, E., H. WIEDENFELD, and M. FRISSE: Pyrrolizidinalkaloide aus Senecio doronicum. Phytochemistry 19, 1275 (1980).

256. RÖDER, E., H. WIEDENFELD, and U. PASTEWKA: Pyrrolizidinalkaloide aus Senecio vernalis. Planta Med. 37, 131 (1979).

257. RODRIGUEZ, F. D., and A. G. GONZALEZ: Canary Plant Alkaloids. XI. Secondary Alkaloids of Senecio kleinia. Farm. Nueva 36, 803 (1971) [Chem. Abstr. 76, 83573 (1972)].

258. — — Canary Plant Alkaloids. XII. Alkaloids of Senecio antieuphorbium. Farm. Nueva 36, 810 (1971) [Chem. Abstr. 76, 83572 (1972)].

259. ROTHSCHILD, M., R. T. APLIN, P. A. COCKRUM, J. A. EDGAR, P. FAIRWEATHER, and R. LEES: Pyrrolizidine Alkaloids in Arctiid Moths (Lep.) with a Discussion on Host Plant Relationships and the Role of these Secondary Plant Substances in the Arctiidae. Biol. J. Linn. Soc. 12, 305 (1979).

260. ROUFFIAC, R., and J. PARELLO: Étude Chimique des Alkaloides du Phyllanthus niruri L. (Euphorbiaceés). Présence de l'Antipode Optique de la Norsécurinine. Plant. Med. Phytother. 3, 220 (1969).

261. SADRITINOV, F. S.: Pharmacology of Alkaloids of 1-Methylpyrrolizidine and their Derivatives. Farmacol. Prirod. Soedin. Tashkent 1979, 29 [Chem. Abstr. 93, 18681 (1980)].

262. SAITO, S., T. TANAKA, K. KOTERA, H. NAKAI, N. SUGIMOTO, Z. HORII, M. IKEDA, and Y. TAMURA: Structures of Norsecurinine and Dihydronorsecurinine. Chem. Pharm. Bull. (Japan) 13, 786 (1965).

263. SASAKI, K., and Y. HIRATA: The Structures of Two New Zwitterionic Alkaloids from Anodendron affine Druce. Tetrahedron Lett. 1969, 4065; — — The Alkaloids of Anodendron affine Druce. Tetrahedron 26, 2119 (1970).

264. SAWHNEY, R. S., and C. K. ATAL: Genus Crotalaria. X. Methylethers of Supinidine and Heliotridine from C. medicaginea Lam. J. Indian Chem. Soc. 47, 741 (1970).

265. — — Genus Crotalaria XII. Identity of Crotalaburnine as Anacrotine. J. Indian Chem. Soc. 48, 887 (1971).

266. — — Croalbidine, a New Pyrrolizidine Alkaloid from Crotalaria albida Heyne ex Roth. Indian J. Chem. 11, 88 (1973).

267. SAWHNEY, R. S., C. K. ATAL, C. C. J. CULVENOR, and L. W. SMITH: Genus Crotalaria. XVII. The Stereochemistry of Croalbinecine. Aust. J. Chem. 27, 1805 (1974).

268. SAWHNEY, R. S., R. N. GIROTRA, C. K. ATAL, C. C. J. CULVENOR, and L. W. SMITH:

Phytochemical Studies on Genus *Crotalaria*: Part VII – Major Alkaloids of *C. mucronata, C. brevifolia* and *C. laburnifolia*. Indian J. Chem. **5**, 655 (1967).

269. Schoental, R.: Pancreatic Islet-Cell and other Tumors in Rats Given Heliotrine, a Monoester Pyrrolizidine Alkaloid, and Nicotinamide. Cancer Res. **35**, 2020 (1975); — Biochemical Basis of Liver Necrosis Caused by Pyrrolizidine Alkaloids and other Hepatotoxins. Biochem. Soc. Trans. **3**, 292 (1975).

270. — Alkylation of Coenzymes and the Acute Effects of Alkylating Hepatotoxins. FEBS Lett. **61**, 111 (1976).

271. Sedmera, P., A. Klásek, A. M. Duffield, and F. Šantavý: Pyrrolizidine Alkaloids. XIX. Structure of the Alkaloid Erucifoline. Collect. Czech. Chem. Commun. **37**, 4112 (1972).

272. Segall, H. J.: Pyrrolizidine Alkaloids from *Senecio jacobaea*. Toxicol. Lett. **1**, 279 (1978).

273. — Reverse Phase Isolation of Pyrrolizidine Alkaloids. J. Liq. Chromatogr. **2**, 429 (1979); — Preparative Isolation of Pyrrolizidine Alkaloids Derived from *Senecio vulgaris*. J. Liq. Chromatogr. **2**, 1319 (1979).

274. Segall, H. J., and R. J. Molyneux: Identification of Pyrrolizidine Alkaloids *(Senecio longilobus)*. Res. Commun. Chem. Pathol. Pharmacol. **19**, 545 (1978).

275. Sethi, M. L., and C. K. Atal: Phytochemical Studies on the Genus *Crotalaria*. Indian J. Pharm. **25**, 159 (1963) [Chem. Abstr. **59**, 4206 (1963)].

276. Sheveleva, G. P., N. V. Plekhanova, and D. S. Sargazakov: Alkaloids from Plants of the Borage, Thistle, and Lily Families of Kirghiz Flora. Mater. Nauch. Konf., Posvyashch. 100-Sto-Letiyu Period. Zakona D. I. Mendeleeva, 1969, 107 [Chem. Abstr. **76**, 23067 (1972)].

277. Siddiqi, M. A., K. A. Suri, O. P. Suri, and C. K. Atal: Genus *Crotalaria*: Part XXXIV – Cronaburmine, a New Pyrrolizidine Alkaloid from *Crotalaria nana* Burm. Indian J. Chem. Sect. B **16**, 1132 (1978).

278. — — — — A New Pyrrolizidine Alkaloid from *Caccinia glauca*. Phytochemistry **17**, 2049 (1978).

279. — — — — Novel Pyrrolizidine Alkaloid from *Crotalaria nana*. Phytochemistry **17**, 2143 (1978).

280. — — — — New Pyrrolizidine Alkaloids from *Crotalaria candicans*. Phytochemistry **18**, 1413 (1979).

281. Soto, J. P.: Decorticasine. Acta Cient. Compostelana **6**, 37 (1969) [Chem. Abstr. **73**, 109943 (1970)].

282. Stermitz, F. R., and J. A. Adamovics: Alkaloids of *Caltha leptosepala* and *Caltha biflora*. Phytochemistry **16**, 500 (1977).

283. Stermitz, F. R., and T. R. Suess: Pyrrolizidine Alkaloids in *Castilleja rhexifolia* (Scrophulariaceae). Phytochemistry **17**, 2142 (1978).

284. Stillman, A. E., R. Huxtable, P. Consroe, P. Kohnen, and S. Smith: Hepatic Veno-occlusive Disease Due to Pyrrolizidine *(Senecio)* Poisoning in Arizona. Gastroenterology **73**, 349 (1977).

285. Stoeckli-Evans, H.: Monocrotaline. A Pyrrolizine Alkaloid. Acta Crystallogr., Sect. B **35**, 231 (1979).

286. — Bulgarsenine (R:R)-(+)-bitartrate: a Pyrrolizidine Alkaloid. Acta Crystallogr. Sect. B **36**, 3150 (1980).

287. Stoeckli-Evans, H., and D. H. G. Crout: The Crystal Structure of Axillarine Hydrobromide Ethanol Solvate: A Pyrrolizidine Alkaloid. Helv. Chim. Acta **59**, 2168 (1976).

288. Subramanian, S. S., S. Nagarajan, and M. N. Ghosh: The Major Alkaloid of *Crotalaria paniculata* Seeds. Indian J. Pharm. **30**, 153 (1968).

289. Suri, K. A., R. S. Sawhney, and C. K. Atal: Pyrrolizidine Alkaloids from *Cynoglossum lanceolatum, C. glochidiatum*, and *Lindelofia angustifolia*. Indian J. Pharm. **37**, 69 (1975).

290. — — — Genus *Crotalaria*: Part XXV — Secopyrrolizidine Alkaloids of *Crotalaria walkeri* Arnott. Indian J. Chem., Sect. B **14**, 471 (1976).

291. Suri, K. A., O. P. Suri, K. L. Dhar, and C. K. Atal: Chemical Components of *Lapulla glochidiata* and *Crotalaria anagyroides*. Indian J. Chem. Sect. B **16**, 78 (1978).

292. Suri, O. P., R. S. Jamwal, K. A. Suri, and C. K. Atal: Ehretinine, a novel pyrrolizidine alkaloid from *Ehretia aspera*. Phytochemistry **19**, 1273 (1980).

293. Suri, O. P., R. S. Sawhney, and C. K. Atal: Pyrrolizidine Alkaloids from *Heliotropium eichwaldii* and *Lindelofia spectabilis*. Indian J. Chem. **13**, 505 (1975).

294. Suri, O. P., R. S. Sawhney, M. S. Bhatia, and C. K. Atal: Novel Secopyrrolizidine Alkaloids from *Crotalaria verrucosa*. Phytochemistry **15**, 1061 (1976).

295. Sussman, J. L., and S. J. Wodak: The Crystal Structure of Fulvine: A Pyrrolizidine Alkaloid. Acta Crystallogr. Sect. B **29**, 2918 (1973).

296. Tandon, H. D., B. N. Tandon, R. Tandon, and N. C. Nayak: A Pathological Study of the Liver in an Epidemic Outbreak of Veno-occlusive Disease. Indian J. Med. Res. **65**, 679 (1977).

297. Tandon, H. D., B. M. Tandon, and A. R. Mattocks: An Epidemic of Veno-occlusive Disease of the Liver in Afghanistan. Amer. J. Gastroenterol. **70**, 607 (1978).

298. Tanino, H., S. Inoue, K. Nishikawa, and Y. Hirata: Syntheses of Tetraacetyl Malaxin and Kuramerine. Tetrahedron **25**, 3033 (1969).

299. Tashkhodzhaev, B., M. V. Telezhenetskaya, and S. Yu. Yunusov: Crystal and Molecular Structure of the Macrocyclic Pyrrolizidine Alkaloid Incanine. Khim. Prir. Soedin. 1979, 363 [Chem. Abstr. **92**, 111199 (1980)].

300. Tashkhodzhaev, B., M. R. Yagudaev, and S. Yu. Yunusov: Crystal and Molecular Structure of the Macrocyclic Pyrrolizidine Alkaloid Trichodesmine. Khim. Prir. Soedin. 1979, 368 [Chem. Abstr. **92**, 111194 (1980)].

301. Tittel, G., H. Hinz, and H. Wagner: Quantitative Bestimmung der Pyrrolizidin-Alkaloide in *Symphyti Radix* durch HPLC. Planta Med. **37**, 1 (1979).

302. Tomczyk, H., and S. Kohlmuenzer; Isolation of a New Pyrrolizidine Alkaloid from *Emilia flammea* (Compositae). Herba Pol. **17**, 226 (1971) [Chem. Abstr. **77**, 19848 (1972)].

303. Tsuda, Y., and L. Marion: The Alkaloids of *Eupatorium maculatum* L. Can. J. Chem. **41**, 1919 (1963).

304. Tufariello, J. F., and G. E. Lee: Functionalized Nitrones. A Highly Stereoselective and Regioselective Synthesis of dl-Retronecine. J. Am. Chem. Soc. **102**, 373 (1980).

305. — — Synthesis in the Pyrrolizidine Class of Alkaloids. dl-Supinidine. J. Org. Chem. **40**, 3866 (1975).

306. Ulubelen, A., and S. Doğanca: Anadoline, a New *Senecio* alkaloid from *Symphytum orientale*. Tetrahedron Lett. 1970, 2583.

307. — — Boraginaceae. Alkaloidal and Other Constituents of *Symphytum orientale*. Phytochemistry **10**, 441 (1971).

308. Ulubelen, A., and F. Öcal: Alkaloids and other Compounds of *Symphytum tuberosum*. Phytochemistry **16**, 499 (1977).

309. Valverde, S., and F. M. Panizo: Chemical Study of the Peninsula *Senecio*. II. Basic Components of *Senecio tournefortii*. Anales de Quim. **67**, 425 (1971) [Chem. Abstr. **75**, 110484 (1971)].

310. Warren, F. L.: The Pyrrolizidine Alkaloids. I. Fortschr. Chem. org. Naturstoffe **12**, 198 (1955).

311. — The Pyrrolizidine Alkaloids. II. Fortschr. Chem. org. Naturstoffe **24**, 329 (1966).

312. — Senecio Alkaloids. In: The Alkaloids (R. H. F. Manske, ed.), Vol. XII. New York: Academic Press. 1970.

313. Wang, S.-T.: The Crystal Structure of Monocrotaline. K'o Hsueh T'ung Pao **23,** 670 (1978) [Chem. Abstr. **90,** 187191 (1979)].

314. White, E. P.: Alkaloids of Some Herbaceous *Senecio* Species in New Zealand. New Zealand J. Sci. **12,** 165 (1969).

315. White, E. P., and F. L. Warren: The *Senecio* Alkaloids. Sarracine *N*-oxide and Sarracine from *Senecio sylvaticus* L. Anales de Quim. **68,** 723 (1972).

316. Wiedenfeld, H., and E. Röder: Das Pyrrolizidinalkaloid Senecionin aus *Senecio fuchsii.* Phytochemistry **18,** 1083 (1979).

317. Willette, R. E., and L. V. Cammarato: Phytochemical Study of Connecticut. I. Isolation of Monocrotaline from *Crotalaria sagittalis* Fruit. J. Pharm. Sci. **61,** 122 (1972).

318. Wright, W. G., and F. L. Warren: Rhizophoraceae Alkaloids. Part I. Four Sulphur-containing Bases from *Cassipourea* spp. J. Chem. Soc. (C) 1967, 283.

319. Wunderlich, J. A.: The Molecular Structures of Retusamine, Otosenine, Renardine and Onetine. Chem. Ind. (London) 1962, 2089; — The Crystal Structure of Retusamine, α'-Bromo-D-camphor-*trans*-π-sulphonate Monohydrate. Acta Crystallogr. **23,** 846 (1967).

320. Yakhontova, L. D., M. G. Pimenov, and L. A. Sapunova: Alkaloids of *Adenostyles alliariae.* Khim. Prir. Soedin. 1976, 122 [Chem. Abstr. **85,** 59575 (1976)].

321. Yamada, K., H. Tatematsu, M. Suzuki, Y. Hirata, M. Haga, and I. Hirono: Isolation and the Structures of Two New Alkaloids, Petasitenine and Neopetasitenine from *Petasites japonicus* Maxim. Chem. Lett. 1976, 461; Yamada, K., H. Tatematsu, Y. Hirata, M. Haga, and I. Hirono: Stereochemistry of Petasitenine, the Carcinogenic Alkaloid from *Petasites japonicus* Maxim. and Transformation of Petasitenine to Senkirkine. Chem. Lett. 1976, 1123.

322. Yamada, K., H. Tatematsu, R. Unno, and Y. Hirata: Petasinine and Petasinoside, Two Minor Alkaloids Possessing a New Necine Isolated from *Petasites japonicus* Maxim. Tetrahedron Lett. 1978, 4543.

323. Yates, S. G., and H. L. Tookey: Festucine, an Alkaloid from Tall Fescue (*Festuca arundinaceae* Schreb.): Chemistry of the Functional Groups. Aust. J. Chem. **18,** 53 (1965).

324. Zalkow, L. H., L. Gelbaum, and E. Keinan: Isolation of the Pyrrolizidine Alkaloid Europine *N*-oxide from *Heliotropium maris-mortui* and *H. rotundifolium.* Phytochemistry **17,** 172 (1978).

325. Zalkow, L. H., S. Bonetti, L. Gelbaum, M. M. Gordon, B. B. Patil, A. Shani, and D. van Derveer: Pyrrolizidine Alkaloids from Middle Eastern Plants. J. Natural Prod. **42,** 603 (1979).

Addendum

Devlin, J. A., and D. J. Robins: Synthesis and Stereochemistry of Dicrotaline, a Macrocyclic Pyrrolizidine Alkaloid. J. Chem. Soc., Chem. Commun. 1981, in the press.

Drewes, S. E., and A. T. Pitchford: Synthesis of New Macrocycles. Part 6. Pyridine-retronecate, a New Synthetic Alkaloid. J. Chem. Soc., Perkin Trans I 1981, 408.

Edgar, J. A., C. C. J. Culvenor, P. A. Cockrum, L. W. Smith, and M. Rothschild: Callimorphine: Identification and Synthesis of the Cinnabar Moth "Metabolite". Tetrahedron Lett. **1980,** 1383.

Frahn, J. L., C. C. J. Culvenor, and J. A. Mills: Preparative Separation of the Pyrrolizidine Alkaloids, Intermedine and Lycopsamine, as their Borate Complexes. J. Chromatogr. **195,** 379 (1980).

Grue-Sørensen, G., and I. D. Spenser: Biosynthesis of Retronecine. J. Am. Chem. Soc. **103,** 3208 (1981).

HERZ, W., P. KULANTHAIVEL, P. S. SUBRAMANIAN, C. C. J. CULVENOR, J. L. FRAHN, and J. A. EDGAR: Alkaloids of *Conoclidium coelestinum* (L.) DC., *Eupatorium compositifolium* Walt. and *E. altissimum* L.: Occurrence of Crystalline Intermedine in *C. coelestinum*. Experientia **37**, 683 (1981).

HIKICHI, M., Y. ASADA, and T. FURUYA: Lycopsamine and O^9-Angelylretronecine, Pyrrolizidine Alkaloids from *Messerschmidia sibirica*. Planta Med. (Suppl. Vol.) 1980, 1.

HUANG, J., and J. MEINWALD: Synthesis of Crobarbatine Acetate. A Macrocyclic Pyrrolizidine Alkaloid Ester. J. Am. Chem. Soc. **103**, 861 (1981).

LÜTHY, J., U. ZWEIFEL, B. KARLHUBER, and C. SCHLATTER: Pyrrolizidine Alkaloids of *Senecio alpinus* L. and their Detection in Feedingstuffs. J. Agric. Food Chem. **29**, 302 (1981).

MATTOCKS, A. R.: A Simple Preparation of Dehydroretronecine using Potassium Nitrosodisulphonate. Chem. Ind. (London) 1981, 251.

— Relation of Structural Features to Pyrrolic Metabolites in Livers of Rats given Pyrrolizidine Alkaloids and Derivatives. Chem.-Biol. Interact. **35**, 301 (1981).

MOHANRAJ, S., P. KULANTHAIVEL, P. S. SUBRAMANIAN, and W. HERZ: Helifoline, a New Pyrrolizidine Alkaloid from *Heliotropium ovalifolium*. Phytochemistry **20**, 1991 (1981).

PASTEWKA, U., H. WIEDENFELD, and E. RÖDER: The Synthesis of Racemic 5-Hydroxy-3,4-dimethyl-1-hexene-2,5-dicarboxylic acid (Senecivernic Acid). Arch. Pharm. (Weinheim) **313**, 785 (1980).

ROBINS, D. J., and J. R. SWEENEY: Pyrrolizidine Alkaloid Biosynthesis. Incorporation of ^{14}C-Labelled Precursors into Retronecine. J. Chem. Soc., Perkin Trans I 1981, in the Press.

RÖDER, E., H. WIEDENFELD, and P. STENGL: Das Pyrrolizidinalkaloid O^7-Angelylheliotridin aus *Senecio ovirensis*. Planta Med. (Suppl. Vol.) 1980, 182.

SMITH, L. W., and C. C. J. CULVENOR: Plant Sources of Pyrrolizidine Alkaloids. Lloydia **44**, 129 (1981).

SUBRAMANIAN, P. S., S. MOHANRAJ, P. A. COCKRUM, C. C. J. CULVENOR, J. A. EDGAR, J. L. FRAHN, and L. W. SMITH: The Alkaloids of *Heliotropium curassavicum*. Aust. J. Chem. **33**, 1357 (1980).

SULLIVAN, G.: Detection of Pyrrolizidine-Type Alkaloids in Matarique *(Cacalia decomposita)*. Vet. Human Toxicol. **23**, 6 (1981).

VEDEJS, E., and G. R. MARTINEZ: Stereospecific Synthesis of Retronecine by Imidate Methylide Cycloaddition. J. Am. Chem. Soc. **102**, 7993 (1980).

WRÓBEL, J. T., and J. A. GLIŃSKI: Stereospecific Total Synthesis of ($\pm$)-Cassipurine, an Alkaloid from Rhizophoraceae. Can. J. Chem. **59**, 1101 (1981).

(Received July 9, 1981)

Alkaloids of Neotropical Poison Frogs (Dendrobatidae)

By J. W. DALY, Laboratory of Bioorganic Chemistry, National Institute of Arthritis, Diabetes, and Digestive and Kidney Diseases, National Institutes of Health, Bethesda, Maryland, U.S.A.

With 27 Figures

Contents

I. Introduction

Poisons for arrows and blow darts have been derived from a wide spectrum of sources in both the plant and animal kingdoms. One unique source of such poisons is the skin secretion of certain brightly colored frogs native to the rain forests of Western Colombia. The Noanamá and Emberá Indians of this region undoubtedly used secretions from these frogs to poison blow darts even in pre-Colombian times, but the first account of dart envenomation with poison frogs did not appear until 1825 (67). Secretions from a single frog were purported to be sufficient for envenomation of at least twenty blow darts [(67, 214, 221, 270, 271); see (176) for review of early literature]. Only three species of neotropical frogs can be stated with assurance to have been used to poison blow darts (191). All of these frogs occur only in western Colombia where the practice of poisoning blow darts with frog secretions still persists today in spite of the inroads of civilization. The poison-dart frog (Phyllobates bicolor) from the headwaters of the Río San Juan is called "neará" by the Indians, while lower in the same drainage the poison-dart frog (Phyllobates aurotaenia) is called "kokoi". Envenomation of darts with the most toxic species (Phyllobates terribilis) from the Río Saija is done simply by drawing the tip of the dart across the back of a living frog while the other two species are impaled in order to elicit a copious flow of skin secretions.

It is not suprising that the nature and action of the poison from these frogs attracted the attention of toxicologists, pharmacologists and chemists. However, perhaps because of the remoteness of the region and the resultant difficulties in obtaining an ample supply of poison, a sustained investigation of the active principles from neotropical poison dart frogs was not initiated until 1961. Early studies demonstrated that the poison — obtained from the blow darts rather than from the frogs themselves — was highly active and probably owed its toxic effects to actions on nerve and muscle (184, 185, 214, 222). The chemical nature of the active principle was undefined although a chemist, J. Aronhson, had concluded that it was an alcohol-soluble alkaloid (cited in *176*). Such a conclusion, however, was not substantiated by SANTESSON (*222*). Finally, the active principle from one of the poison-dart frogs was established to be an extremely toxic alkaloid (*176*), thus providing the starting point and the impetus for some eighteen years of research on the skin alkaloids from this family of brightly colored poison frogs, the Dendrobatidae.

Batrachotoxins. After five years of research, the structures of the first class of dendrobatid alkaloids were reported in 1968-9 (*251, 252*). The parent alkaloid of this first class of dendrobatid alkaloid was named batrachotoxin from the Greek *batrachos* meaning frog. Four *major* alkaloids had been isolated from skin extracts from the Colombian poison-dart frog, *Phyllobates aurotaenia*, and all were closely related complex steroidal compounds whose presence in nature remains unique to this one group of neotropical frogs. Indeed, the presence of batrachotoxins has been used as one character for the definition of the true poison-dart frogs as a monophyletic genus, *Phyllobates* (*191*). This genus, *Phyllobates*, contains five species ranging from Costa Rica through Panama and Western Colombia.

Pumiliotoxin-C. Extension of these investigations to other poison frogs of the family Dendrobatidae led to the isolation of simpler alkaloids (*77*). The structure of the first of the simpler alkaloids was reported in 1969 (*87*). The compound proved to be a *cis*-decahydroquinoline and was named pumiliotoxin C after the specific name of the Panamanian frog *Dendrobates pumilio* from which it was first isolated. Although termed a toxin, the compound has relatively low toxicity, unlike the fairly toxic pumiliotoxin A and B which had been isolated from the same Panamanian frog (*77*).

Histrionicotoxins. Another poison frog, *Dendrobates histrionicus* occurred sympatric with the poison-dart frog *Phyllobates aurotaenia* in Western Colombia. Preliminary studies based primarily on mass spectrometry of compounds isolated by thin-layer chromatography, indicated the presence of another class of relatively simple alkaloids in extracts from a few specimens of this species. Three major alkaloids were later isolated from skin extracts of a population of *Dendrobates histrionicus* from southwestern Colombia and the structures of two of these were reported in 1971 (*81*). The compounds were unique spiropiperidine alkaloids with remarkable acetylenic and allenic centers of unsaturation in the side chain substituents. The parent compound was named histrionicotoxin after the specific name of the frog *Dendrobates histrionicus* from which it was first isolated. A number of histrionicotoxins were subsequently isolated and structurally defined (*86, 253*). Like pumiliotoxin C, the histrionicotoxins exhibit relatively low toxicity to mammals.

Gephyrotoxins. One of the major alkaloids from *Dendrobates histrionicus* proved to be not a histrionicotoxin but instead a tricyclic alkaloid which after elucidation of its structure was named gephyrotoxin (*86*). The name is derived from the Greek *gephyra* meaning bridge and literally refers to the "bridge", which has been presumed to be formed biosynthetically by addition of the nitrogen to one of two side chains of a proposed parent 2,6-disubstituted piperidine. This bridge forms a bicyclic indolizidine ring system. A further biosynthetic cyclization — analogous to the cyclization which has been presumed to form the decahydroquinoline pumiliotoxin C from a 2,6-disubstituted piperidine — would then form the tricyclic gephyrotoxin. The term gephyrotoxin has been applied both to simple bicyclic indolizidines and to the tricyclic perhydrobenzoindolizidines (*80*). Gephyrotoxin itself exhibits relatively low toxicity to mammals.

Recently, the proposed structure of one of the simpler bicyclic gephyrotoxins designated alkaloid **223AB**, from a population of the Colombian frog *Dendrobates histrionicus* (*80*), was established by comparison with synthetic compounds (*244*). The natural alkaloid was shown to be a simple 3,5-disubstituted indolizidine which, because of the presence of the above mentioned "bridge" and the requisite apparent biosynthetic lineage to a presumed precursor 2,6-disubstituted piperidine has come to be termed gephyrotoxin **223AB** (*175, 244*).

Pumiliotoxin A. After some eighteen years of research on dendrobatid alkaloids over one hundred compounds had been characterized or partially characterized. The simpler alkaloids were given numerical designations and placed for the purpose of classification into various classes or categories (*80*), viz. the batrachotoxin class, the pumiliotoxin-C class, the histrionico-toxin class, the gephyrotoxin class, and the pumiliotoxin-A class. The structural nature of alkaloids of the pumiliotoxin-A class remained an elusive challenge for many years. Pumiliotoxin A and pumiliotoxin B first isolated from the Panamanian frog *Dendrobates pumilio* (*77*) were the parent members of the class, all of which exhibit mass spectra containing major fragment ions of C_4H_8N (m/z 70), and either $C_{10}H_{16}NO$ (m/z 166) or $C_{10}H_{16}NO_2$ (m/z 182). The structure of a relatively simple member of the class, designated alkaloid **251D** from the Ecuadorian frog *Dendrobates tricolor* was finally elucidated in 1980 (*84*). The compound proved to be an indolizidine but, unlike pumiliotoxin C, the histrionicotoxins or the gephyrotoxins, it was not an alkaloid which could be derived directly by biosynthetic cyclizations of a presumed precursor 2,6-disubstituted pipe-ridine. Its structure provided the key to the structures of pumiliotoxin A, pumiliotoxin B, and a few further alkaloids of the pumiliotoxin-A class. Pumiliotoxin A and B, representatives of the second class of dendrobatid alkaloids to be documented (*77*), are relatively toxic compounds, although at least two orders of magnitude less toxic than the batrachotoxins.

Other Alkaloids. There remain a large number of dendrobatid alkaloids of unknown or poorly defined structures. Many of these are only trace constituents or occur in species of frogs which are rare or difficult to obtain in adequate numbers. Many were proposed based on limited data to be members of the pumiliotoxin-C class or to be related hydroxy-compounds (*80*). However, it now appears likely that further classes of dendrobatid alkaloids will be defined as research continues and that some of these trace constituents will be representatives of new classes. Indeed, both indole alkaloids and a pyridine alkaloid have recently been isolated from dendrobatid frogs (*254*). For the present review, all compounds which are not unambigously members of the present five major classes of dendrobatid alkaloids have been included as "Other Alkaloids".

Pharmacology and Biology. The dendrobatid alkaloids have proved to have remarkable pharmacological activities, a circumstance lending impetus to synthetic studies and to further research on the structures of the remaining undeciphered compounds from this amazing family of frogs. The pharmacological activities undoubtedly owe their derivation from evolutionary development of these skin alkaloids as part of a chemical defense system designed to protect the brightly colored dendrobatid frogs from predators. The skin secretions from species of *Phyllobates* and *Dendrobates* are indeed quite effective deterents against certain predators (*191, 234, 235*). The alkaloids are presumably stored in cutaneous granule glands and secreted under duress (*196*). Deterent or noxious activities at buccal tissue

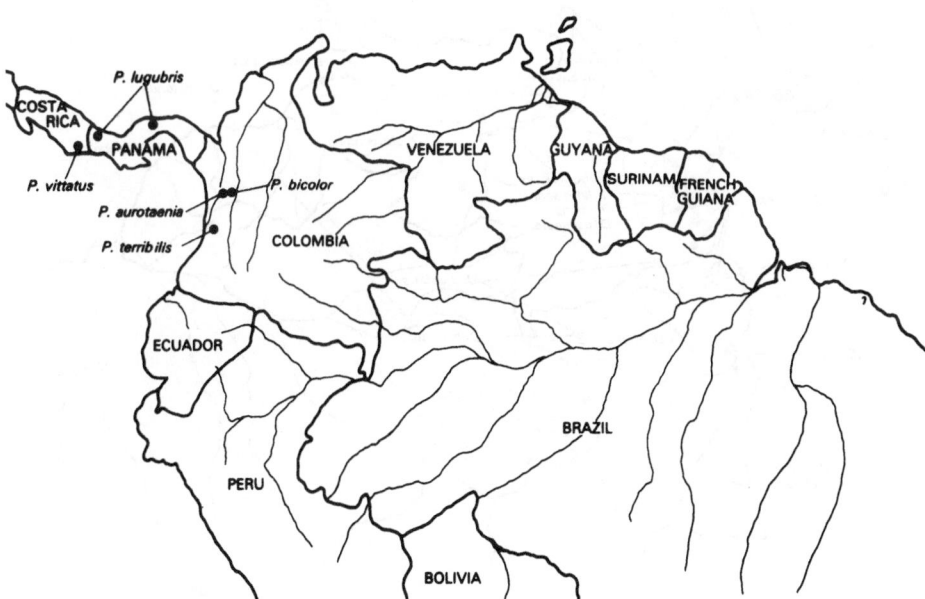

Fig. 1. Sites of collection of poison-dart frogs (Genus *Phyllobates*) for analysis of skin extracts
(*191*)

rather than toxicity *per se* have undoubtedly been the main selective factors involved in the evolutionary development of such alkaloids. This hypothesis perhaps explains why certain alkaloids such as pumiliotoxin C, the histrionicotoxins, and gephyrotoxin are relatively nontoxic.

At the present time, extracts from some forty-five species of dendrobatid frogs have been examined, but detailed analyses have been carried out and published for only about one-half of these species. Sites of collection of various species of *Phyllobates* and *Dendrobates* for analysis of skin extracts are shown in Figs. 1 and 2. The five species of the genus *Phyllobates* all contain batrachotoxins (*191*, see Table 3). Certain Phyllobates species also contain pumiliotoxins and histrionicotoxins (*191*, unpublished results). The some thirty species of the genus *Dendrobates* whose skin extracts have been examined contain only the simpler dendrobatid alkaloids, that it is the pumiliotoxins, the histrionicotoxins, the gephyrotoxins, and related compounds (*80, 187 – 190*, unpublished results; see Table 22). As yet, dendrobatid alkaloids have not been positively identified in skin extracts from ten species of a third genus of dendrobatid frogs, the usually cryptically colored *Colostethus*.

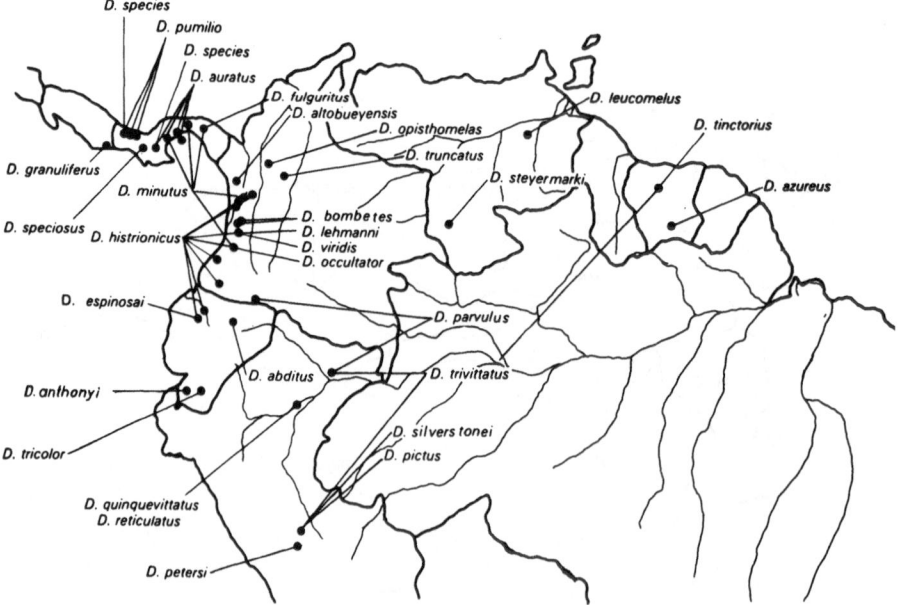

Fig. 2. Sites of collection of poison frogs (Genus *Dendrobates*) for analysis of skin extracts (*80, 187—190*, and unpublished data)

Biosynthetically, little is known of the origin of these alkaloids. In a preliminary study, radioactive acetate and mevalonate were incorporated into skin steroids of frogs, while incorporation of these potential precursors into skin alkaloids was not detected (*150*). Neither radioactive cholesterol nor serine was significantly incorporated into the batrachotoxins. Three species, *Phyllobates aurotaenia, Dendrobates pumilio* and *Dendrobates auratus* were used in these exploratory biosynthetic studies.

The present review summarizes current knowledge of the properties, occurrence, synthesis and biological activity of the dendrobatid alkaloids.

II. Batrachotoxins

A. Structures

The initial studies by Märki and Witkop (*176*) on poison-dart frogs established a pattern for subsequent studies. During the summer of 1962, the poison-dart frogs were obtained in the field in western Colombia, sacrificed and the skins placed in methanol. Preliminary field evaluation of extracts by Dr. Märki had established that at least ninety percent of the toxic principles of a poison-dart frog were in the skin and could be efficiently extracted into methanol. In these initial studies, the methanol extracts were evaporated *in vacuo* at a field station and the residues transported to the National Institutes of Health for isolation of active principles. The active principles appeared more stable in the dry residue than in the crude methanol extract. A bioassay was also established during these initial studies. It was found that the time of death of white mice was related to the amount of crude extract injected subcutaneously (*176*). Fortunately, the active principles from the poison-dart frog were very toxic so that only a small amount, actually only about 200 ng of pure toxin, was needed for a single bioassay. Ultimately, a total of a "few hundred" micrograms of apparently pure toxin was isolated from extracts of 330 frog skins.

Briefly, the most satisfactory isolation procedure was as follows: 1) extraction of minced skins twice with 70% aqueous methanol, followed by evaporation *in vacuo* to yield a dry tan residue; 2) trituration with 0.9% NaCl, adjustment of pH to 2 with HCl, followed by extraction of lipids into chloroform; 3) adjustment of pH to 8.5 with aqueous NH_3, followed by extraction of toxic principles into chloroform; 4) after evaporation to dryness *in vacuo* the active principles were purified by thin-layer chromatography on silica gel with chloroform-methanol (6:1). Two iodoplatinate-positive alkaloid spots were detected; the higher R_f material accounted for virtually all of the toxicity. In view of subsequent studies this higher R_f

material was undoubtedly a mixture of two very similar compounds, batrachotoxin and homobatrachotoxin, while the lower R_f material was undoubtedly the much less toxic batrachotoxinin A. With only a "few hundred micrograms", little chemical characterization was possible. The material had basic properties and an apparent carbonyl absorption at about 1690 cm⁻.

In order to accomplish structure elucidation, larger numbers of frogs were clearly necessary. Fortunately, the frog occurs over a vast area in the Pacific drainage of Colombia and is fairly common in relatively accessible regions of the middle and upper Rio San Juan drainage. From extracts of 2400 frogs a total of 30 mg of "batrachotoxin", 15 mg of batrachotoxinin A, 6 mg of a compound referred to as batrachotoxinin B, and 1 mg of a compound referred to as batrachotoxinin C were isolated (85). The "batrachotoxin" undoubtedly still represented a mixture of two very similar compounds, batrachotoxin and homobatrachotoxin. The batracho-toxin B and C may have represented artefacts formed by oxidation and/or hydrolysis during isolation and have not isolated in subsequent studies. The extraction, partition and thin-layer isolation procedure used in this study (85) was similar to that employed by Märki and Witkop (176). Extractions and partitions were carried out at room temperatures and large losses were incurred during preparative thin-layer chromatography. The major com-pounds were partially characterized both by physical and chemical methods (85). The "batrachotoxin" appeared to have a molecular ion in its mass spectrum corresponding to $C_{24}H_{33}NO_4$ (m/z 399). This apparent molecular ion was subsequently found to represent a pyrolysis fragment with the true molecular ion being $C_{31}H_{42}N_2O_6$ (m/z 538) for batrachotoxin and $C_{32}H_{44}N_2O_6$ (m/z 552) for homobatrachotoxin. Batrachotoxinin A affor-ded a true molecular ion corresponding to $C_{24}H_{35}NO_5$ (m/z 417). The two minor alkaloids, batrachotoxinin B and C which had low R_f values and may have represented isolation artefacts (vide supra) appeared, based on mass spectra, to be isomers of batrachotoxinin A. The isolated alkaloids all afforded positive reactions with iodoplatinate. A key observation was that only the batrachotoxin fraction gave a positive Ehrlich reaction, red with p-dimethylaminobenzaldehyde and blue with p-dimethylaminocinnam-aldehyde. This extremely sensitive reaction was suggestive of a pyrrole moiety. However, the absence of an apparent ultraviolet chromophore argued against a pyrrole. Furthermore, only one nitrogen was present in the apparent molecular ion of the batrachotoxin and "the nitrogen" of batrachotoxin was basic in character and capable of being converted under mild conditions to a methiodide. Thus the data appeared to disfavor the actual presence of a pyrrole nucleus, but appeared to suggest formation of a pyrrole under the acid conditions of the Ehrlich reaction. This proved to be an incorrect conclusion and the pure batrachotoxin and homobatrachotoxin do contain the requisite ultraviolet chromophore and two rather than one nitrogen atom. The infrared spectrum of the batrachotoxin fraction showed a carbonyl absorption at about 1690 cm^{-1}. Proton nuclear magnetic resonance spectra were reminiscent of a steroid spectrum and showed a methyl resonance singlet at 0.85 ppm and an N-methyl resonance peak at 2.3 – 2.5 ppm for both the batrachotoxin fraction and for the batrachotoxinins.

Chemical properties of the batrachotoxin fraction were assessed on a microscale. In retrospect it appears that most reactions resulted in loss of the pyrrole entity and hence a product which did not afford a positive Ehrlich reaction. These chemical reactions included catalytic hydrogenation with palladium on charcoal, reduction with lithium aluminium hydride, treatment with acidic methanol, oxidation with manganese dioxide, treatment with acid, reaction with 2,4-dinitrophenylhydrazine, and exhaustive methylation with methyl iodide. An Ehrlich-positive methiodide could be obtained under milder conditions with methyl iodide. Acetylation of the batrachotoxin fraction with acetic anhydride and pyridine afforded two Ehrlich-positive O-acetyl derivatives. Reaction with methoxyamine afforded an Ehrlich-positive O-methyloxime. Reduction with sodium borohydride afforded an Ehrlich-positive dihydro-derivative. This product apparently isomerizes to other dihydro-compounds (251). Autoxidation, a serious problem during isolation of batrachotoxin, led to Ehrlich-negative products.

It should be mentioned that the initial studies (85, 176) were reported to be on extracts from *Phyllobates bicolor*. This was due to an error in identification and the extracts were actually from the species *Phyllobates aurotaenia* (191). *Phyllobates aurotaenia* is a black frog with yellow dorsal lateral stripes and occurs at lower to middle elevations in the Río San Juan drainage. The collection site for poison-dart frogs was at Playa de Oro, a village not far from highest elevations at which *Phyllobates aurotaenia* occurs commonly. *Phyllobates bicolor,* a somewhat larger frog, has a solid gold dorsum and occurs at higher elevations in the same river drainage. A few specimens of *Phyllobates bicolor* were brought by Indians down to the collection site at Playa de Oro, but were not included in the large sample for extraction and purification.

At this stage it had been determined that isolation of batrachotoxins would best be carried out at low temperatures and that thin-layer chromatography needed to be replaced with column chromatography to avoid large losses by autoxidation. It was also decided that efforts should be concentrated on the preparation of a crystalline derivative from batrachotoxinin A, the most stable alkaloid present in the extracts. Extracts from 5000 frogs finally afforded after silica gel column chromatographies five alkaloids: batrachotoxin (11 mg), homobatrachotoxin (16 mg), batracho-

toxinin A (46 mg) and a previously unsuspected labile compound pseudobatrachotoxin (1 mg) (252). The latter compound spontaneously converted
to batrachotoxinin A during isolation. Pseudobatrachotoxin represented a
significant portion of the alkaloids in the initial alkaloid fraction when care
was taken to maintain low temperatures during preparation and storage of
extracts and during fractionation. In earlier studies, batrachotoxin and
homobatrachotoxin were not separated from each other and the batrachotoxin fraction of the earlier studies undoubtedly contained some pseudobatrachotoxin and/or its conversion product batrachotoxinin A.
Pseudobatrachotoxin has no ultraviolet chromophore and its presence or
that of its product batrachotoxinin A probably contributed to the low
ultraviolet extinction coefficients of early batrachotoxin fractions.
Homobatrachotoxin, when first isolated, was thought to be an isomer of
batrachotoxin because of an apparent mass spectral parent ion identical to
the apparent parent ion of batrachotoxin. It was termed isobatrachotoxin
(252), but this term was corrected to homobatrachotoxin in a subsequent
paper (251). Chromatography of the batrachotoxins was carried out on
silica gel columns with cyclohexane, chloroform, triethylamine and methanol (16:4:1:1). Certain fractions were further purified on silica gel
columns with chloroform-methanol mixtures (9:1 and 94:6).

A crystalline derivative for x-ray analysis was now needed. Possible
derivatives of the most stable and abundant alkaloid batrachotoxinin A
were investigated. Ultimately, batrachotoxinin A was acylated with *p*-
bromobenzoic anhydride under Schotten-Baumann conditions (252). The
resultant O-*p*-bromobenzoate was purified by column chromatography on
silica gel and crystallized from acetone. X-ray diffraction analysis of a single
crystal afforded the structure of the batrachotoxinin A *p*-bromobenzoate
(3α,9α-epoxy-14β,18β-(epoxyethano-N-methylimino)-5β-pregna-7,16-
diene-3β,11α,20α-triol) (252) and provided the key to the structures of
batrachotoxin and homobatrachotoxin. The structure of the labile
pseudobatrachotoxin remains unknown due to paucity of material for
characterization. The most plausible hypothesis is that it is a highly labile
ester which hydrolyses to batrachotoxinin A. Further x-ray analysis of
batrachotoxinin A *p*-bromobenzoate (112, 153) led to the absolute configuration of this steroid. The absolute configuration was identical with that
of cholesterol at carbons 3, 5, 9, 10 and 13 but different from that of
cholesterol at carbon 14 and reminiscent of the cardenolides. The structure
of batrachotoxinin A is shown in Fig. 3.

A knowledge of the structure of batrachotoxinin A led to a re-
evaluation of the spectral properties of the much more toxic batrachotoxin
and homobatrachotoxin. It soon became apparent that the ultraviolet
absorption peaks (λ_{max} 234, 264 nm), the infrared red absorption band at
1690 cm^{-1}, and additional nuclear magnetic resonance absorption peaks in

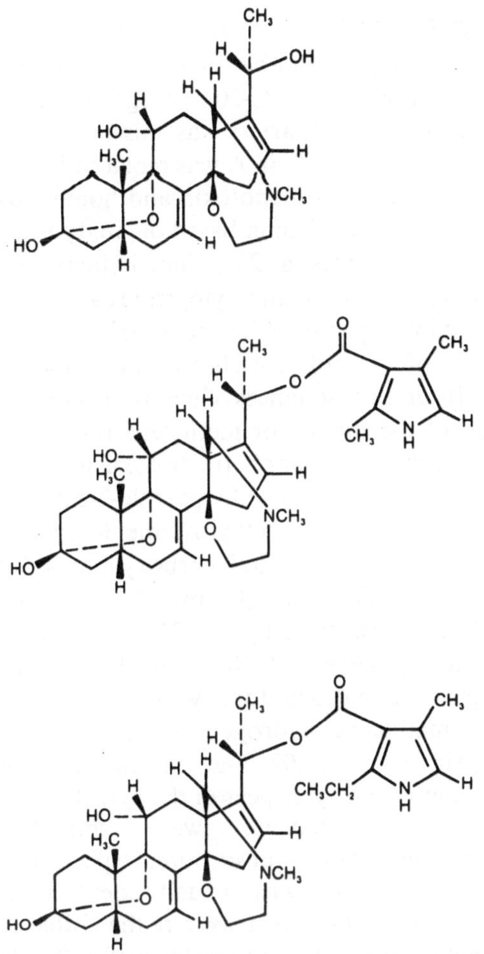

Fig. 3. Structures for batrachotoxin class alkaloids. **A** Batrachotoxinin A. **B** Batrachotoxin.
C Homobatrachotoxin

the spectra of batrachotoxin and homobatrachotoxin ("isobatrachotoxin")
were impossible to accomodate by an apparent molecular ion of
$C_{24}H_{33}NO_4$ (m/z 399) for these two alkaloids. The molecular ion of
batrachotoxinin A was $C_{24}H_{35}NO_5$ (m/z 417). Instead, it was concluded
that batrachotoxin and homobatrachotoxin must contain an additional
moiety that was responsible for the carbonyl absorption band, the
ultraviolet chromophore, the positive Ehrlich reaction, a downfield singlet
corresponding to one hydrogen in the nuclear magnetic resonance spectra
at about 6.35 ppm units and resonance peaks corresponding to two methyl
groups (singlets, 2.24 and 2.47 ppm) in batrachotoxin and to a methyl group
(singlet, 2.24 ppm) and an ethyl group in homobatrachotoxin. Otherwise,

the nuclear magnetic spectra of batrachotoxin and homobatrachotoxin contained most of the identifiable elements seen in the spectrum of batrachotoxinin A. However, the C-20 hydrogen resonance of batrachotoxinin A at about 4.6 ppm (quartet) was missing and as in the 20-α-*p*-bromobenzoate of batrachotoxinin A was replaced by a downfield quartet at about 5.9 ppm for both batrachotoxin and homobatrachotoxin.

The most reasonable conclusion based on analysis of the spectral data was that batrachotoxin was a 20-α-dimethylpyrrole carboxylate and homobatrachotoxin a 20-α-ethylmethylpyrrole carboxylate of batrachotoxinin A. The ultraviolet spectra of batrachotoxin and homobatrachotoxin resembled those of pyrrole-3-carboxylates rather than pyrrole-2-carboxylates. Shifts of the resonances of methyl groups on model pyrrole-3-carboxylates and of the methyl or methylene resonances on the pyrrole moiety of batrachotoxin and homobatrachotoxin led to the conclusion that batrachotoxin contains a 2,4-dimethylpyrrole-3-carboxylate moiety and homobatrachotoxin a 2-ethyl-4-methylpyrrole-3-carboxylate moiety. A 2-methyl substituent in model pyrrole-3-carboxylates is shifted downfield in $CDCl_3$ relative to its position in C_6D_6 by $+0.07$ to 0.08 ppm, a 5-methyl substituent is shifted downfield by $+0.26$ to 0.30 ppm and a 4-methyl substituent is shifted upfield by -0.18 to -0.24 ppm. The shifts for the two pyrrole methyl groups of batrachotoxin were $+0.10$ and -0.11 ppm, while the shift of the methylene hydrogens of the pyrrole ethyl group of homobatrachotoxin was $+0.02$ and of the pyrrole methyl group -0.12 ppm. The methylene hydrogens of the pyrrole ethyl group in ethyl 2-ethyl-4-methylpyrrole-3-carboxylate were shifted downfield by $+0.04$ ppm. The ultraviolet absorption spectrum of ethyl 2,4-dimethylpyrrole-3-carboxylate was virtually identical with those of batrachotoxin and homobatrachotoxin, as were the visible absorption spectra of the Ehrlich reaction products obtained from ethyl 2,4-dimethylpyrrole-3-carboxylate and the two toxins.

The mass spectra of batrachotoxin and homobatrachotoxin could now be readily interpreted. Indeed, it even proved possible to detect the molecular ion at 538 for batrachotoxin, but its intensity even under the most favorable conditions was only about two percent of the apparent molecular ion at 399. Furthermore, the true molecular ion at 538 could never be observed except transiently probably due to pyrolytic elimination of the pyrrole carboxylate moiety. The pyrrole carboxylate moiety was responsible for major ions of $C_7H_9NO_2$ (m/z 139) and C_6H_9N, C_6H_8N (m/z 95, 94) in batrachotoxin and for major ions of $C_8H_{11}NO_2$ (m/z 153), $C_7H_8NO_2$, $C_7H_{11}N$ (m/z 109) and C_6H_8N in homobatrachotoxin (see Table 1 at the end of this section).

Hydrolysis of the hindered pyrrole ester moiety of batrachotoxin proved possible with 2 N NaOH at 60° for 16 hr. Under these conditions a partial hydrolysis to a compound identical to batrachotoxinin A was attained. Reaction of the mixed anhydride of ethyl chloroformate and 2,4-dimethylpyrrole-3-carboxylate with batrachotoxinin A under Schotten-Baumann conditions (Scheme I) afforded a product identical in properties to batrachotoxin, thereby confirming the proposed structure.

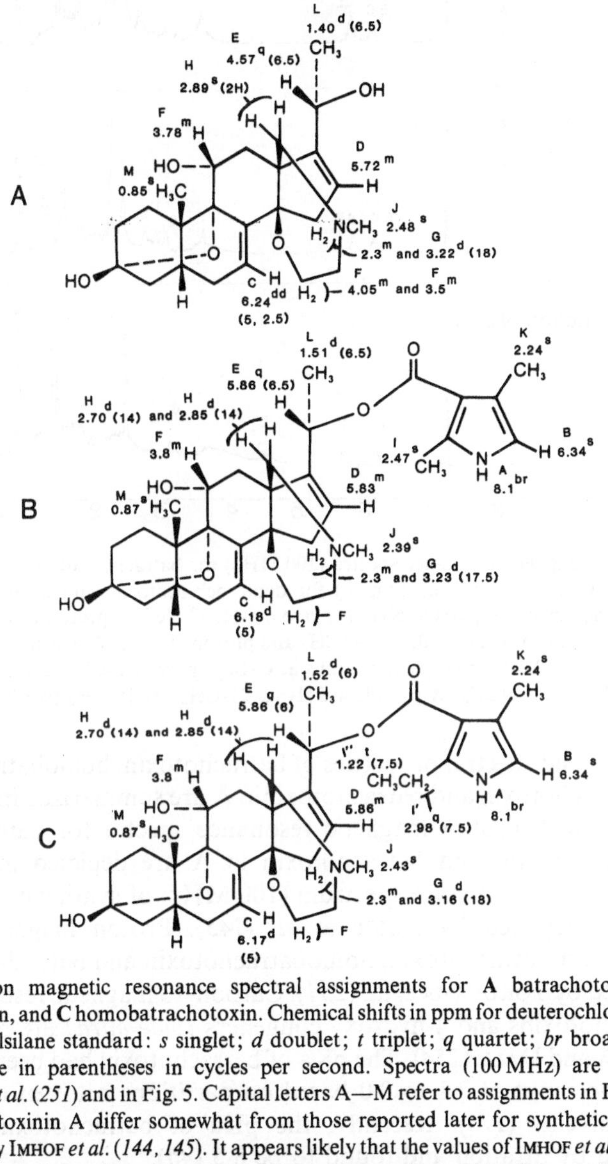

Fig. 4. Proton magnetic resonance spectral assignments for **A** batrachotoxinin A, **B** batrachotoxin, and **C** homobatrachotoxin. Chemical shifts in ppm for deuterochloroform with a tetramethylsilane standard: *s* singlet; *d* doublet; *t* triplet; *q* quartet; *br* broad. Coupling constants are in parentheses in cycles per second. Spectra (100 MHz) are depicted by TOKUYAMA *et al.* (*251*) and in Fig. 5. Capital letters A—M refer to assignments in Fig. 5. Values for batrachotoxinin A differ somewhat from those reported later for synthetic and natural compound by IMHOF *et al.* (*144, 145*). It appears likely that the values of IMHOF *et al.* correspond to the free base and that the earlier values of TOKUYAMA *et al.* were for mixtures of free base and cationic form present in the CDCl₃. Certain earlier assignments (*251*) have been revised in light of the detailed examination of the spectrum of batrachotoxinin A by IMHOF *et al.* (*145*)

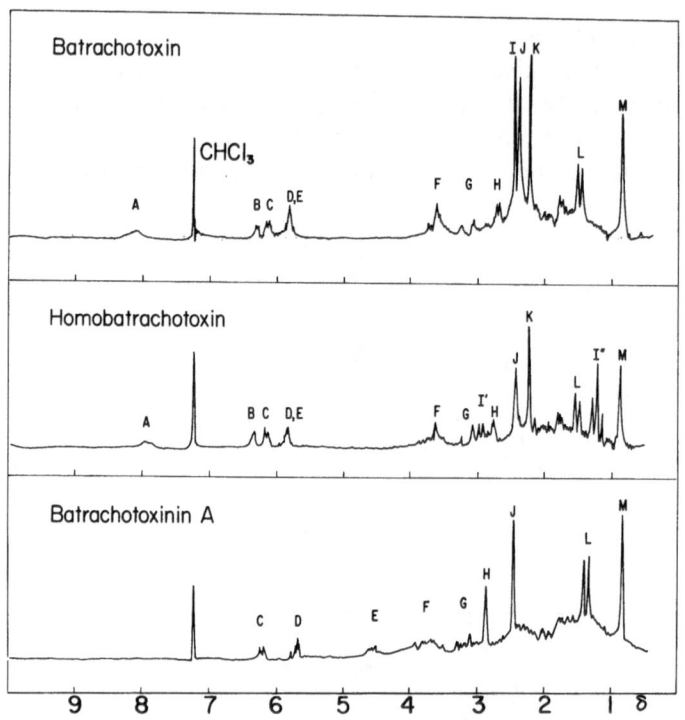

Fig. 5. Proton magnetic resonance spectra (100 MHz) for batrachotoxin class alkaloids and assignments. Chemical shifts in ppm (δ) for deuterochloroform with a tetramethylsilane standard. Assignments *A* pyrrole NH; *B* pyrrole 5-H; *C* olefinic proton at C-7; *D* olefinic proton at C-16; *E* proton at C-20; *F* 14-OCH₂ and proton at C-11; *G* one proton at C-15, the other appears at δ 2.3; *H* methylene protons at C-18; *I* pyrrole 2-CH₃; *I′* and *I″* pyrrole 2-CH₂CH₃; *J* NCH₃; *K* pyrrole 4-CH₃; *L* 21-CH₃; *M* 19-CH₃ (see Fig. 4)

Physical and spectral properties of batrachotoxin, homobatrachotoxin, pseudobatrachotoxin and batrachotoxinin A are summarized in Fig. 4 and Tables 1 and 2. Proton magnetic resonance spectra for batrachotoxin, homobatrachotoxin and batrachotoxin in A are depicted in Fig. 5. A proton magnetic resonance spectrum (100 MHz) of synthetic batrachotoxinin A is depicted by IMHOF *et al.* (*145*). Proton magnetic spectra (100 MHz) for batrachotoxin, homobatrachotoxin and batrachotoxinin A are depicted by TOKUYAMA *et al.* (*251*). Carbon-13 magnetic resonance data for batrachotoxins and 4-hydroxy congeners *(vide infra)* are reported by TOKUYAMA and DALY (*254*). The pKa of batrachotoxin had been estimated to be in the range of 7.1 to 8.0 based on partitions between buffers and organic solvents (*176*). Recently, the pKa of batrachotoxinin A was determined by titration and found to be 8.5 (*46*).

The toxicity of batrachotoxin and the sensitivity of the Ehrlich reaction provide two sensitive probes for occurrence of batrachotoxin and homobatrachotoxin in extracts. Positive identification of (homo)batrachotoxin has been made with skin extracts from five species of dendrobatid frogs (Table 3). It should be noted that not all populations of *Phyllobates lugubris* have contained levels of batrachotoxin sufficient for positive identification. Furthermore, captive-raised specimens of a normally highly toxic frog, *Phyllobates terribilis*, did not contain levels of batrachotoxin sufficient for positive identification (83). Batrachotoxin was not identified in tadpoles of *Phyllobates aurotaenia* hatched in captivity (cited in 150). Batrachotoxin has not been detected in any of the other dendrobatid frogs *(Dendrobates, Colosthetus)* or in extracts from a rather extensive spectrum of nondendrobatid frogs (see 103, for partial listing of nondendrobatid frogs which have been examined).

Table 1. *Mass Spectra of Batrachotoxins* (259)
Only the most diagnostic peaks are presented with the intensity relative to the base peak set equal to 100 in parentheses. Intensities of fragments due to the pyrrole moiety (m/z 153, 138, 122, 121, 120, 109, 95, 94) vary due to thermolysis. Parent peaks of batrachotoxin and homobatrachotoxin can be observed only briefly if at all. For example, the parent ion of batrachotoxin could be seen only briefly, and even then was only 1.7% of the intensity of m/z 399. Mass spectra are depicted and fragmentation pathways are discussed in (259).

Formulae	Batrachotoxin	Homobatracho-toxin	Batracho-toxinin A	Pseudobatra-toxin[a]
$C_{24}H_{35}NO_5$			417 (2)	
$C_{23}H_{32}NO_5$			402 (6)	
$C_{24}H_{33}NO_4$	399 (3)	399 (6)	399 (11)	399 (6)
$C_{23}H_{32}NO_3$	370 (6)	370 (10)	370 (6)	370 (6)
$C_{20}H_{26}O_4$			330 (100)	
$C_{20}H_{24}O_3$	312 (13)	312 (25)	312 (30)	312 (17)
$C_{20}H_{22}O_2$	294 (10)	294 (20)	294 (8)	294 (15)
$C_{18}H_{22}O_3$	286 (10)	286 (22)	286 (9)	286 (6)
$C_{13}H_{14}O_2$	202 (3)	202 (3)	202 (15)	202 (3)
$C_{13}H_{12}O$	184 (30)	184 (60)	184 (11)	184 (48)
$C_{11}H_{10}O$			158 (14)	
$C_8H_{11}NO_2$		153 (90)		
$C_7H_9NO_2$	139 (65)	139 (22)		
$C_7H_8NO_2$	138 (24)	138 (100)		
C_7H_8NO	122 (20)	122 (12)		
C_7H_7NO	121 (13)	121 (15)		
C_7H_6NO	120 (10)	120 (26)		
$C_7H_{11}N$	109 (15)	109 (60)		
C_6H_9N	95 (100)	95 (23)		
C_6H_8N	94 (100)	94 (94)		
$C_4H_{10}NO$	88 (34)	88 (72)	88 (60)	88 (100)
C_4H_9N	71 (26)	71 (28)	71 (8)	71 (45)

[a] The spectrum for pseudobatrachotoxin contained ions at m/z 342 ($C_{22}H_{32}NO_2$, 6) and at 166 ($C_{10}H_{16}NO$, 12), but these were proposed to be due to an impurity.

Table 2. *Other Physical and Spectral Properties of the Batrachotoxins*[a]

Batrachotoxin
  R_f 0.45 (SiO_2 thin-layer plates, $CH_3OH - CHCl_3$, 9:1)   *(191)*
  Ultraviolet: λ_{max} 234 nm, ε 9,800   *(251)*
                262 nm, ε 5,000
  Infrared: 1,690 cm^{-1} *(251)*
  Optical rotation: *(85)*
      (c 0.23, CH_3OH) $[\alpha]^{24}_{584}$ −5 to −10°
                                $[\alpha]^{24}_{300}$ −260°

Homobatrachotoxin   *(251)*
  R_f 0.50
  Ultraviolet: λ_{max} 233 nm, ε 8,900
                264 nm, ε 5,000
  Infrared: 1,690 cm^{-1}

Batrachotoxinin A
  mp 160 − 2° (synthetic)   *(145)*
  R_f 0.28   *(191)*
  Ultraviolet: end absorption   *(251)*
  Optical rotation (synthetic)   *(145)*
      (c 0.45, CH_3OH) $[\alpha]^{20}_D$ −42°
  pKa 8.5   *(45)*

Pseudobatrachotoxin   *(251)*
  R_f 0.25
  Ultraviolet: end absorption

---

[a] Ultraviolet spectra are depicted in *(251)*.

Table 3. *Distribution of Batrachotoxins in Dendrobatid Frogs of the Genus Phyllobates*
(191, MYERS *and* DALY, *unpublished results)*[a]

Species	Locale	Microgratoxin Batracho-toxin	Micrograms per frog Homo-batrachotoxin	Batracho-toxinin A
P. terribilis	Rio Saija, Cauca, Colombia	500	300	200
P. aurotaenia	Rio San Juan, Choco, Colombia	20	10	50[b]
P. bicolor	Rio San Juan, Risaralda, Colombia	24	12	60
P. vittatus	Palmar Norte, Costa Rica	0.2	0.2	2
P. lugubris	Almirante, Panama		not detected	
	Cope, Panama	0.2	0.1	0.5

[a] As yet batrachotoxins have only been detected in these five species.
[b] Including pseudobatrachotoxin initially present at 20 µg per frog. This alkaloid undergoes very facile conversion to batrachotoxinin A and has been detected only in *P. aurotaenia*.

Recently, two additional batrachotoxin class alkaloids were isolated as minor constituents from skin extracts of *Phyllobates terribilis*. Mass spectra and proton and carbon-13 magnetic resonance spectra allowed assignment of structures as 4β-hydroxybatrachotoxin and 4β-hydroxyhomobatrachotoxin (*254*).

B. Syntheses

The structure of batrachotoxin was confirmed by a partial synthesis from batrachotoxinin A. This proved feasible using a mixed anhydride prepared from 2,4-dimethylpyrrole-3-carboxylic acid and ethyl chloroformate (Scheme I). A variety of analogs of batrachotoxin were prepared in a similar manner. These compounds are documented in Table 4. A dihydrobatrachotoxin was prepared by sodium borohydride reduction (Scheme II) and apparently is subject to allylic rearrangement to other dihydroproducts (*251*).

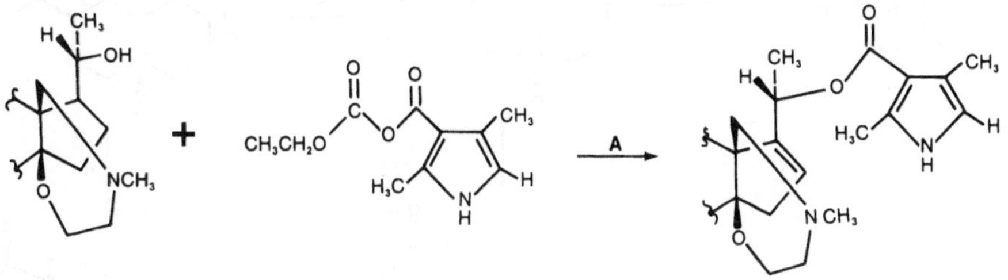

Scheme I. Partial synthesis of batrachotoxin from batrachotoxinin A (*251*). **A** i) NaOH, H₂O/HCCl₃

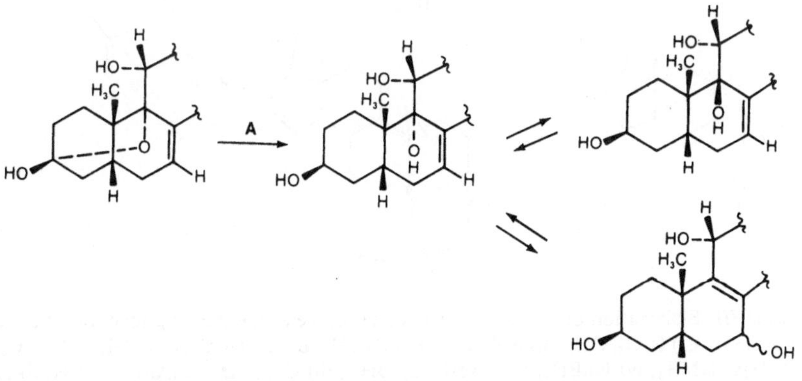

Scheme II. Formation of dihydrobatrachotoxin and postulated conversion to lower R_f isomers (*251*). **A** i) NaBH₄

The elaboration of the steroidal portion of a molecule as complex as
batrachotoxin represented a formidable challenge. This challenge was
accepted by scientists at Zurich and led over a period of two years to the
synthesis of batrachotoxinin A and its dihydro- and tetrahydro-congeners
from steroid precursors. These synthetic studies have been reviewed by
IMHOF et al. (145). Initially efforts were devoted to the development of
synthetic routes for the elaboration of the homomorpholine bridge in a
model steroid, but were directed to the AB rather than the CD ring junction
(Scheme III) (35). After this initial study, attention was directed towards
application of such routes to the elaboration of the homomorpholine ring at
the CD ring junction of a model steroid (Scheme IV) (36). The starting
material for the latter studies was a readily available steroid (20R)-3-β-
acetoxy-20-hydroxy-5α-pregnane, a compound ultimately used for the
synthesis of batrachotoxinin A itself. Synthesis of a homomorpholone

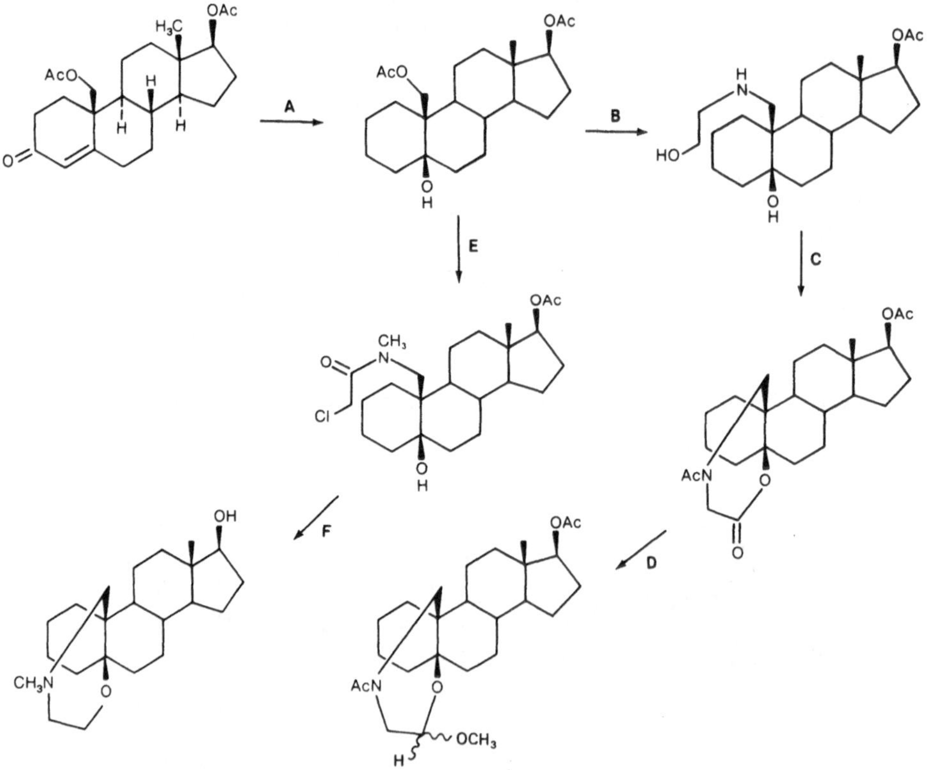

Scheme III. Elaboration of the homomorpholine bridge at the AB ring junction of a model
steroid (35). **A** i) H$_2$O$_2$, OH$^-$, ii) NH$_2$NH$_2$, iii) PtO$_2$, H$_2$. **B** i) NaHCO$_3$, CH$_3$OH, ii) Ag$_2$CO$_3$, iii)
HOCH$_2$CH$_2$NH$_2$, iv) NaBH$_4$. **C** i) Ac$_2$O, ii) OH$^-$, iii) CrO$_3$. **D** i) LiAl(t-BuO)$_3$H, ii) HCl,
CH$_3$OH. **E** i) NaHCO$_3$, CH$_3$OH, ii) Ag$_2$CO$_3$, iii) CH$_3$NH$_2$, iv) NaBH$_4$, v) ClCH$_2$COCl. **F** i)
NaH, ii) Ac$_2$O, iii) LiAlH$_4$

bridge at the CD ring junction from 3-β-acetoxy-16α,17α-epoxy-5β-pregnane-20-one by another group (*158*) has been briefly described (Scheme V).

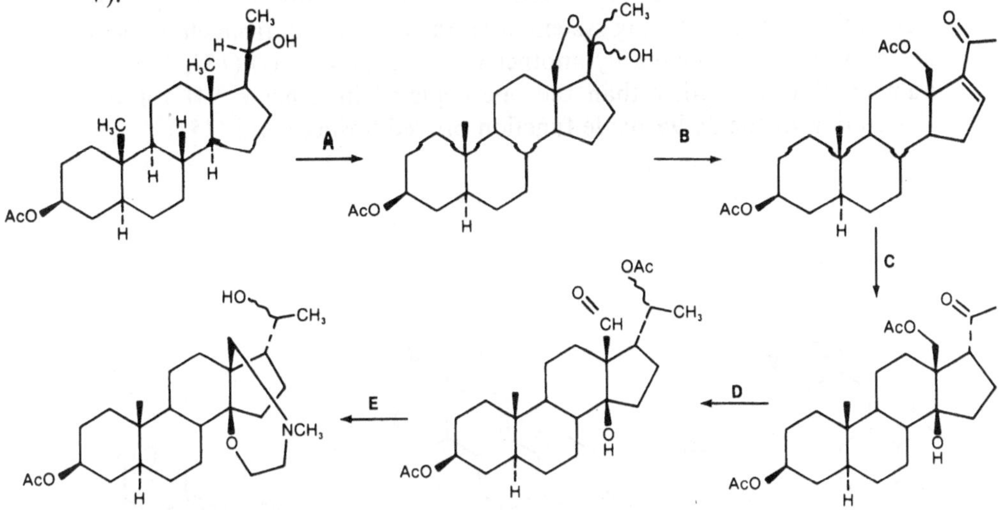

Scheme IV. Elaboration of the homomorpholine bridge at the CD ring junction of a model steroid (*36*). **A** i) Pb(OAc)₄, I₂, ii) CrO₃, iii) AgOAc. **B** i) Ac₂O, pyridine, ii) Pyridinium hydrobromide perbromide, iii) Heat, dimethylformamide. **C** i) N-Bromosuccinimide, ii) NaI, iii) p-Nitroperbenzoic acid, iv) Pd/BaSO₄, H₂. **D** i) LiAl(t-BuO)₃H, ii) Ac₂O, pyridine, iii) NaHCO₃, iv) CrO₃. **E** i) CH₃NH₂, ii) NaBH₄, iii) ClCH₂COCl, iv) NaH, v) LiAlH₄

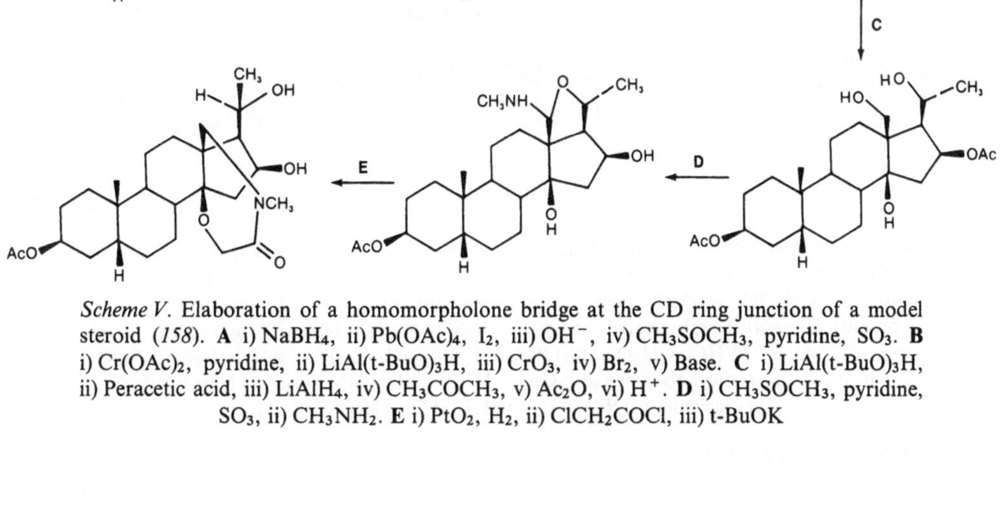

Scheme V. Elaboration of a homomorpholone bridge at the CD ring junction of a model steroid (*158*). **A** i) NaBH₄, ii) Pb(OAc)₄, I₂, iii) OH⁻, iv) CH₃SOCH₃, pyridine, SO₃. **B** i) Cr(OAc)₂, pyridine, ii) LiAl(t-BuO)₃H, iii) CrO₃, iv) Br₂, v) Base. **C** i) LiAl(t-BuO)₃H, ii) Peracetic acid, iii) LiAlH₄, iv) CH₃COCH₃, v) Ac₂O, vi) H⁺. **D** i) CH₃SOCH₃, pyridine, SO₃, ii) CH₃NH₂. **E** i) PtO₂, H₂, ii) ClCH₂COCl, iii) t-BuOK

Routes for the elaboration of the requisite hemiketal function and 7,8-double bond of batrachotoxinin A in the AB ring of the steroid precursor were now developed by the Zurich scientists (Scheme VI) (*146*). The elaboration of the ABC ring system of batrachotoxinin A from cholic acid has also been investigated by another group (Scheme VII) (*157, 228*). A number of routes other than the one depicted in Scheme VII for the elaboration of the 3α,9α-oxide function proved unsuccessful (*157*).

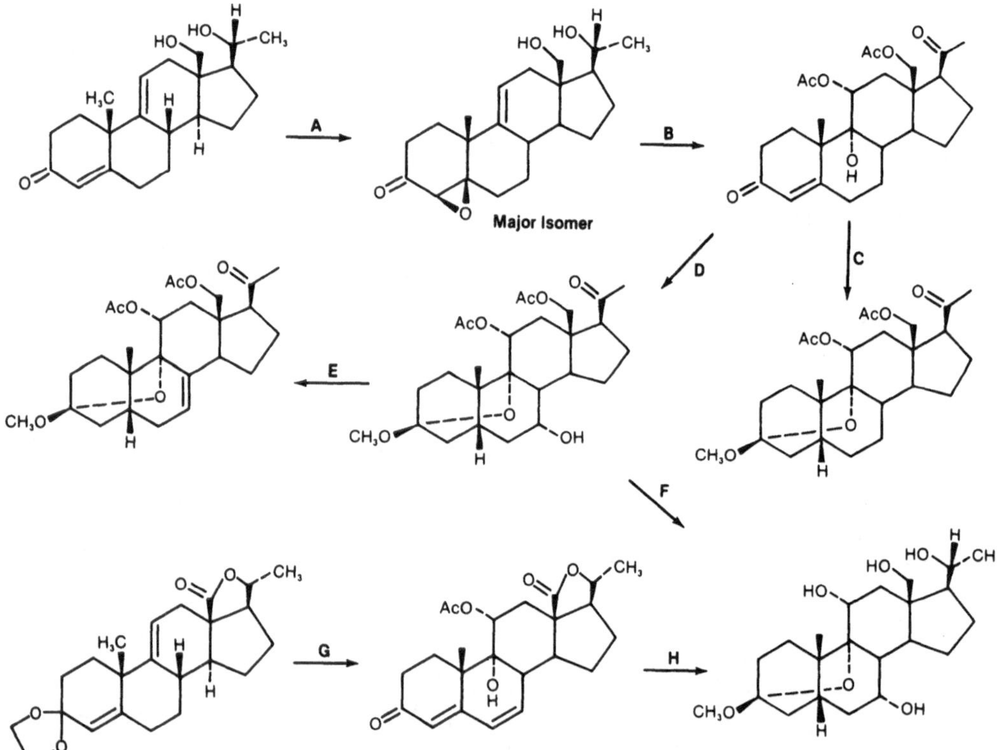

Scheme VI. Elaboration of the ABC ring system of batrachotoxin (*146*). **A** i) H_2O_2, OH^-. **B** i) Ac_2O, pyridine, ii) CrO_3, iii) a. OsO_4, b. H_2S, iv) Ac_2O, pyridine, v) Zn, AcOH. **C** i) Pd/C, H_2, ii) HCl, CH_3OH. **D** i) Dichlorodicyano-p-benzoquinone (Dehydrogenation to Δ-6,7 compound), ii) p-Nitroperbenzoic acid, iii) Pd/BaSO₄, cyclohexene (Reduction of 6,7-epoxide), iv) Pd/C, H_2, v) HCl, CH_3OH. **E** i) $POCl_3$. **F** i) $LiAlH_4$. **G** i) a. OsO_4, b. H_2S, ii) Ac_2O, pyridine, iii) Dichlorodicyano-p-benzoquinone. **H** i) p-Nitroperbenzoic acid, ii) Pd/BaSO₄, cyclohexene, iii) Pd/C, H_2, iv) HCl, CH_3OH, v) $LiAlH_4$

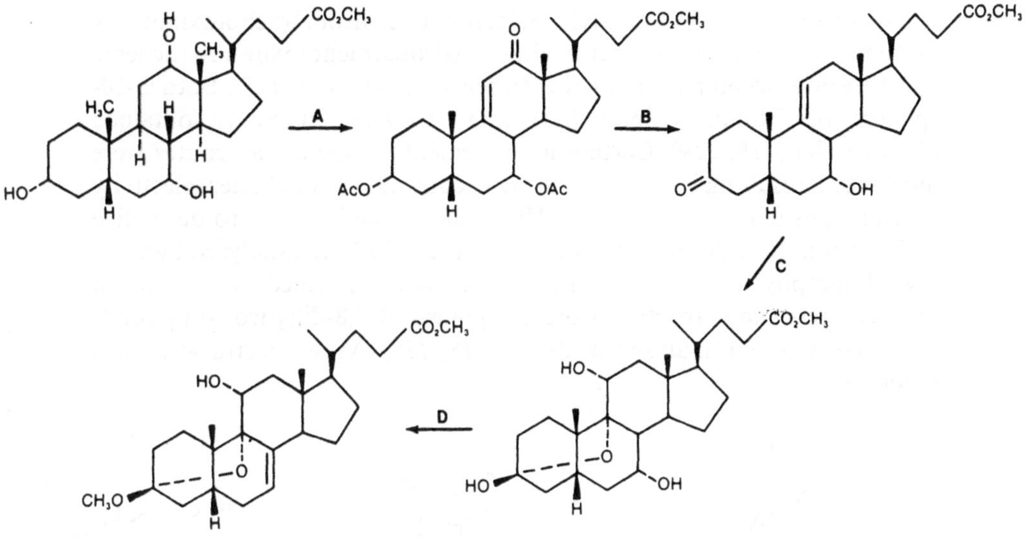

Scheme VII. Elaboration of the ABC ring system of batrachotoxin (*157, 228*). **A** i) Ac₂O, ii) CrO₃, iii) SeO₂. **B** i) HSCH₂CH₂SH, ii) Raney nickel, iii) K₂CO₃, CH₃OH, iv) CH₂N₂, v) CrO₃, vi) CH₃ONa. **C** i) a. OsO₄, b. H₂S. **D** i) H⁺, CH₃OH, ii) Ac₂O, pyridine, iii) POCl₃

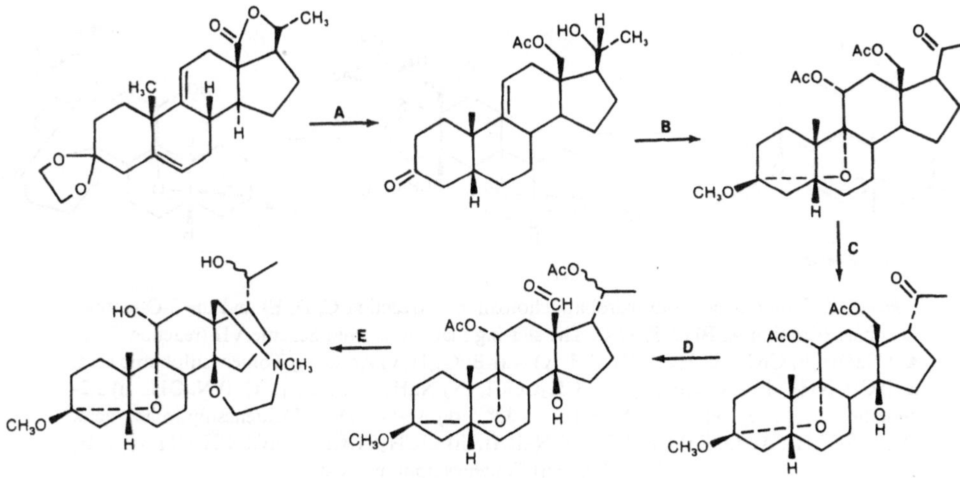

Scheme VIII. Synthesis of a 3-O-methyl-7,8,15,16-tetrahydrobatrachotoxinin A (*115, 117*). **A** i) LiAlH₄, ii) H⁺, iii) Pd, H₂, iv) Ac₂O, pyridine. **B** i) CrO₃, ii) OsO₄, iii) H₂S, iv) HCl, CH₃OH, v) Ac₂O, pyridine. **C** i) N-Bromosuccinimide, ii) LiBr, Li₂CO₃, iii) N-Bromosuccinimide, iv) LiBr, Li₂CO₃, v) p-Nitroperbenzoic acid, vi) Pd/BaSO₄, cyclohexene (Reduction of 16,17 epoxide), vii) Pd/C, H₂. **D** i) NaBH₄, ii) Ac₂O, pyridine, iii) NaHCO₃, iv) CrO₃. **E** i) CH₃NH₂, ii) NaBH₄, iii) (ClCH₂CO)₂O, iv) NaH, v) LiAlH₄

Syntheses of 3-O-methyl-7,8,15,16-tetrahydrobatrachotoxinin A (Scheme VIII) and of 3-O-methyl-7,8-dihydrobatrachotoxinin A (Scheme IX) were now accomplished by the Zurich scientists (*115, 117*). Both C-20-epimers of 7,8-dihydrobatrachotoxinin A were ultimately obtained (Scheme IX) (*118, 119*). Certain improvements in the earlier routes were introduced, resulting in more quantitative conversions and better selectivity in reductions to desired 20S and 20R alcohols. The 7,8-dihydro-derivative with the unnatural 20-β-hydroxy configuration (20S) was analyzed by x-ray crystallography and its structural assignment confirmed (*152*). Proton magnetic resonance spectra of the 20-epimers of 7,8-dihydro compounds were depicted and analyzed in detail (*118, 119*). Mass spectra were also reported.

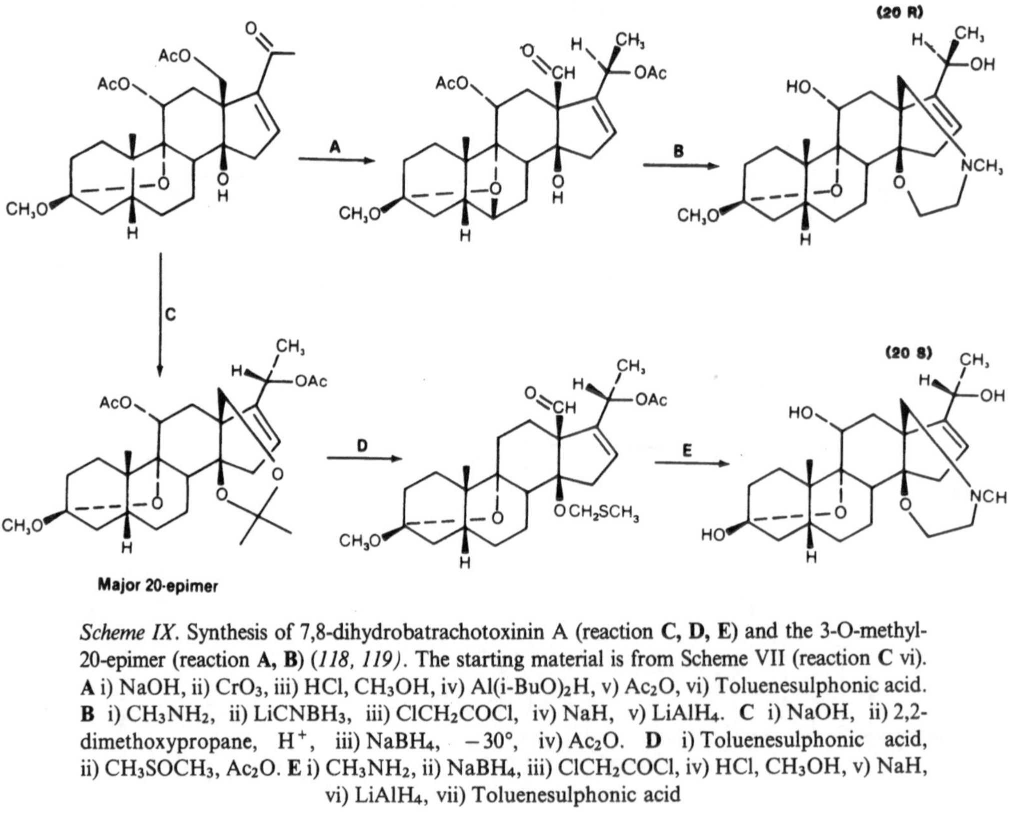

Major 20-epimer

Scheme IX. Synthesis of 7,8-dihydrobatrachotoxinin A (reaction **C, D, E**) and the 3-O-methyl-20-epimer (reaction **A, B**) (*118, 119*). The starting material is from Scheme VII (reaction **C** vi). **A** i) NaOH, ii) CrO₃, iii) HCl, CH₃OH, iv) Al(i-BuO)₂H, v) Ac₂O, vi) Toluenesulphonic acid. **B** i) CH₃NH₂, ii) LiCNBH₃, iii) ClCH₂COCl, iv) NaH, v) LiAlH₄. **C** i) NaOH, ii) 2,2-dimethoxypropane, H⁺, iii) NaBH₄, −30°, iv) Ac₂O. **D** i) Toluenesulphonic acid, ii) CH₃SOCH₃, Ac₂O. **E** i) CH₃NH₂, ii) NaBH₄, iii) ClCH₂COCl, iv) HCl, CH₃OH, v) NaH, vi) LiAlH₄, vii) Toluenesulphonic acid

Synthesis of batrachotoxinin A could now be realized based on the developed routes in thirty-six steps beginning with the readily available (20R)-3-β-acetoxy-20-hydroxy-5-α-pregnane (Scheme X) (*144, 145*). The overall yield for this impressive synthesis was 0.12%.

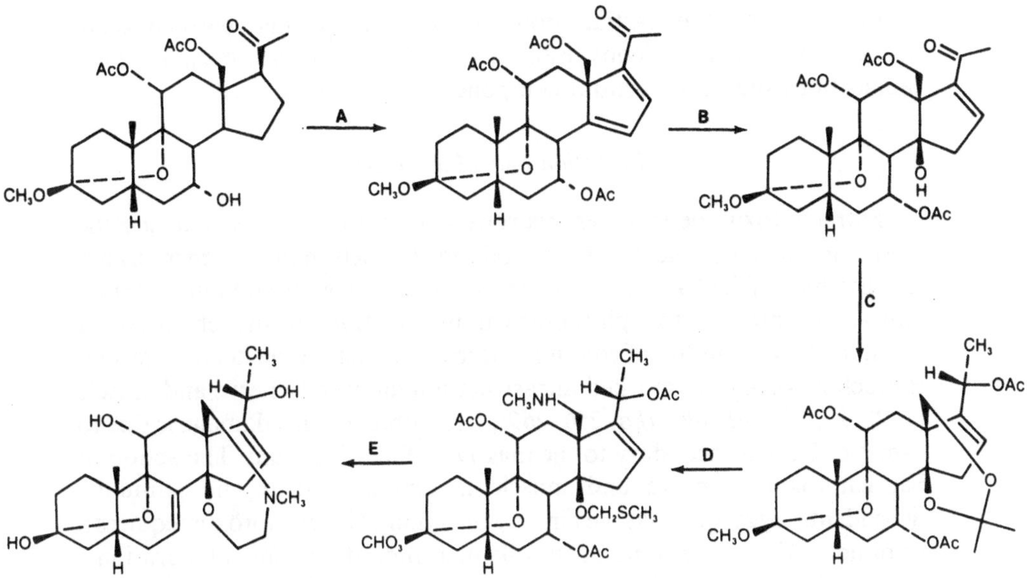

Scheme X. Synthesis of batrachotoxinin A (*144, 145*). The starting material is from Scheme VI (reaction **D**). **A** i) Ac₂O, pyridine, ii) N-Bromosuccinimide, iii) LiBr, Li₂CO₃, iv) N-Bromosuccinimide, v) LiBr, Li₂CO₃. **B** i) p-Nitroperbenzoic acid, ii) Pd/BaSO₄, cyclohexene. **C** i) OH⁻, ii) 2,2-Dimethoxypropane, H⁺, iii) NaBH₄, −30°, iv) Ac₂O. **D** i) Toluenesulphonic acid, ii) CH₃SOCH₃, Ac₂O, iii) CH₃NH₂, iv) NaBH₄. **E** i) ClCH₂COCl, ii) HCl, CH₃OH, iii) NaH, iv) CH₃ONa, v) Ac₂O, vi) SOCl₂, pyridine, vii) LiAlH₄, viii) Toluenesulphonic acid

The unique pyrrole-3-carboxylate moiety of batrachotoxin has not proven to be essential to high activity in these alkaloids since the corresponding benzoate is very active (*46*). 2,4,5-Trimethylpyrrole carboxylates of codeine, ephedrine, jervine, scopoline and methylreserpate have been prepared using trifluoroacetic in a mixed anhydride procedure (*272*). The scopoline derivative showed modest analgesic activity.

C. Biological Activity

1. Toxicity

Batrachotoxin is among the most toxic substances known to man (see *78* for comparison with other toxins). A lethal dose in mouse is only about 100 nanograms and it has been estimated that in man a lethal dose would be much less than 200 µg (*191*). Undoubtably cytotoxic effects on heart leading to arrhythmias and cardiac arrest play a dominant role in the toxicity of this agent. However, the partial protection of mice against batrachotoxin afforded by certain anticonvulsants (R. Cadenas Carrera, personal communication) suggests a central component to toxicity as well.

2. Action and Mechanism

Batrachotoxin depolarizes neurons and muscle cells *via* a specific interaction with voltage dependent sodium channels in plasma membranes (*5*). Binding of batrachotoxin to sites associated with sodium channels appears to prevent the physiological inactivation of the channels. A resultant massive influx of sodium ions leads to membrane depolarization. Batrachotoxin causes marked ultrastructural damage to nerve and muscle (*1, 22, 23, 25, 110, 180, 186, 266, 269*). Undoubtedly, this damage is due to osmotic changes secondary to the massive influx of sodium. The action of batrachotoxin in nerve and muscle is time-dependent and stimulus-dependent, suggesting that the binding or action of batrachotoxin requires a prior activation or opening of the channel. Indeed, in some preparations, batrachotoxin has no effect unless channels are opened by stimulation (*32*). Certain other alkaloids, such as aconitine and veratridine, and the diterpene grayanotoxin appear to interact with the same site on the sodium channel as batrachotoxin, but these compounds are less potent and efficacious. All cause depolarization in a similar manner, that is by preventing inactivation or closing of sodium channels. The interactions of batrachotoxin with the site on the sodium channel can be antagonized by the "partial agonists" aconitine and veratridine (*59*). However, the mode of binding of batrachotoxin may differ slightly from the mode of binding of the "partial agonists". Thus, muscles of the frogs which produce batrachotoxin are virtually insensitive to its action, while still retaining some sensitivity to the action of veratridine and grayanotoxin (*25, 83*). Muscles of other dendrobatid frogs, such as *Dendrobates histrionicus*, are sensitive to batrachotoxin (cited in *5, 83*). The potencies for batrachotoxin-elicited effects in different preparations extend over nearly a 50-fold range, but it is not known whether this is due to actual changes in affinity for binding sites on sodium channels or to other factors (*5*).

Tetrodotoxin and saxitoxin, which block the voltage-dependent sodium channels, prevent and actually reverse batrachotoxin-elicited depolarizations. The blockade by tetrodotoxin occurs at a different channel site than that at which batrachotoxin acts. Classical anesthetics, however, would appear to antagonize the action of batrachotoxin through competition for the binding site (*50* and references therein). Local anesthetics have been shown to block but perhaps not reverse the action of batrachotoxin in squid axon (*19*), rat diaphragm (*11*), eel electroplax (*33*), frog nerve (*160, 163*), synaptosomes (*106, 179*), and neuroblastoma cells (*50, 135, 136*).

Batrachotoxin will not cause depolarization in the absence of sodium ions, consonant with its proposed mechanism of action. In denervated muscle, sodium channels develop which are not sensitive to tetrodotoxin but do retain sensitivity to batrachotoxin (8). Batrachotoxin causes a massive release of acetylcholine in neuromuscular preparations. This effect, undoubtedly due to depolarization of the presynaptic terminal, is prevented by tetrodotoxin (23, 149, 269) or by botulinus toxin (236).

In addition to the tetrodotoxin-site and the batrachotoxin-site, the sodium channel contains another site which interacts with certain polypeptides, including scorpion toxin and anemone toxin (59). There is an apparent cooperative interrelationship between binding of agents at the batrachotoxin-site and binding of polypeptides at the peptide-site. Thus, polypeptides increase the apparent affinity of batrachotoxin and other agents for the batrachotoxin-site (248). In the presence of scorpion toxin, the affinity of the "partial agonists" veratridine, aconitine, and grayanotoxin for the batrachotoxin-site not only increases, but these agents now also appear to be full agonists with respect to activation of sodium channels. The converse cooperative interrelationship is also true, with batrachotoxin, veratridine, and aconitine increasing the affinity of polypeptides for their site. The activation of sodium channels by batrachotoxin has also been reported to be enhanced by a toxin from the dinoflagellate, *Ptychodiscus brevis* (63). This toxin does not appear to act *via* the polypeptide-site.

Pretreatment of lobster axons with sulfhydryl reagents such as *p*-chloromercuribenzene sulphonic acid and dithiothreitol prevent or reduce respectively depolarization elicited by batrachotoxin (17). Such treatment does not prevent stimulus-evoked activation of sodium channels. The results with such sulfhydryl reagents provides evidence for the importance of a protein to the action of batrachotoxin.

The effects of batrachotoxin in nerve and muscle preparations are often relatively irreversible. This apparent irreversibility probably reflects a slow removal of the alkaloid from tissues because of lipid solubility and because only a small percentage of sodium channels ($< 5\%$) need to be activated to cause and maintain complete depolarization in most electrogenic membranes. The effects of batrachotoxin are readily reversible in neuroblastoma cells (52).

Batrachotoxin has no effect on a calcium channel (18), or on potassium channels (193). However, batrachotoxin does appear to antagonize the increase in conductance elicited by nicotinic agonists in striated neuromuscular preparations (111) and adrenal glands (164, 165). The mechanism involved in inhibition of nicotinic receptor-controlled conductances by batrachotoxin remains unclear but actually might represent another site of action for the alkaloid. Batrachotoxin, however, has no effect on binding of α-bungarotoxin or histrionicotoxin to the acetylcholine receptor-channel

complex in *Torpedo* electroplax (*96*). It has been proposed that batrachotoxin either may prevent tetrodotoxin-insensitive sodium channels in the
muscle endplate from being coupled to and activated by carbamylcholine-
activated acetylcholine channels or may convert such sodium channels to a
tetrodotoxin-sensitive state (*111*). Alternatively, it was proposed that
batrachotoxin may directly interfere with function of the acetylcholine
receptor-channel complex (*164*). The inhibitory effects of batrachotoxin on
nicotinic responses are mimicked by aconitine and veratridine and, of
course, can be observed only in the presence of tetrodotoxin.

Batrachotoxin is extremely potent in antagonizing axonal transport
(*198*). It would appear that the basis of the effect of batrachotoxin on axonal
transport is dependent on interactions with sodium channels and the
resultant influx of sodium ions. The effect is blocked by tetrodotoxin. In the
mollusc, *Aplysia californica*, it has been proposed that the inhibition of
axonal transport by batrachotoxin is not due to interactions with sodium
channels (*169*), but this interpretation has been questioned (*105*). Blockade
of axonal transport by batrachotoxin reduced uptake and transport of
nerve growth factor at distal terminals (*41*), increased activity of certain
muscle lysosomal enzymes (*40*), and altered uptake of calcium in muscle
sarcoplasmic reticulum (*264, 265*). Batrachotoxin inhibits saltatory movements in neuroblastoma cells (*105*). This inhibitory effect is blocked by
tetrodotoxin.

Batrachotoxin is an extremely potent cardiotonic agent (*129, 132, 133,
156, 183, 233*). Its actions lead ultimately to arrhythmias and cardiac arrest.
The basis for the action of batrachotoxin in cardiac preparations is linked to
activation of sodium channels and can be antagonized by tetrodotoxin.
Unlike the cardiotonic cardiac glycosides, batrachotoxin has little effect on
Na^+-K^+-ATPase, causing only a slight inhibition of the enzyme at a
concentration (60 μM) much higher than that usually employed for
depolarization of cells (*79*).

The effect of batrachotoxin on sodium fluxes has been studied
extensively in neuroblastoma cells (see Table 5 at the end of this section for
references). This system has been invaluable for delineating interactions of
batrachotoxin with polypeptides. Such interactions involve allosteric
modifications of the voltage-dependent sodium channel (*57, 59*). It has been
proposed that batrachotoxin binds selectively to an active state of the
channel and that partial agonists such as aconitine and veratridine bind to
both active and inactive states (*168*). Polypeptides such as scorpion toxin
have been proposed to inhibit the conversion of active to inactive state,
thereby decreasing the amount of inactive state available to the alkaloids.

It appears that the alkaloid batrachotoxin changes the voltage dependency of sodium channels such that the active form is now stable at
resting membrane potentials (*162* and ref. therein). The batrachotoxin-

activated channel shows differences not only in voltage dependency but also in interactions with local anesthetics and in pore size apparently being larger than the stimulus-activated channel (134, 148, 162, 173).

Structure activity correlations for analogs of batrachotoxin have been studied to a limited extent. Toxicity in white mice (Table 4 at the end of this section) (251), effects on nerve-striated muscle preparations (268), cardiac preparations (233), ATPase (79), and eel electroplax (cited in 5) show similar profiles with the different analogs. The substitution pattern on the pyrrole moiety is important. The relative order of activity of certain substituted pyrrole carboxylates in a neuromuscular preparation as depolarizing agents was as follows: Batrachotoxin > homobatrachotoxin = 2,4,5-trimethylpyrrole analog > 5-Acetyl-2,4-dimethylpyrrole analog > 4,5-dimethylpyrrole analog. It appeared that replacement of a pyrrole-3-carboxylate with another ester would result in marked reduction in activity. For example, the 20-α-p-bromobenzoate and the pyrrole-2-carboxylate analogs showed very low toxicity (Table 4). However, recently it was found that the 20-α-benzoate analog was virtually equipotent with batrachotoxin (46). The steroid configuration of batrachotoxin appeared quite important since reduction of the $3\alpha, 9\alpha$-hemiketal function with borohydride yields a relatively inactive dihydro-derivative. The quarternary derivative of batrachotoxin showed low activity. This might be due to lack of penetration of this positively charged compound. Batrachotoxin is more effective applied internally to squid axons than when present in the external media (193).

Batrachotoxin was most effective in neuromuscular preparations at pH 9, at which pH much of the alkaloid would be in the unionized form (268). It also has been suggested that a membrane constituent ionized at pH values > 6.0 is essential to the action of batrachotoxin (5). Binding studies support this interpretation (45, vide infra). Obviously, further research is required to define the active form of batrachotoxin and the nature of its binding site.

A speculative model has been advanced for interaction of batrachotoxin with polypeptides proposed to be components of the sodium channel (237—240). Structure activity relationships for batrachotoxin analogs have been discussed (102).

3. Binding of a Radioactive Batrachotoxin Analog

Batrachotoxinin-A 20α-p-[³H]benzoate ([³H]-BTX-B) has proven to be a satisfactory ligand for the investigation of batrachotoxin-binding sites in brain membranes and synaptosomes (45, 46, 66). The affinity constant for [3H]-BTX-B is approximately 50 nM in the presence of scorpion toxin and approximately 700 nM in the absence of a polypeptide. The density of sites is similar to that noted for tetrodotoxin, saxitoxin and scorpion toxin. Binding of [^{3}H] BTX-B is antagonized by batrachotoxin and analogs and by veratridine and grayanotoxin (46, 66). Local anesthetics also antagonize binding of [^{3}H] BTX-B (unpublished results). The specific binding of [^{3}H]-BTX-B is greatest at pH 8.5 and is markedly reduced at pH < 8 and pH > 9. It would appear that pH dependence relates to two groups with pKa of 7.7 and 8.8. The pKa of batrachotoxinin A is 8.5. Two interpretations are possible: Either i) the protonated form of batrachotoxin might be the most active form, accounting for reduction in binding at pH > 8.8; or ii) both protonated and nonprotonated alkaloid might be active, and deprotonation of a pKa 8.8 group in the channel proteins might be unfavorable to binding of batrachotoxin. In either case, protonation of a group with pKa of 7.7, perhaps representing a histidine residue in the sodium channel, appears unfavorable, accounting for the reduction in binding at pH < 7.7. Photochemical or chemical treatments which should selectively modify histidine residues did reduce binding of [^{3}H]-BTX-B.

4. Summary

Batrachotoxin has proven to be an invaluable tool for the mechanistic study of voltage-dependent sodium channels and for investigation of effects of depolarization and/or influx of sodium ions on physiological functions. A summary of the scope of investigations with batrachotoxin is provided in Table 5. A number of general and selective reviews on the pharmacology of batrachotoxin are available (4, 5, 12, 59, 78, 159, 176, 192).

Table 4. *Toxicity of Batrachotoxin and Related Compounds*
Toxicity determined by subcutaneous injection into 20 g mice (46, 251, and unpublished data). Dihydrobatrachotoxin was formed by NaBH$_4$ reduction of the 3α,9α-oxide function

	LD$_{50}$, μg/kg
Batrachotoxin	2
Homobatrachotoxin	3
Batrachotoxinin A	1000
Batrachotoxinin A 20-(2,5-dimethylpyrrole-3-carboxylate)	2.5
Batrachotoxinin A 20-(4,5-dimethylpyrrole-3-carboxylate)	260
Batrachotoxinin A 20-(2,4,5-trimethylpyrrole-3-carboxylate)	1
Batrachotoxinin A 20-(2,4-dimethyl-5-ethylpyrrole-3-carboxylate)	8
Batrachotoxinin A 20-(2,4-dimethyl-5-acetylpyrrole-3-carboxylate)	280
Batrachotoxinin A 20-(N,2,4,5-tetramethylpyrrole-3-carboxylate)	> 1000
Batrachotoxinin A 20-(pyrrole-2-carboxylate)	> 1000
Batrachotoxinin A 20-p-bromobenzoate	> 1000
Batrachotoxinin A 20-benzoate	2
Dihydrobatrachotoxin	250
Batrachotoxin methiodide	500
3-O-Methylbatrachotoxin	30

Table 5. *Effects of Batrachotoxins on Biological Systems*

Preparation	Species	Comments	References
Whole animal	Dog, cat, rabbit, mouse, frog, toad, sheep	Cardiac effects appear to play a major role in toxicity: LD_{50} $2-5\,\mu g/kg$ in mammals	(*1, 78, 156, 176, 184, 185, 191, 214, 222, 223, 251, 252*)
		Tissue Distribution	(*5*)
Neuromuscular preparation (striated)	Rat, frog, toad	Depolarization of striated muscle more rapid than nerve. Depolarization of nerve terminal rapid and results in an initial increase and then blockade of transmission	(*8, 11, 23, 25, 46, 111, 149, 176, 183–185, 222, 223, 236, 265, 269*)
	Lobster, crayfish	No effect on calcium channels	(*18*)
Cardiac preparations	Cat, rabbit, dog, guinea pig, toad, mouse	Initial increase contractile force, depolarization, arrhythmias, cardiac arrest	(*129, 132, 133, 156, 183–185, 222, 223, 233*)
		Blockade of iso-proterenol-elicited accumulation of cyclic AMP	(*47*)
Heart cells	Chick	Increases rate beating followed by arrhythmias and arrest	(*220*)
Smooth muscle	Rabbit, guinea pig	Rhythmic contractures in ileum, blocked by tetrodotoxin. No effect uterus.	(*183–185*)
Nerve (axons)	Rat, rabbit, sheep, mudpuppy, frog, lobster, squid, mollusc, garfish	Depolarization; blocked by tetrodotoxin. Batra-chotoxin-activated channels show alterations in properties	(*1, 17, 19, 23, 68, 126, 149, 160–163, 169, 193, 194, 212, 218, 219, 249*)
	Rat, cat, mouse, mollusc	Reduction in axonal transport and resultant distal effects on nerve and muscle; axonal degeneration	(*22, 39–43, 110, 169, 180, 186, 198, 264, 265, 266*)

Table 5 *(continued)*

Preparation	Species	Comments	References
Superior cervical ganglion	Rabbit	Depolarization and blockade of ganglionic transmission	*(156)*
Adrenal medulla	Bovine	Release of catecholamines; inhibition of nicotinic response in presence of tetrodotoxin	*(164, 165)*
Electroplax	Eel	Depolarization; no effect in unstimulated preparations	*(32, 33, 174)*
	Malapterurus	Depolarization	*(97)*
Skin	Frog	No effect on short circuit current	Cited in *(5)*
Neuro-blastoma cells	Mouse	Increased sodium fluxes; no effect in strains which lack voltage-dependent sodium channels. Cooperative interactions with polypeptide toxins	*(50 – 53, 55 – 57, 60 – 63, 65, 134 – 136, 148, 173, 216, 274)*
Muscle cells	Chick, rat	Increased sodium fluxes; cooperative interactions polypeptide toxins	*(54, 55, 58, 171, 172, 173)*
Glial cells	Mudpuppy	No effect on membrane potential	*(249)*
Pituitary cells	Rat	Enhanced release of luteinizing hormone	*(69)*
Erythrocytes	Human	No effect on sodium permeability	Gardner, J., cited in *(5)*
Brain	Rat, guinea pig, mouse	Depolarization; linked to cyclic AMP and cyclic GMP formation in brain slices and microsacs	*(82, 137, 138, 182, 229 – 232, 263)*
	Rat, guinea pig, mouse	Depolarization of microsacs and synaptosomes and of neurons in fetal brain cell cultures	*(64, 76, 106, 113, 168, 178, 179, 181, 213, 215, 217, 248, 261)*

Table 5 *(continued)*

Preparation	Species	Comments	References
Brain	Rat, cat	Inhibition of uptake and increase in release of neurotransmitters	*(127, 131)*
	Rat, mouse, guinea pig	Binding of radioactive batrachotoxin analog	*(45, 46, 66)*
ATP levels	Rat, cat, mouse	Little effect on levels of ATP in nerve-muscle; reduces phosphocreatine. Reduction of ATP in brain slices and atria	*(47, 79, 198)*
Na^+-K^+ ATPase	Eel	Slight inhibition at high concentrations	*(79)*
Cyclic AMP-phosphodiesterase	Guinea pig	Slight inhibition at high concentrations	*(229)*
Tyrosine hydroxylase	Rat	Lowers enzyme levels in superior ganglia	*(199)*

III. Pumiliotoxin-C Class (cis-decahydroquinolines)

A. Structures

In 1965, the investigation of active principles from skins of dendrobatid frogs was extended to *Dendrobates pumilio,* a small, brightly-colored, extremely variable frog of Panama. Batrachotoxin was not detected, but instead three simpler alkaloids were found *(77, 87)*. The simplest of these, after isolation of 16 mg from skin extracts of 250 frogs, crystallized as the hydrochloride salt. This compound, pumiliotoxin C, was a 2,5-dialkyl-*cis*-decahydroquinoline (Fig. 6). The absolute stereochemistry of pumiliotoxin C ([2S,4aS,5R,8aR]-5-methyl-2-n-propyl-*cis*-decahydroquinoline) was determined by x-ray crystallography *(87)*. It should be noted that the absolute stereochemistry at C-2 is the same in pumiliotoxin C and gephyrotoxin (C-3a), but differs from that of C-2 of (dihydroiso)-histrionicotoxin *(vide infra)*. The mass spectrum of pumiliotoxin C was extremely simple, consisting, as would be expected, of a base peak due to loss of the n-propyl substituent at $C_{10}H_{18}N$ (m/z 152). The parent ion and other fragment ions were very weak.

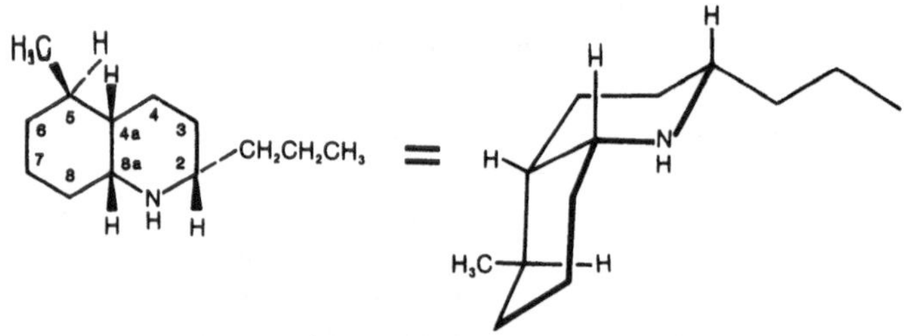

Fig. 6. Structure of pumiliotoxin C. Configuration depicted at the right is based on x-ray analysis of the hydrochloride salt (87)

The entry below for pumiliotoxin C is in the format adopted by DALY et al. (1978) for dendrobatid alkaloids. 1) The trivial name is listed if any and a numerical designation based on molecular weight and for isomers a letter or letters are given in bold face. This is followed by 2) the empirical formula based on high resolution mass spectrometry; 3) a R_f value from thin layer chromatography on silica gel with methanol : chloroform, 9 : 1; 4) an emergent temperature on a 1.5% OV-1 packed gas chromatographic column programmed from 150—280°C at 10° per minute; 5) the electron impact mass spectral ions (m/z) followed for each ion by the intensity in parentheses relative to base peak set equal to 100. Only the peaks diagnostically most useful are reported; 6) the perhydrogenation derivative obtained with palladium on carbon catalyst in methanol and 30 psi hydrogen (Ho = no addition of hydrogen) and mass spectral data; 7) the result of treatment with acetic anhydride-sodium acetate in acetone at room temperature and any other pertinent comments. Most of the other dendrobatid alkaloids will be tabulated in this format in appropriate sections of this review.

Pumiliotoxin-C Class

195A. Pumiliotoxin C, $C_{13}H_{25}N$, 0.20; 157°, m/z 195(3), 194(5), 180(1), 152(100), 109(8). H_0-derivative. N-acetyl derivative.

Further properties of pumiliotoxin C are documented in Figs. 7 and 8 and Table 6. These include nuclear magnetic resonance spectral assignments and optical rotations. Natural pumiliotoxin C is the l-enantiomer.

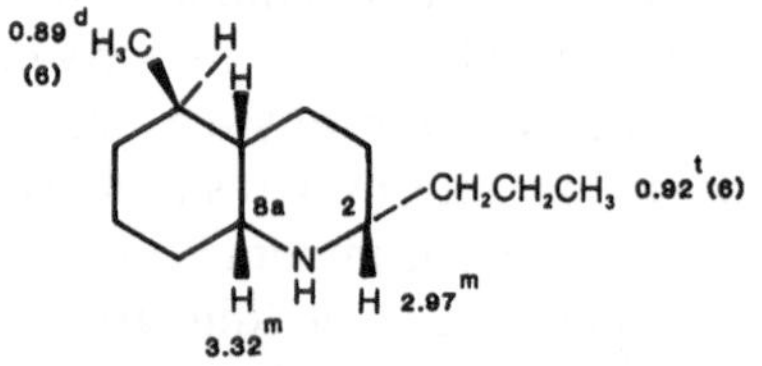

Fig. 7. Proton magnetic resonance spectral assignments for pumiliotoxin C hydrochloride. Chemical shifts in ppm for deuterochloroform with coupling constants in parentheses (data from 87, 143, 200, 209). Chemical shifts for the free base are 2-H, 2.64; 8a-H, 2.95. The spectrum of the HCl salt (100 MHz) is given by DALY et al. (87)

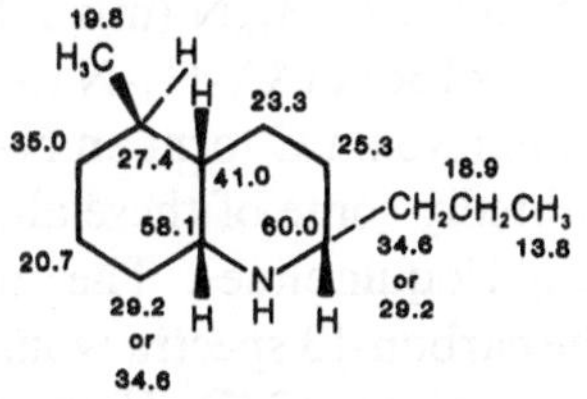

Fig. 8. Carbon-13 magnetic resonance spectral assignments for pumiliotoxin C, hydrochloride: Chemical shifts are in ppm for deuterochloroform (data from *209*; similar data reported in *122*)

Table 6. *Other Physical and Spectral Properties of Pumiliotoxin C*

Natural l-Pumiliotoxin C HCl (from *D. auratus*)		(*122, 200*)
m.p. 230–40°		
Optical Rotation	$[\alpha]_D^{20}$ −13.1°	
(c 1.0, CH₃OH)		
	$[\alpha]_{436}^{20}$ −27.6°	
Synthetic l-Pumiliotoxin C HCl		(*122, 200*)
m.p. 277–8°		
Optical Rotation	$[\alpha]_D^{20}$ −14.5°	
(c 1.0, CH₃OH)		
	$[\alpha]_{436}^{20}$ −25.5°	
Synthetic d-Pumiliotoxin C HCl		(*200*)
m.p. 286–8°		
Optical Rotation	$[\alpha]_D^{20}$ +16.4°	
(c 1.0, CH₃OH)		
	$[\alpha]_{436}^{20}$ +28.1°	
Synthetic dl-Pumiliotoxin C HCl		(*140, 202, 203,*
m.p. 243–4°		*209*)
Infrared data		

As yet no other *cis*-decahydroquinolines besides pumiliotoxin C itself have been *positively* identified in extracts from dendrobatid frogs (*80*). Pumiliotoxin C itself is not widely distributed in dendrobatids occuring in only one fourth of the species examined (see Table 22). A large number of alkaloids are present in extracts from dendrobatid frogs which are bicyclic compounds exhibiting mass spectra similar to that of pumiliotoxin C. The spectra have a very small parent ion and exhibit only with one major fragmention ion either at $C_{10}H_{18}N$ (m/z 152) as in pumiliotoxin C, or with one less methylene residue at $C_9H_{16}N$ (m/z 138), or with more methylene residues at $C_{11}H_{20}$ N (m/z 166) or $C_{12}H_{22}N$ (m/z 180), etc. (80). Certain of these alkaloids like pumiliotoxin C form an N-acetyl derivative. However, others do not. Undoubtedly some of these alkaloids will prove to be 2,5-disubstituted-*cis*-decahydroquinolines. The nuclear magnetic resonance spectra particularly the carbon-13 spectra would appear diagnostic for 2,5-dialkyl-*cis*-decahydroquinolines (207). However, for the purpose of the present review, the alkaloids which have been previously *tentatively* assigned to a pumiliotoxin-C or hydroxypumiliotoxin-C class (80) are relegated to the section *Other Alkaloids (vide infra)* until their structures are defined with greater assurance. Such definition may prove a difficult task since many of these particular alkaloids are present in skin extract only as trace components or occur in frogs which are difficult to collect in any quantity.

Some of these compounds have been isolated in sufficient quantities for partial characterization by nuclear magnetic resonance spectroscopy. These include three trace compounds designated I, II and III which were isolated from extracts of some 3200 skins of *Dendrobates histrionicus* (86). All three compounds were bicyclic and based on the mass spectra and proton resonance spectra were proposed to be members of the pumiliotoxin-C class (*cis*-decahydroquinolines) of alkaloids.

Compound I (mol. wt. 195, $C_{13}H_{25}N$) was saturated and lost a $-C_4H_9$ fragment to yield a base peak at m/z 138 in its mass spectrum. The data, including proton magnetic resonance spectrum, were compatible with a structural formulation of a 2-n-butyl-*cis*-decahydroquinoline, but a carbon-13 magnetic resonance spectrum was not obtained and this compound is tabulated for the present as **195 C** under *Other Alkaloids*.

Compound II (mol. wt. 223, $C_{15}H_{29}N$) was also a saturated bicyclic alkaloid and lost a $-C_3H_7$ fragment to yield a base peak at m/z 180 in the mass spectrum. It was tentatively formulated as a 2,5-dipropyl-*cis*-decahydroquinoline, but a carbon-13 magnetic resonance spectrum was not obtained, and for the present review it is tabulated as **223 C** under *Other Alkaloids*.

Compound III (mol. wt. 269, $C_{19}H_{27}N$) contained two unsaturated five carbon side chains based on its proton magnetic resonance spectrum and yielded on catalytic reduction a decahydro derivative. Thus, compound III appeared to contain a saturated bicyclic ring system and both a 3,4-pentadienyl and a 2-penten-4-ynyl side chain. The mass spectrum differed from other compounds proposed to belong to the pumiliotoxin-C class, since it exhibited *two* major fragment ions corresponding to loss of C_5H_5 and C_5H_7. The decahydro derivative of compound III differed from an isomeric semisynthetic deoxydodecahydrohistrionicotoxin *(vide infra)* both in mass spectrum and proton magnetic resonance spectrum (86). The structure was tentatively proposed as that of a 2-(3,4-pentadienyl)-5-(2-penten-4-ynyl)-*cis*-decahydroquinoline. A carbon-13 magnetic resonance spectrum was not obtained. There remains the possibility that compound III is a mixture of two isomeric compounds, one with 2-(3',4'-pentadienyl)-5-(2'-penten-4'-ynyl) substituents and the other with 2-(2'-penten-4'-ynyl)-5-(3',4'-pentadienyl) substituents (*80*). Reduction of these proposed isomers would yield a common 2,5-dipentyl-*cis*-decahydroquinoline. In view of the tentative nature of the structural assignment, this compound or compounds is for the present review tabulated as **269A** and **269B** under *Other Alkaloids*.

Another dendrobatid alkaloid designated **219A** was tentatively proposed to be 2,5-diallyl-*cis*-decahydroquinoline (*80*). However, a subsequent proton magnetic resonance spectrum was not compatible with the proposed structure and the perhydro-(H_4)-derivative was separated from synthetic 2,5-dipropyl-*cis*-decahydroquinoline on gas chromatography (unpublished results, see *209* for synthesis). The emergent temperatures of perhydro-**219A** was 168°C on a 1.5% OV-1 column, while the synthetic dipropyl compound emerged at 165°C. Alkaloid **219A** is for the present tabulated under *Other Alkaloids*. It is possible that it represents a deoxyhistrionicotoxin (see p. 273 – 274).

B. Syntheses

Synthesis of pumiliotoxin C has been accomplished by a number of routes. These are summarized in the Schemes XI—XVII. Most of the synthetic routes provided racemic pumiliotoxin C in overall yields of only 1—20%. A short and convenient route (Scheme XVII) has provided racemic pumiliotoxin C in an overall yield of about 50% (*208, 209*). Both enantiomers of pumiliotoxin C have been synthesized using either R-norvaline and S-norvaline as starting materials: The latter afforded the levo-rotatory (2S)-pumiliotoxin C identical to the natural compound (Scheme XII) (*200*). An earlier publication erred in reporting that levorotatory pumiliotoxin C could be synthesized from 1-(2-bromoethyl-1R)-butylamine (Scheme XIII) (*122*). In actuality the 1-(2-bromoethyl-1S)-butylamine had been used and led to levo-rotatory (2S)-pumiliotoxin C. X-ray crystallographic analyses of synthetic pumiliotoxin hydrochloride have been reported (*122, 203*). The early syntheses of pumiliotoxin C have been reviewed (*147*). Earlier literature relevant to synthesis of decahydroquinolines and to synthesis of octahydroquinolines has been cited by OPPOLZER and FROSTL (*201*). Syntheses of *cis*-decahydroquinolines with substituents at the 2- and 5-position other than n-propyl and methyl have been reported (Schemes XVII, XVIII). Certain of these were obtained during development of routes to perhydrogephyrotoxin *(vide infra)*.

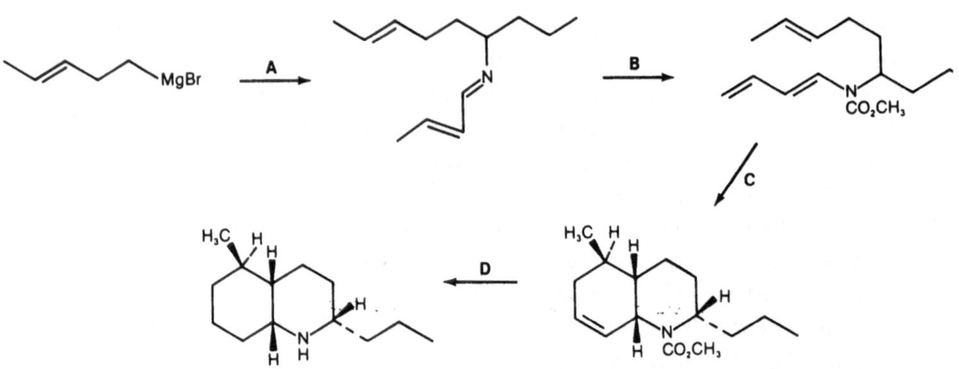

Scheme XI. Synthesis of dl-pumiliotoxin C (*201, 203*). A minor isomeric compound was formed in reaction **D**. **A** i) CH₃(CH₂)₂CN, ii) NH₂OH, iii) LiAlH₄, iv) *trans*-Crotonaldehyde. **B** i) ClCO₂CH₃, NaN(Si(CH₃)₃)₂. **C** i) Heat. **D** i) Pd/C, H₂, ii) HCl

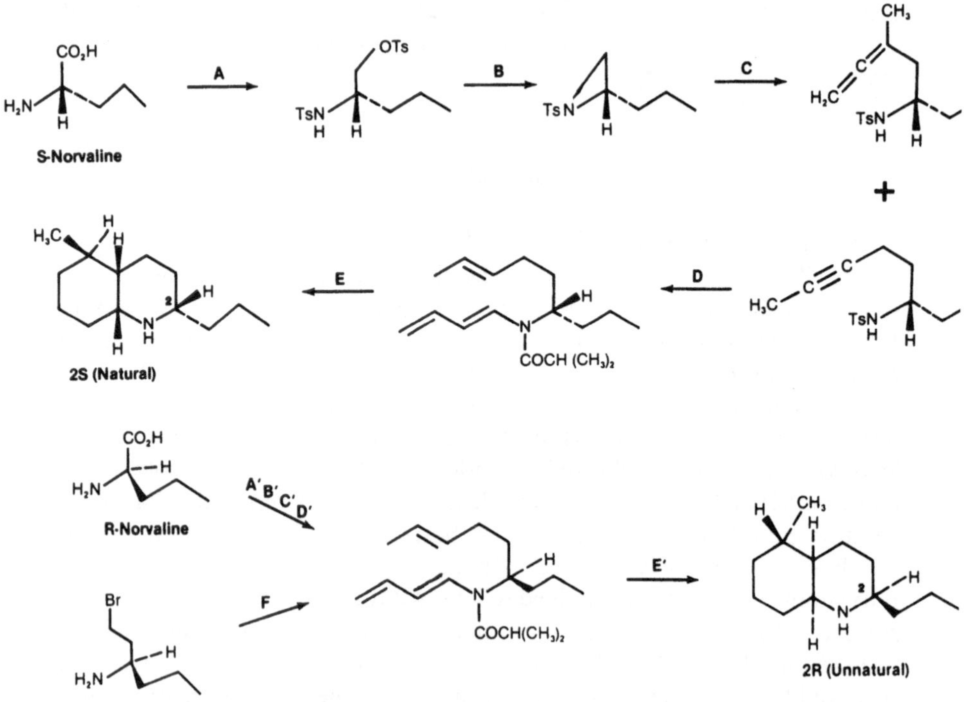

Scheme XII. Synthesis of natural l-pumiliotoxin C (reactions **ABCDE**) and unnatural d-pumiliotoxin C (reaction **A'B'C'D'E'** or **FE'**) (*200*). Reaction **A = A'**, etc. **A** i) LiAlH₄, ii) Tosyl chloride, pyridine. **B** i) KOH, CH₃OH. **C** i) CH₃C≡CCH₂MgBr. **D** i) Na, NH₃, ii) *trans*-Crotonaldehyde, iii) NaH, iv) (CH₃)₂CHCOCl. **E** i) Heat, ii) Pd, H₂, iii) Al(i-BuO)₂H. **F** i) HC≡CCH₂Na, ii) Na, NH₃, iii) *trans*-Crotonaldehyde, iv) NaH, v) (CH₃)₂CHCOCl

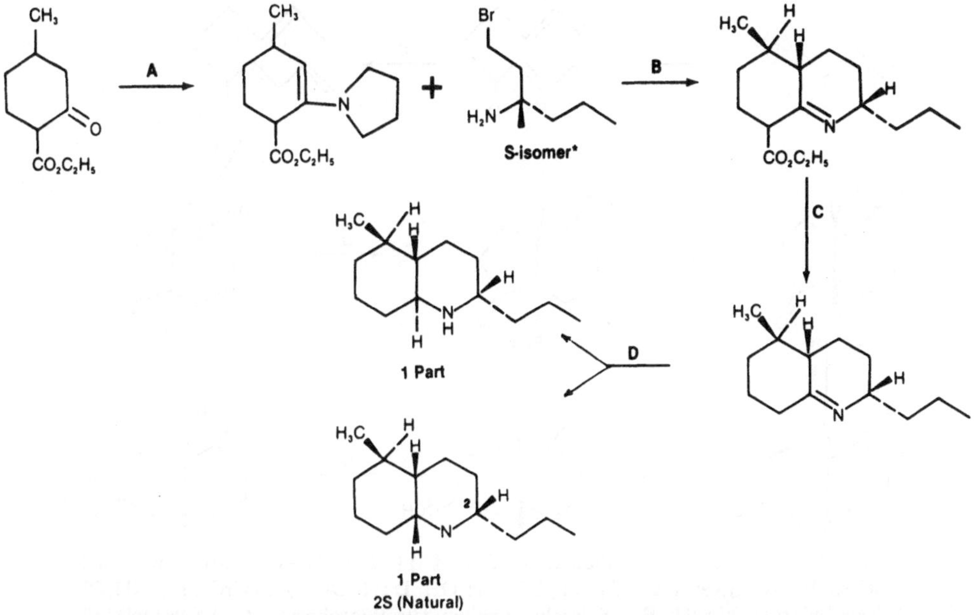

Scheme XIII. Synthesis of natural l-pumiliotoxin C and a *trans*-isomer (*122*). The synthesis of dl-pumiliotoxin C was also reported (*120, 122*). The optically active amine* used in this synthesis was erroneously reported as the R-isomer (*122*). The structure of the *trans*-isomer was not rigorously proven. **A** i) Pyrrolidine, HCO₂H. **B** i) Heat. **C** i) HCl, heat. **D** i) Pd/C, H₂, ii) C₆H₅COCl, iii) LiAlH₄, iv) Pd/C, H₂. (Reactions **D** ii, iii, iv were necessary to facilitate separation of the *cis*- and *trans*-decahydroquinoline products)

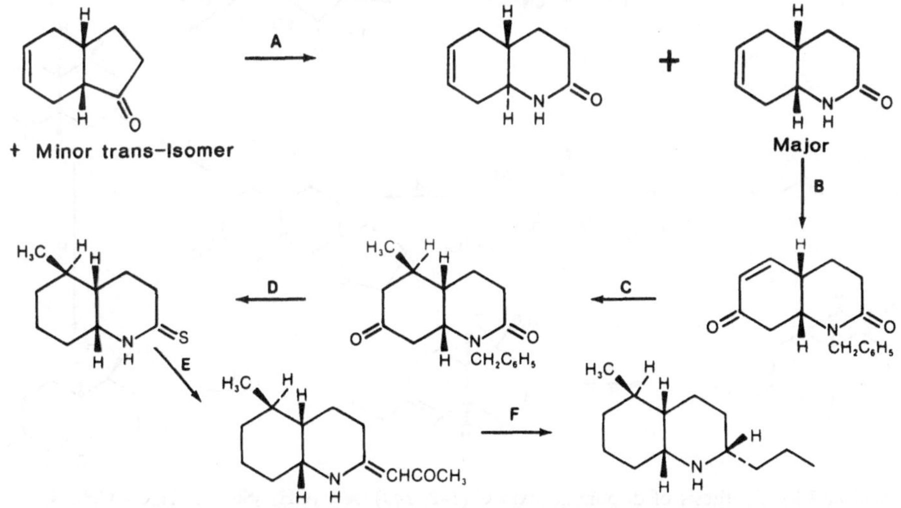

Scheme XIV. Synthesis of dl-pumiliotoxin C (*139, 140*). The starting material contains about 10 percent of the *trans*-isomer. **A** i) NH₂OH, ii) Tosyl chloride, pyridine. **B** i) C₆H₅CH₂Cl, ii) m-Chloroperbenzoic acid, iii) HBr, iv) CrO₃, v) LiBr, Li₂CO₃. **C** i) Li(CH₃)₂Cu. **D** i) HSCH₂CH₂SH, ii) Raney nickel, iii) Na, NH₃, iv) P₂S₅. **E** i) BrCH₂COCH₃, ii) (C₆H₅)₃P. **F** i) PtO₂, H₂, ii) CrO₃, iii) HSCH₂CH₂SH, iv) Raney nickel

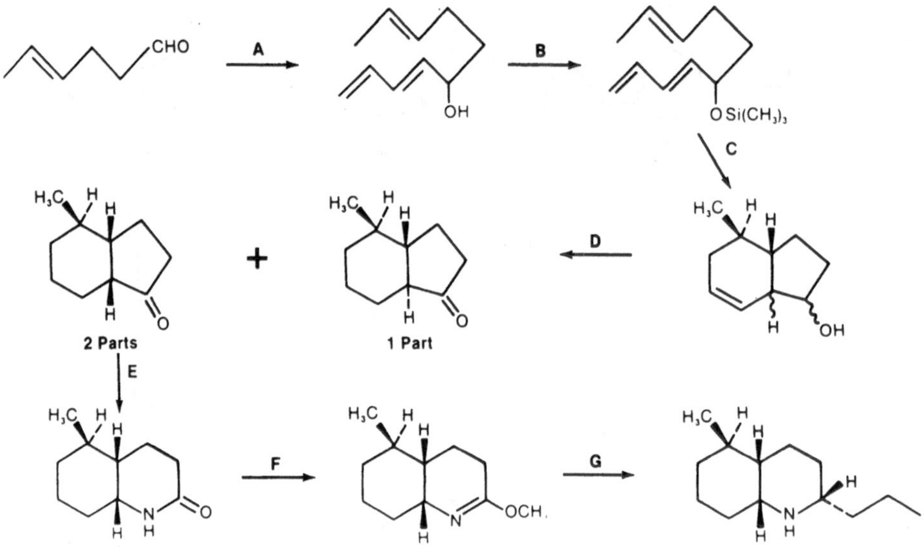

Scheme XV. Synthesis of dl-pumiliotoxin C (*202*). **A** i) CH$_2$=CH−C≡CMgBr, ii) LiAlH$_4$, CH$_3$ONa. **B** i) (CH$_3$)$_3$SiCl. **C** i) 245°, ii) KF, CH$_3$OH. **D** i) PtO$_2$, H$_2$, ii) CrO$_3$. **E** i) NH$_2$OH, ii) Tosyl chloride, NaOH. **F** i) Trimethyloxonium tetrafluoroborate. **G** i) n-PropylMgBr, ii) Pd, H$_2$

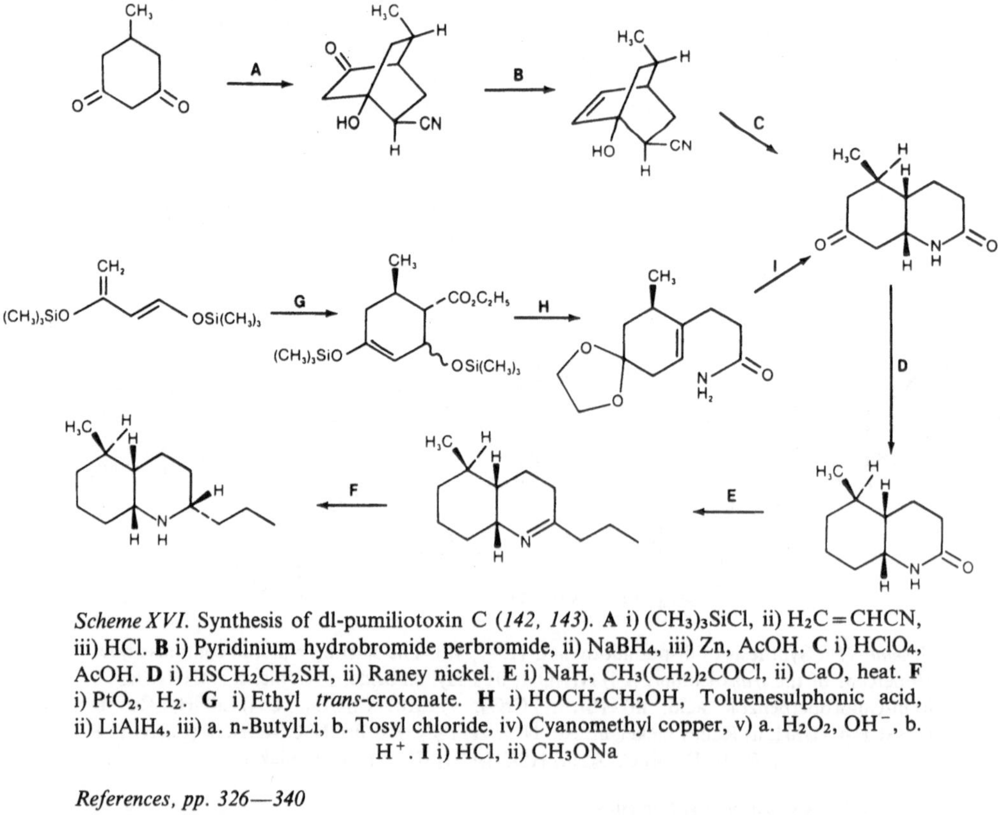

Scheme XVI. Synthesis of dl-pumiliotoxin C (*142, 143*). **A** i) (CH$_3$)$_3$SiCl, ii) H$_2$C=CHCN, iii) HCl. **B** i) Pyridinium hydrobromide perbromide, ii) NaBH$_4$, iii) Zn, AcOH. **C** i) HClO$_4$, AcOH. **D** i) HSCH$_2$CH$_2$SH, ii) Raney nickel. **E** i) NaH, CH$_3$(CH$_2$)$_2$COCl, ii) CaO, heat. **F** i) PtO$_2$, H$_2$. **G** i) Ethyl *trans*-crotonate. **H** i) HOCH$_2$CH$_2$OH, Toluenesulphonic acid, ii) LiAlH$_4$, iii) a. n-ButylLi, b. Tosyl chloride, iv) Cyanomethyl copper, v) a. H$_2$O$_2$, OH$^-$, b. H$^+$. **I** i) HCl, ii) CH$_3$ONa

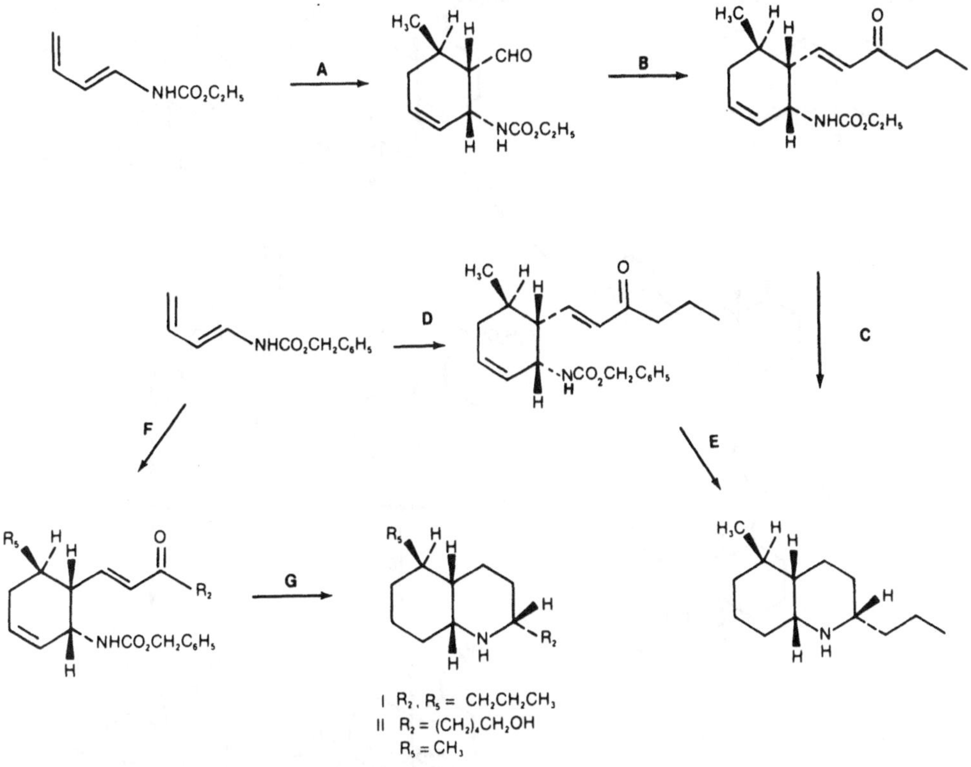

Scheme XVII. Synthesis of dl-pumiliotoxin C and analogs I and II (208, 209). **A** i) trans-Crotonaldehyde. **B** i) Sodium dimethyl-2-oxopentyl phosphonate. **C** i) Pd/C, H$_2$, ii) Cu, HBr, AcOH, iii) PtO$_2$, H$_2$. **D** i) trans-Crotonaldehyde, ii) sodium dimethyl-2-oxopentyl phosphonate. **E** i) Pd/C, H$_2$, HCl. **F** As in reactions **D** with appropriate aldehyde and phosphonate. **G** As in reaction **E**

A trans-isomer (HCl, m.p. 276°) was reported by HABERMEHL et al. (122) as a minor product in the synthesis of pumiliotoxin C (Scheme XIII). The mass spectrum was virtually identical to pumiliotoxin C. The carbon-13 magnetic resonance peaks (D$_2$O) were as follows: 61.5 (C-8a), 57.9 (C-2), 45.7 (C-4a), 36.6, 36.2, 34.8, 30.8, 29.2, 27.1, 24.4, 19.1, 18.7, 13.9. The optical rotation of this presumed (2S)-trans-isomer was [α]$_D^{19}$ −27.5 (c 0.6, CH$_3$OH). No further characterization was reported. A trans-decahydroquinoline, stated to be obtained "as a by-product of the synthesis of pumiliotoxin C" (reference to a 1972 Ph. D. thesis by W. KISSING, University of Darmstadt), was subjected to x-ray analysis and shown to be 2-n-propyl-7-methyl-trans-decahydroquinoline (104).

16*

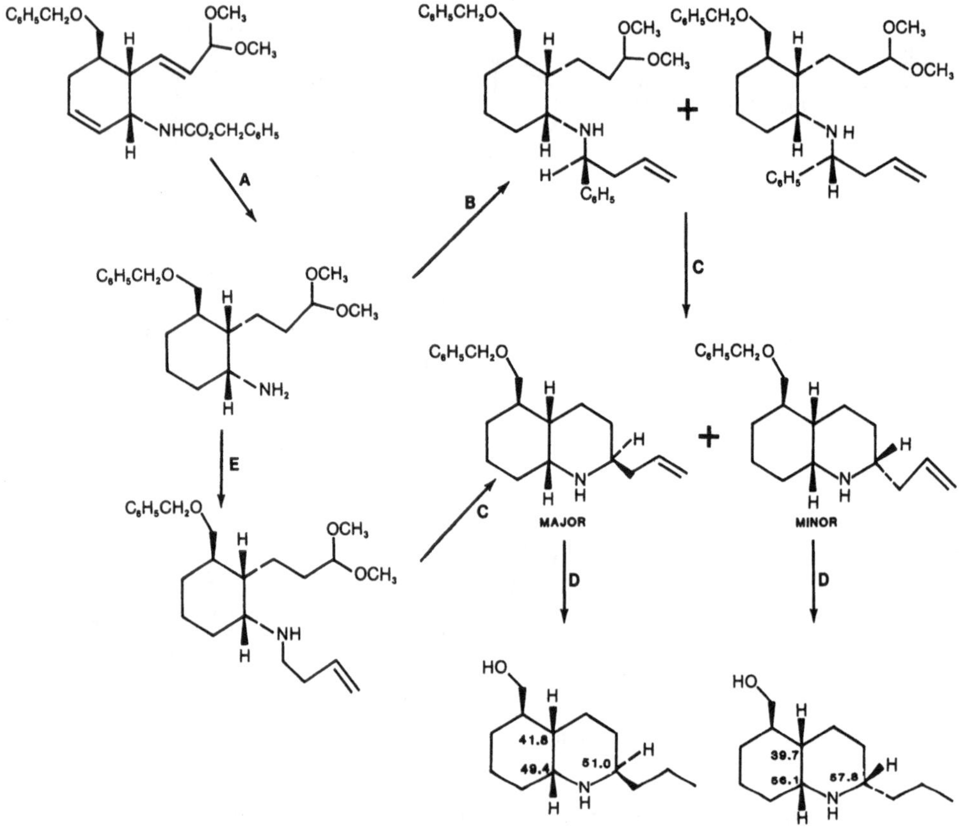

Scheme XVIII. Synthesis of further analogs of pumiliotoxin C (*207, 210*). The starting material was obtained in reactions similar to those of Scheme XVII. Carbon-13 magnetic resonance peaks which are diagnostic for stereochemistry at C-2 in *cis*-decahydroquinolines are given for products of reaction **D**. The ratio of the isomers obtained by reaction **C** (see Scheme XXXIX) is dependent on the precursor. **A** i) Pd/C, H$_2$. **B** i) C$_6$H$_5$CHO, ii) CH$_2$=CHCH$_2$MgBr. **C** i) H$^+$, heat. **D** i) Pd/C, H$_2$. **E** i) CH$_2$=CHCH$_2$CH$_2$Br

An approach to pumiliotoxin-C *via* dihydroquinolines was not successful because of incomplete reduction of such precursors (Scheme XIX) (*34*).

A synthesis of a 2-n-propyl-*cis*-decahydroquinolin-5α-ol has been reported (Scheme XX) (*121*). This compound is of interest since it is related to structures previously and tentatively proposed for certain trace alkaloids of a so-called hydroxypumiliotoxin-C class (*80*). The mass spectrum was as follows: m/z 197 (7); 154 (100); 136 (11); 111 (5); 72 (12). Infrared and proton magnetic resonance data were reported (*121*). The synthesis of the corresponding 2-n-propyl-*cis*-dehydroquinoline was reported in the early part of this century (*262*).

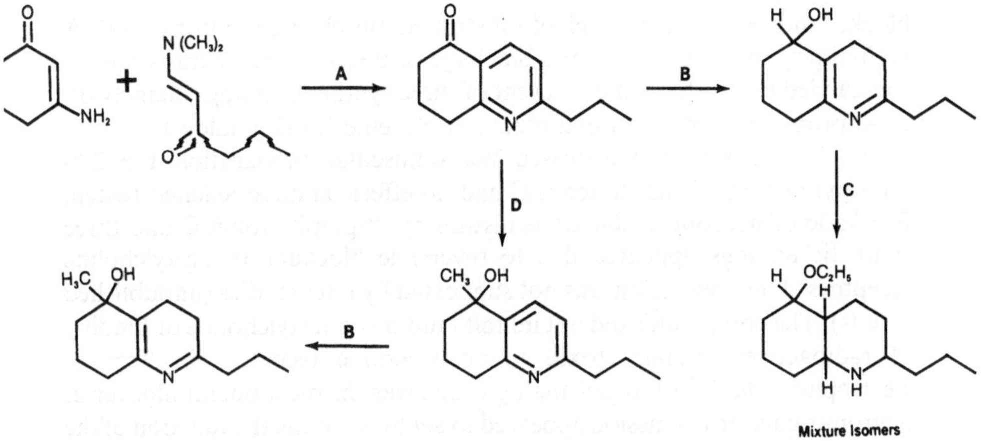

Mixture Isomers

Scheme XIX. Synthesis of hexahydro- and decahydroquinolines related in structure to pumiliotoxin C (*34*). The reductive procedures were not successful. **A** i) AcOH. **B** i) PtO$_2$, H$_2$. **C** i) (C$_2$H$_5$O)$_3$CCH$_3$, toluenesulphonic acid, ii) Pd/C, H$_2$. **D** i) CH$_3$Li

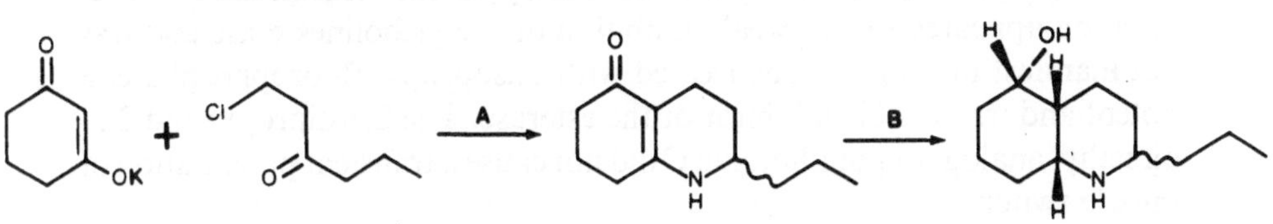

Scheme XX. Synthesis of a 5-hydroxy-*cis*-decahydroquinoline analog of pumiliotoxin C (*121*). **A** i) KOH, ii) NH$_3$, iii) Urushibara nickel. **B** i) PtO$_2$, H$_2$

A synthetic precursor, namely a tricarbonyl compound, RCH$_2$CO(CH$_2$)$_3$CO (CH$_2$)$_3$COCH$_2$R, which might be the basis for syntheses of both pumiliotoxin and histrionicotoxin class alkaloids, has been proposed (*275*). Initial cyclization of such a compound could lead to a cyclohexenone, followed by reductive amination and then imine formation to yield the pumiliotoxin precursor.

C. Biological Activity

Pumiliotoxin C is a relatively nontoxic compound and the toxin designation is, therefore, somewhat a misnomer. Pharmacological effects on biological systems also require relatively high concentrations of this alkaloid. Pumiliotoxin C at concentrations up to 10 µM had no effect on tension of guinea pig ileum nor on acetylcholine-induced contractures (*183*). At these concentrations pumiliotoxin C also had no effect on spontaneous activity of guinea pig atria or on direct or indirect elicited twitch in rat phrenic nerve diaphragm preparations. At higher concentrations of 80—320 µM dl-pumiliotoxin C initially potentiated and then blocked indirect elicited twitch of rat striated muscle preparations (*197*). A preliminary investigation of the blockage of neuromuscular transmission was carried out with the most potent of three synthetic analogs, namely dl-2,5-dipropyl-*cis*-decahydroquinoline (see Scheme XVII, analog I).

In an α-bungarotoxin-treated neuromuscular preparation the 2,5-dipropyl analog of pumiliotoxin C had no effect on direct elicited twitch. Blockade of neuromuscular transmission by dl-pumiliotoxin C and three synthetic analogs appeared due to reversible blockade of acetylcholine receptors. This conclusion was not supported by later studies (unpublished results). The compounds did not inhibit binding of acetylcholine or binding of radioactive α-bungarotoxin to acetylcholine receptors in *Torpedo* electroplax. The 2,5-dipropyl analog which was the most potent blocker of neuromuscular transmission appeared to act by affecting the function of the channel associated with the acetylcholine receptor in the muscle endplate. A transient potentiation of indirect elicited twitch by pumiliotoxin C and by the 2-n-butyl-*cis*-decahydroquinoline analog in the neuromuscular preparation appeared due to weak inhibition of acetylcholinesterase and was not manifest in preparations treated with diisopropyl fluorophosphate, a potent and irreversible inhibitor of the esterase. The 2,5-dipropyl and 2,5-dimethyl analogs of pumiliotoxin C did not cause a transient potentiation of muscle twitch.

Table 7. *Effect of Pumiliotoxin C on Biological Systems*

System	Species	Comments	References
Whole animal	Mouse	Relatively nontoxic, locomotor difficulties at 5 mg/kg; Convulsions, and death in 10 min at 125 mg/kg	(*80, 140*)
Neuromuscular preparation (striated)	Rat	Blockade acetylcholine responses at high concentrations	(*183, 197*)
Cardiac preparations	Guinea pig atrium	No effect at 10 µM	(*183*)
Smooth muscle	Guinea pig ileum	No effect at 10 µM	(*183*)
Electroplax	*Torpedo* membranes	No effect on binding of α-bungarotoxin	Unpublished results
Acetylcholine-esterase	Rat	Inhibition at high concentrations	(*197*)

Another analog, 2-n-propyl-*cis*-decahydroquinoline, had been suggested to have pharmacological properties similar to the plant alkaloid coniine (*262*). The decahydroquinoline caused headaches and stupefication in man. The effects on heart preparations were stated to be similar to coniine. It was stated to have no effect on striated muscle preparations, but no details were reported.

The effects of pumiliotoxin C on biological systems are summarized in Table 7.

IV. Histrionicotoxins

A. Structures

A preliminary investigation of the alkaloids in extracts from another brightly colored, extremely variable species of dendrobatid frog, *Dendrobates histrionicus,* was carried out in tle late 1960's. The results indicated the presence of simple alkaloids differing markedly in mass spectral properties from the pumiliotoxins (J. DALY, unpublished results). The source of these extracts were a population of *Dendrobates histrionicus* sympatric with *Phyllobates aurotaenia* in the Río San Juan drainage of western Colombia, and had been obtained during field work on the latter frog. Since the Río San Juan populations of *Dendrobates histrionicus* were not particularly abundant, further investigations were carried out with an extremely abundant population of the same species, which was known to occur near the town of Guayacana in southwestern Colombia. Methanol extracts from some four hundred specimens of the Guayacana population of *Dendrobates histrionicus* afforded, after partitions and silica gel column chromatography, two major alkaloids, histrionicotoxin (53 mg) and dihydroisohistrionicotoxin (93 mg) (*81*). The dihydroisohistrionicotoxin required further purification which was carried out by preparative thin-layer chromatography on silica gel. In addition, another 40 mg of alkaloid fractions were obtained which contained one major and several minor alkaloids.

The hydrochloride salt of histrionicotoxin was crystallized twice from methanol-acetone and then from isopropanol; the hydrobromide salt of histrionicotoxin from isopropanol and the hydrochloride salt of dihydro-isohistrionicotoxin from isopropanol-acetone. X-ray crystallographic analyses revealed the structures and absolute configurations of histrionico-toxin, [2R,6R,7S,8S]-7-(1-cis-buten-3-ynyl)-2-(cis-2-penten-4-ynyl)-1-aza-spiro-[5.5]-undecan-8-ol, and dihydroisohistrionicotoxin (*81, 151,* Fig. 9). These compounds were unusual spiropiperidine alkaloids with acetylene and/or allenic moieties in the side chains.

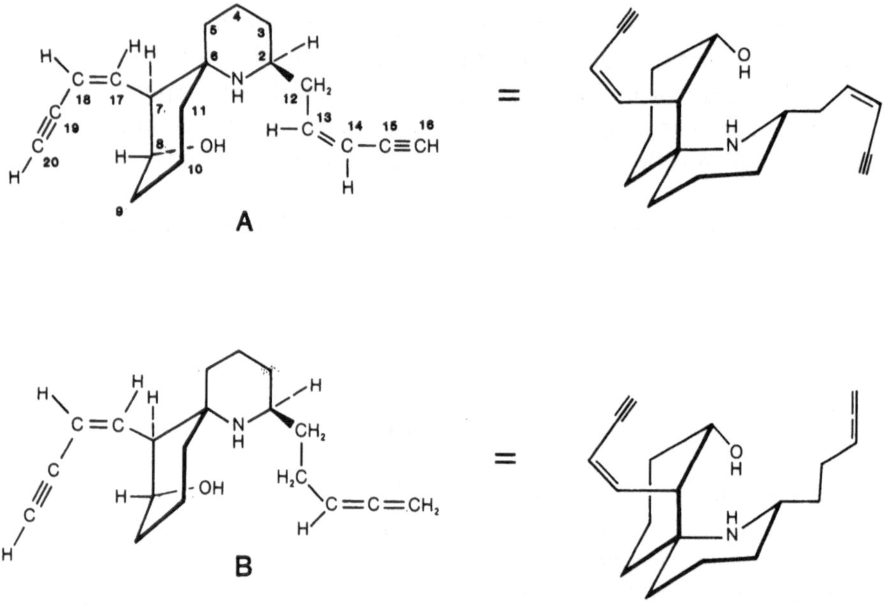

Fig. 9. Structures of **A** histrionicotoxin and **B** isodihydrohistrionicotoxin. The configurations depicted on the right are based on x-ray analysis of hydrochloride salts (*81, 151*)

In a subsequent study, the alkaloids from methanol extracts from some 1100 skins of *Dendrobates histrionicus* were subjected to multiple column chromatographies on silica gel and Sephadex LH 20 (*253*). Histrionicotoxin (226 mg), dihydrohistrionicotoxin (320 mg), and four analogs were isolated. The mass spectra and nuclear magnetic spectra defined the structures of the analogs as neodihydrohistrionicotoxin (19 mg), tetrahydrohistrionicotoxin (2 mg), isotetrahydrohistrionicotoxin (6 mg), and octahydrohistrionico-toxin (9 mg) (Table 8). A fifth compound referred to as HTX-D (47 mg) was obviously not a histrionicotoxin, although it contained a five carbon side chain (*cis*-CH$_2$CH=CHC≡CH) identical with one side chain in histrionicotoxin. The structure of HTX-D was later elucidated by x-ray crystallographic analysis and the tricyclic alkaloid was renamed gephyro-toxin (*vide infra*). A number of minor alkaloids also were detected as mixtures in certain column fractions. Three additional histrionicotoxins were later isolated and characterized from extracts of further large samples of skins of *Dendrobates histrionicus* (*86, 255*). These were allodihydro-histrionicotoxin, allotetrahydrohistrionicotoxin and dihydrohistrionico-toxin (Table 8). The former is a major alkaloid constituent, while the latter two are minor or trace constituents. All of these histrionicotoxins reduce to a common perhydroderivative, dodecahydrohistrionicotoxin, whose struc-ture is shown with proton magnetic resonance assignments in Fig. 10. The proton magnetic resonance spectrum of isodihydrohistrionicotoxin is depicted in Fig. 11. Carbon-13 magnetic resonance spectral assignments for histrionicotoxin and certain congeners are presented in Table 9.

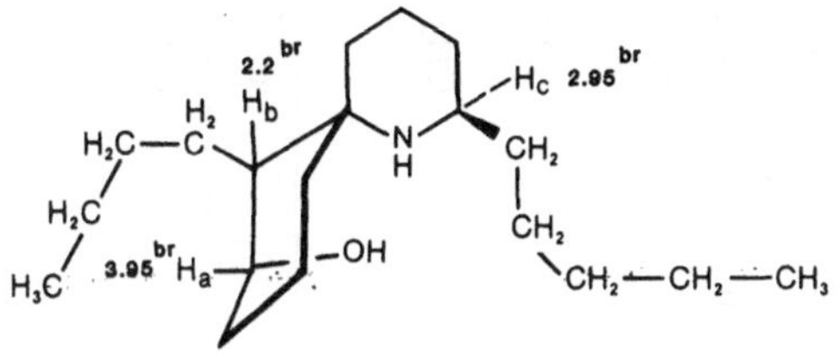

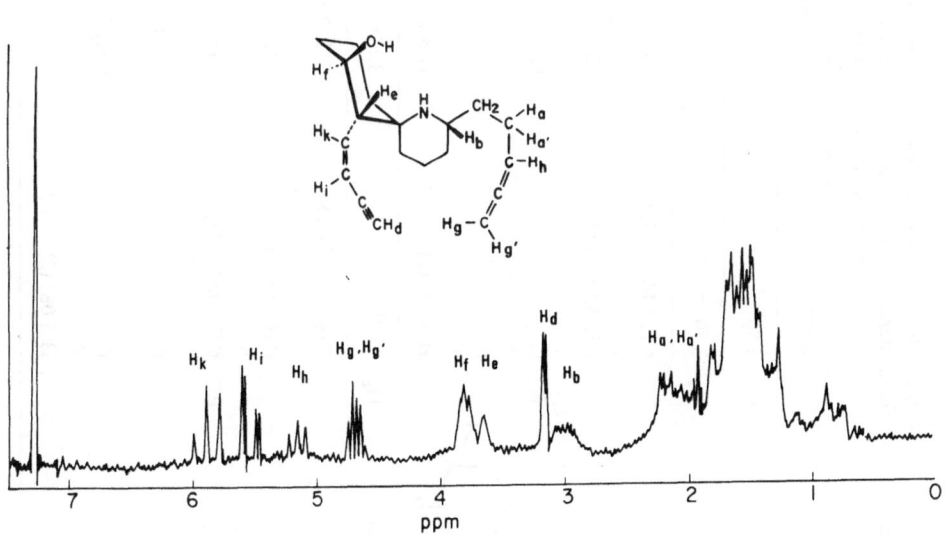

Fig. 10. Structure of perhydrohistrionicotoxin (dodecahydro) and proton magnetic resonance spectral assignments. Chemical shifts are ppm for deuterochloroform (253). The proton magnetic resonance spectrum (100 MHz) of perhydrohistrionicotoxin is depicted in TOKUYAMA et al. (253). See Table 8 for chemical shifts of Ha, Hb and Hc in various naturally occurring histrionicotoxins

Fig. 11. Proton magnetic resonance spectrum (100 MHz) and assignments for dihydroiso-histrionicotoxin. Chemical shifts in ppm for deuterochloroform with a tetramethylsilane standard (TOKUYAMA, unpublished spectrum)

Table 8. *Proton Magnetic Resonance Spectral Assignments for Histrionicotoxin (HTX) and Congeners*

Chemical shifts in ppm for deuterochloroform with a tetramethylsilane standard: Coupling constants (*J*) in parentheses: br, broad; m, multiplet; s, singlet; d, doublet; t, triplet; d,d doublet of doublets; etc. (86, 253, 255)

	HTX	DihydroHTX[1]	IsodihydroHTX[2]	NeodihydroHTX
H_a	3.80^{br}	3.78^{br}	3.82^{br}	3.71^{br}
H_b	$3.70^{d,br}$ (11)	$3.67^{d,br}$ (11)	$3.73^{d,br}$ (10)	$3.41^{d,br}$ (10)
R'	CH $5.88^{d,d}$ (11,11) =C CH $5.55^{d,d}$ (11,2) —C≡C H 3.19^{d} (2)	CH $5.83^{d,d}$ (11,11) =C CH $5.56^{d,d}$ (11,2) —C≡C H 3.17^{d} (2)	CH $5.89^{d,d}$ (10,10) =C CH $5.54^{d,d}$ (10,2) —C≡C H 3.17^{d} (2)	CH $5.28^{d,d}$ (10,10) = CH $6.04^{d,d}$ (10,10) — CH $6.76^{d,d,d}$ (10,10,16) = CH$_2$ 5.16^{d} (10) / 5.20^{d} (16)
H_c	3.04^{br}	3.15^{br}	3.00^{br}	3.15^{br}
R''	CH$_2$ $2.40^{d,d}$ (8,8) — CH $6.06^{d,t}$ (8,11) = CH $5.60^{d,d}$ (11,2) —C≡C H 3.08^{d} (2)	CH$_2$ $2.37^{d,d}$ (7,7) — CH $5.45^{d,t}$ (7,11) = CH $6.12^{d,d}$ (11,11) — CH $6.64^{d,d,d}$ (11,10,16) = CH$_2$ 5.13^{d} (10) / 5.20^{d} (16)	CH$_2$ — CH$_2$ 2.18^{m} — CH $5.16^{t,t}$ (7,7) =C— CH$_2$ $4.69^{d,t}$ (7,3.5)	CH$_2$ $2.39^{d,d}$ (8,8) — CH $6.06^{d,t}$ (8,11) = CH $5.56^{d,d}$ (11,2) —C≡C H 3.07^{d} (2)

AllodihydroHTX[3]	TetrahydroHTX	IsotetrahydroHTX	AllotetrahydroHTX	OctahydroHTX[4]
3.78^{br}	3.76^{br}	3.76^{br}	3.68^{br}	3.94^{br}
3.66^{d,d}	3.45^{d,br} (11)	3.46^{d,br} (11)	3.88^{d,br} (11)	~2.1
CH 5.82^{d,d} (11,11) =	CH 5.29^{d,d} (11,11) =	CH 5.33^{d,d} (11,11) =	CH 5.24^{d,d} (11,11) =	CH₂ —
CH 5.46^{d,d} (9,2) —	CH 6.08^d (11,11) —	CH 6.08^{d,d} (11,11) —	CH 5.98^{d,d} (11,11) —	CH₂ 2.13^m —
C ≡	CH 6.80^{d,d,d} (11,11,17) =	CH 6.80^{d,d,d} (11,11,17) =	CH 6.93^{d,d,d} (11,11,16) =	CH 5.80^{d,d,t} (8,11,17) =
C —	CH₂ 5.19^d (11) / 5.24^d (17)	CH₂ 5.20^d (11) / 5.24^d (17)	CH₂ 5.10^d (11) / 5.16^d (16)	CH₂ 4.99^d (11) / 5.07^d (17)
H 3.14^d (2)				
3.00^{br}	3.11^{d,br} (11)	3.04^{br}	3.00^{br}	3.00^{br}
CH₂ —	CH₂ 2.30^{d,d} (7,7) —	CH₂ —	CH₂ —	CH₂ —
CH₂ —	CH₂ 5.49^{d,t} (7,11) =	CH₂ 2.1^m —	CH₂ —	CH₂ —
CH₂ 2.18^{d,t} (2,7) —	CH 6.14^{d,d} (11,11) —	CH 5.15^{t,t} (7,7) =	CH₂ 2.18^{d,t} (2,7) —	CH₂ 2.04^m —
C ≡	CH 6.64^{d,d,d} (11,11,17) =	C	C	CH 5.77^{d,d,t} (8,11,17) =
C —	CH₂ 5.15^d (11) / 5.21^d (17)	C ≡	C ≡	CH₂ 4.29^d (11) / 4.96^d (17)
H 1.91^t (2)		CH₂ 4.68^{d,t} (7,3.5)	CH 1.90^t (2)	

[1] DihydroHTX was previously referred to as syn-dihydroHTX (253). It was first prepared by partial reduction of histrionicotoxin, but has now been isolated as a trace alkaloid from *Dendrobates histrionicus*, Guayacana, Narino, Colombia (255). [2] The spectrum (100 MHz) of iso-dihydrohistrionicotoxin is depicted in Fig. 11. [3] The spectrum (100 MHz) of allodihydroHTX is depicted in (86). [4] Assignments of the protons to side chains of octahydroHTX are arbitrary with respect to R and R' and may be reversed.

Table 9. *Carbon-13 Magnetic Resonance Assignments for Histrionicotoxin (HTX) and Congeners (CDCl₃) (255)*

Resonance peaks designated with a superscript a may be interchanged as may peaks designated with a superscript b

Carbon number	HTX	Dihydro-HTX	Isodihydro-HTX	Neodihydro-HTX	Allodihydro-HTX
2	49.7	50.1	49.2	50.0	49.6
3	36.8	35.3	36.0	36.8	36.3
4	19.5	19.7	18.9	19.5	19.2
5	29.0	29.2	28.5	28.7	28.8
6	54.1	54.4	53.8	54.4	54.3
7	41.4	41.7	40.9	38.8	41.2
8	71.3	71.6	70.8	72.4	71.2
9	32.8	32.6	32.4	32.9	32.7
10	15.0	15.3	14.7	15.2	15.0
11	37.8[a]	37.1	37.5	37.5	37.8
5-Carbon side chain					
12	37.9[a]	38.2	36.2	38.0	36.5
13	141.6	128.1	23.9	141.7	18.3
14	110.0[b]	131.6	89.2	110.6	24.5
15	80.3	132.0	207.9	80.5	83.9
16	81.7	117.8	74.7	81.9	68.5
4-Carbon side chain					
17	142.9	143.1	142.3	129.8	142.6
18	110.3[b]	110.2	109.6	131.8	110.0
19	79.7	80.0	79.3	130.8	79.6
20	82.6	82.7	82.3	118.6	82.6

A protocol for the analysis of constituent alkaloids from dendrobatid frogs by gas-chromatography-mass spectrometry had been developed and refined (*80, 188*). Such a protocol allowed the characterization of alkaloid profiles in extracts from a single frog skin and has in some instances provided sufficient data for structural assignment. A satisfactory protocol for such analyses is as follows:

Frogs were skinned in the field, and skins stored in methanol (1 part skins to from 2 to 100 parts methanol), when possible at $-5°$C, otherwise at ambient temperature. Skins were macerated twice with methanol to extract alkaloids. Skin samples with a total wet weight of 50 mg to 500 mg were macerated twice each time with 5 to 10 ml of methanol, skin samples from 1 to 4 g wet weight were macerated three times each time with 10—20 ml of methanol. Larger skin samples were macerated twice each time with about 2—4 volumes of methanol (weight/volume). The methanol extracts were then diluted with an equal volume of water. In some instances for larger samples the methanol extract was first concentrated *in vacuo* to a smaller volume before dilution. The aqueous methanol was extracted twice each time with 2 volumes of chloroform. The basic chloroform-soluble alkaloids were then extracted three times from the combined chloroform fractions, each time into one-half volumes of 0.1 N HCl. The combined 0.1 N HCl fractions were adjusted to pH 9 with 1 N aqueous ammonia followed by re-extraction into chloroform: twice, each time with an equal volume of chloroform. The combined chloroform fractions were dried over anhydrous sodium sulfate and then evaporated *in vacuo* to dryness. Certain of the alkaloids have appreciable volatility and evaporations *in vacuo*, therefore, must be done carefully. The resulting alkaloid residue was dissolved in methanol so that 10 μl corresponds to 10 mg of the original wet weight of the skins. This alkaloid fraction contains primarily alkaloids although traces of steroids and environmental artefacts such as phthalates sometimes are present as contaminents.

Such alkaloid fractions were analyzed by thin-layer chromatography, gas chromatography and mass spectrometry (*80*). The studies have been carried out in a manner designed to facilitate quantitative and qualitative comparisons of alkaloid profiles between populations and species of dendrobatid frogs and other amphibians.

Thin-layer chromatographic analyses were routinely carried out with alkaloid fraction equivalent to 10 mg wet weight skin on silica gel plates with chloroform-methanol (9 : 1). Subsequent detection was routinely performed with iodine vapor. In many instances preparative thin-layer chromatography on a microscale, followed by analysis by gas-chromatography-mass spectrometry was used to determine R_f values of specific alkaloids.

Gas chromatographic analyses were routinely carried out with alkaloid fraction equivalent to 2 mg wet weight skin on a 1.5% OV-1 column. The column was programmed from 150° to 280° at 10° per min with a flow rate of 20—25 cm³/min of nitrogen. Analysis with a flame ionization detector provided a quantitative profile of alkaloid components (see *80* for further details). The second step in analysis was combined gas chromatography-chemical ionization mass spectrometry with nitrogen carrier gas and ammonia as the ionizing gas. Ammonia chemical ionization provided virtually exclusively the protonated parent ion of the various dendrobatid alkaloids. Computer-assisted analysis of ammonia chemical ionization mass spectra for the gas chromatographic run provided the number, elution sequence, and parent ions of the alkaloids corresponding to flame ionization profiles. Often an apparent single flame ionization peak was found to represent two or more alkaloids. For further characterization gas chromatography-mass spectral analysis was repeated with methane as the ionizing gas. Under such conditions typical fragments are seen for the various classes of dendrobatid alkaloids. Recently, "pseudoelectron impact spectra" have been determined using N_2-NO in a chemical ionization mode (unpublished data). Such spectra appear to be suitable for detailed characterization and identification of individual dendrobatid alkaloids. After the chemical ionization measurements, the gas chromatography-mass spectral analysis was repeated by the electron impact method. Such electron impact spectra were usually sufficiently detailed to allow characterization and identification of individual alkaloids. Gas chromatographic profiles are presented in Fig. 12 for alkaloid fractions from two populations of a *Dendrobates bombetes* with various peaks indentified by numerical designations (*80, 190*). Further examples of gas chromatographic profiles of dendrobatid alkaloid fractions are reproduced in other publications (*187—191*).

Finally, empirical formulae for parent ion (or protonated parent ion) and fragment ions were determined for many alkaloids by high resolution gas chromatography-mass spectrometry. Additional data on the alkaloids could be obtained after perhydrogenation, acetylation, or trimethyl-silylation. In such instances, alkaloid fractions corresponding to 100 to 300 μg wet weight skin were either reduced in methanol with 10% palladium on charcoal catalyst and 30 psi hydrogen gas overnight, or acetylated in acetone with acetic anhydride and sodium acetate at room temperature overnight, or trimethylsilylated in pyridine with *bis*-(trimethylsilyl)-fluoroacetamide for 1 hr at room temperature (*80, 84*). The resultant derivatized alkaloid fraction was subjected to analysis by gas chromatography-mass spectrometry.

Such a protocol has been used to characterize a large number of dendrobatid alkaloids. Two additional alkaloids of the histrionicotoxin class were detected and characterized. These were histrionicotoxins **235A** and **259** (see Fig. 13). The structure of histrionicotoxin **259** has been confirmed by proton and carbon-13 magnetic resonance spectroscopy (*255*). Histrionicotoxins as a class are characterized by a major fragment ion ($C_6H_{10}N$) at m/z 96. All form O-acetyl derivatives and all of the eleven

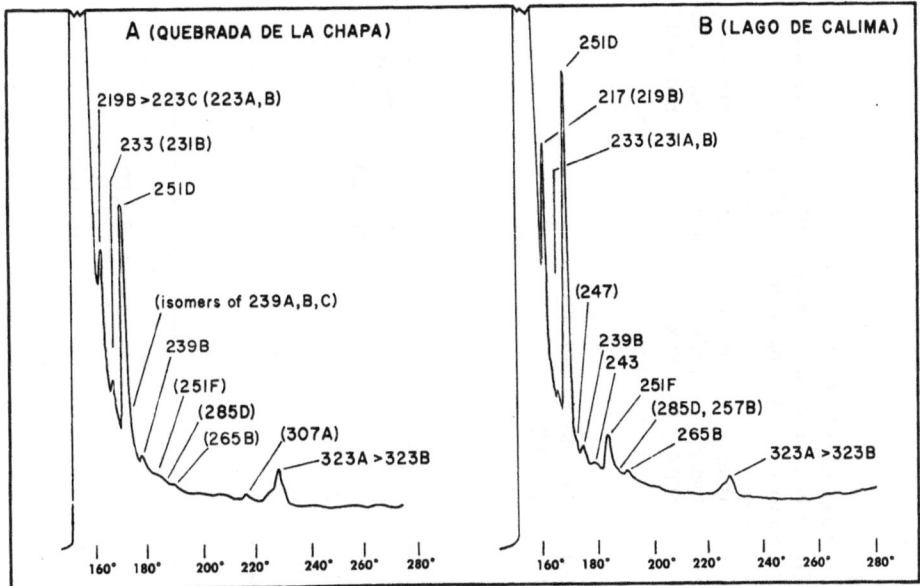

Fig. 12. Gas chromatographic profiles for alkaloids from two populations of a small Colombian poison frog, *Dendrobates bombetes A*. From Quebrada de la Chapa (7 specimens, February, 1974). *B* From Lago de Calima (10 specimens, November, 1976). Chromatography was with 2 µl of methanolic alkaloid fraction corresponding to 2 mg of wet skin. Alkaloids are designated by molecular weights and in most cases a letter: Parentheses indicate trace compounds. See listings of compounds in text and Table 22. The gas chromatographic column was 1.5% OV-1 and was programmed 150°—280° at 10°/min. For methodology see DALY *et al.* (*80*) (data from *190*)

natural histrionicotoxins have unsaturated side chains. The detailed electron impact mass spectra of the histrionicotoxins with a five carbon side chain at C-2 and a four carbon side chain at C-7 are depicted in Table 10. These spectra were obtained with a different mass spectrometer than the spectra reported for the eleven histrionicotoxins in the following section which is in the previously detailed format (see p. 236).

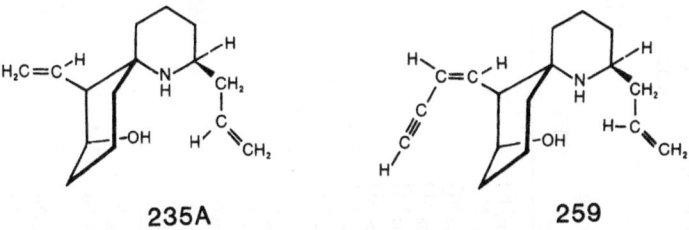

235A   259

Fig. 13. Structures of histrionicotoxins **235A** and **259**. Recent nuclear magnetic resonance spectra support the structure of histrionicotoxin **259** (*255*)

Table 10. *Mass Spectra of Histrionicotoxin (HTX) and Congeners (78, 86, 253)*

Major fragmentation pathways are discussed in (253). Loss of hydroxyl and of the five carbon side chain are major pathways. OctahydroHTX has a major loss of $-CH_2CH=CH_2$. (Each mass entry is followed by its relative intensity in parentheses, see Table 1)

Fragment	HTX	Dihydro HTX	Iso-dihydro HTX	Neo-dihydro HTX	Allo-dihydro HTX	Tetra-hydro HTX	Isotetra-hydro HTX	Allotetra-hydro HTX	Octa-hydro HTX	Dodeca-hydro (Per-hydro) HTX
$C_{19}H_{37}NO$										295 (22)
$C_{19}H_{33}NO$									291 (28)	
$C_{19}H_{29}NO$						287 (20)	287 (28)	287 (29)		
$C_{19}H_{28}NO$						286 (8)	286 (14)			
$C_{19}H_{27}NO$		285 (24)	285 (24)	285 (60)	285 (36)					
$C_{19}H_{26}NO$		284 (12)	284 (18)	284 (18)						
$C_{19}H_{25}NO$	283 (19)									
$C_{19}H_{24}NO$	282 (11)									
$C_{19}H_{36}N$										278 16
$C_{19}H_{32}N$									274 (21)	
$C_{19}H_{28}N$						270 (8)	270 (24)	270 (16)		
$C_{19}H_{26}N$		268 (35)	268 (26)	268 (20)	268 (23)					
$C_{19}H_{24}N$	266 (14)									
$C_{17}H_{24}NO$			256 (8)							
$C_{19}H_{22}NO$	254 (5)									
$C_{16}H_{30}NO$		252 (26)[a]								252 (25)
$C_{16}H_{28}NO$									250 (88)	
$C_{16}H_{24}NO$			246 (14)							
$C_{16}H_{22}NO$	244 (5)								244 (10)[b]	

Formula	Observed ions m/z (rel. int.)
$C_{17}H_{24}N$	240 (6)
$C_{17}H_{22}N$	242 (11), 238 (8)
$C_{17}H_{20}N$	238 (15)
$C_{15}H_{26}NO$	236 (15)
$C_{16}H_{26}N$	232 (13)
$C_{16}H_{24}N$	230 (14)
$C_{16}H_{22}N$	228 (16)
$C_{16}H_{20}N$	226 (12)
$C_{14}H_{26}NO$	224 (68)
$C_{14}H_{24}NO$	222 (50)
$C_{14}H_{22}NO$	220 (100), 220 (80), 220 (30), 220 (24)
$C_{14}H_{20}NO$	218 (100), 218 (58), 218 (46), 218 (26)
$C_{14}H_{20}N$	202 (36), 202 (35), 202 (35), 202 (32), 202 (15), 202 (10)
$C_{14}H_{18}N$	200 (84), 200 (21), 200 (18), 200 (16)
$C_{13}H_{24}N$	196 (36), 194 (36)
$C_{13}H_{22}N$	192 (22), 192 (14)
$C_{13}H_{20}N$	190 (27), 190 (27), 190 (15), 190 (10)
$C_{13}H_{18}N$	188 (10)
$C_{13}H_{16}N$	
$C_{12}H_{22}N$	180 (100)
$C_{12}H_{20}N$	178 (100)
$C_{12}H_{18}N$	176 (50), 176 (50), 176 (45), 176 (38), 176 (20), 176 (15), 176 (13)
$C_{12}H_{16}N$	174 (14)
$C_{12}H_{14}N$	172 (12)

Table 10 *(continued)*

Fragment	HTX	Dihydro HTX	Iso-dihydro HTX	Neo-dihydro HTX	Allo-dihydro HTX	Tetra-hydro HTX	Isotetra-hydro HTX[b]	Allotetra-hydro HTX	Octa-hydro HTX	Dodeca-hydro (Per-hydro) HTX
$C_{11}H_{21}N$										167 (40)
$C_{11}H_{19}N$									165 (36)	
$C_{11}H_{16}N$			162 (37)	160 (65)	162 (31)		162 (40)	162 (49)		
$C_{11}H_{14}N$	160 (28)									
$C_{10}H_{14}N$	146 (10)		148 (21)					148 (21)		
$C_{10}H_{12}N$		145 (43)								
$C_{10}H_{11}N$					144 (33)					
$C_{10}H_{10}N$										
$C_{10}H_{14}$			134 (24)					134 (70)		
$C_{10}H_{12}$	132 (12)			132 (25)						
$C_8H_{14}N$			122 (27)		122 (36)			122 (24)		
$C_8H_{12}N$			120 (21)					120 (38)		
$C_7H_{10}N$	108 (16)		108 (34)					106 (40)		
C_7H_8N			106 (30)							
$C_6H_{10}N$	96 (100)	96 (78)	96 (100)	96 (50)	96 (100)	96 (100)	96 (100)	96 (100)	96 (34)	96 (65)

[a] The ion for dihydroHTX is $C_{18}H_{22}N$.
[b] The ion for octahydroHTX is $C_{17}H_{26}N$.

Histrionicotoxin Class

235A. $C_{15}H_{25}NO$, 0.36, 176°, m/z 235(5), 234(2), 218(15), 194($C_{12}H_{20}NO$, 76), 176($C_{12}H_{18}N$, 25), 150(8), 96($C_6H_{10}N$, 100). H_4-derivative, m/z 239, 196, 178, 96. O-acetyl derivative.

259. $C_{17}H_{25}NO$, 0.36, 190°, m/z 259(4), 242(2), 218($C_{14}H_{20}NO$, 18), 200(6), 96($C_6H_{10}N$, 100). H_8-derivative, m/z 267(10), 250(13), 224(39), 196(15), 168(19), 152(100), 96(68). O-acetyl derivative.

283A. Histrionicotoxin, $C_{19}H_{25}NO$, 0.50, 210°, m/z 283(9), 282(2), 266(5), 250(2), 218(48), 200(27), 160(22), 96(100). H_{12}-derivative, 0·36, 214°, m/z 295(12), 294(2), 278(13), 252(18), 224(73), 196(27), 180(100), 168(39), 96(68). O-acetyl derivative.

285A. Isodihydrohistrionicotoxin, $C_{19}H_{27}NO$, 0.39, 215°, m/z 285(7), 284(2), 268(8), 252(12), 238(3), 218(6), 200(9), 190(4), 176(24), 162(18), 96(100). H_{10}-derivative, m/z see H_{12}-derivative of **283A**. O-acetyl derivative.

285B. Neodihydrohistrionicotoxin, $C_{19}H_{27}NO$, 0.46, 211°, m/z 285(4), 284(1), 268(3), 250(2), 220(37), 202(9), 160(20), 96(100). H_{10}-derivative, m/z see H_{12}-derivative of **283A**.

285C. Allodihydrohistrionicotoxin, $C_{19}H_{27}NO$, 0.40, 211°, m/z 285(4), 284(1), 268(2), 252(2), 218(5), 200(3), 190(4), 176(15), 162(17), 96(100). H_{10}-derivative, m/z see H_{12}-derivative of **283A**.

285E. Dihydrohistrionicotoxin, $C_{19}H_{27}NO$, 0.50, 212°, m/z 285(13), 268(10), 218(100), 200(84), 176(13), 96(78). H_{10}-derivative, m/z see H_{12}-derivative of **283A**.

287A. Isotetrahydrohistrionicotoxin, $C_{19}H_{29}NO$, 0.42, 216°, m/z 287(12), 286(4), 270(3), 220(30), 202(34), 176(45), 162(60), 148(24), 96(100). H_8-derivative, m/z see H_{12}-derivative of **283A**.

287B. Tetrahydrohistrionicotoxin, $C_{19}H_{29}NO$, 0.43, 213°, m/z 287(13), 286(2), 270(2), 220(43), 202(18), 176(6), 162(4), 148(4), 96(100). H_8-derivative, m/z see H_{12}-derivative of **283A**.

287D. Allotetrahydrohistrionicotoxin, $C_{19}H_{29}NO$, 0.35, 215°, m/z 287(14), 270(8), 220(24), 202(36), 176(38), 162(49), 96(100). H_8-derivative, m/z see H_{12}-derivative of **283A**.

291A. Octahydrohistrionicotoxin, $C_{19}H_{33}NO$, 0.35, 212°, m/z 291(12), 290(2), 274(14), 250(54), 222(24), 194(18), 192(12), 178(100), 96(52). H_4-derivative, m/z see H_{12}-derivative of **283A**. O-acetyl derivative.

Further physical and spectral properties of histrionicotoxins are presented in Table 11. The natural histrionicotoxins have all been levorotatory. The pKa values for histrionicotoxin and perhydrohistrionicotoxin as determined by titration were 9.0—9.3 (unpublished results). A Δ17,18-*trans*-isomer of histrionicotoxin has recently been isolated as a trace constituent from extracts of *Dendrobates histrionicus* (*255*). The occurrence of histrionicotoxins in dendrobatid frogs and in various populations of *Dendrobates histrionicus* has been discussed in terms of a biological character (*80, 188*). Histrionicotoxins are fairly widely distributed in frogs of the genus *Dendrobates*, occurring in half of the twenty species of Table 22. Often they occur at relatively high levels.

Table 11. *Other Physical and Chemical Properties of Histrionicotoxins* (*86, 246* — unless otherwise noted). Physical and spectral properties of (+) and (−) desamyl analogs of perhydrohistrionicotoxin are reported in (*246*)

Histrionicotoxin	
m.p. (HCl) 225 – 228°	
Ultraviolet (C₂H₅OH)	λ_{max} 224 nm, ε 22,300
Infrared (HCCl₃)	2100 cm⁻¹ (acetylene)
	1664 cm⁻¹ (*cis*-olefin)
Optical rotation (HCl)	$[\alpha]_D^{25} = -96.3°$
(c 1.0, C₂H₅OH)	
Dihydrohistrionicotoxin	
Ultraviolet (C₂H₅OH)	λ_{max} 226 nm, ε 24,700
Infrared (HCCl₃)	2100 cm⁻¹ (acetylene)
	1670 cm⁻¹ (diene)
Isodihydrohistrionicotoxin	
m.p. (HCl) 240 – 243°	
Ultraviolet (C₂H₅OH)	λ_{max} 225 nm, ε 8100
	λ_{max} 235 nm, ε 7200
Infrared (HCCl₃)	2100 cm⁻¹ (acetylene)
	1598 cm⁻¹ (allene)
Optical rotation	$[\alpha]_D^{25} = -35.3°$
(c 0.5, C₂H₅OH)	
Neodihydrohistrionicotoxin	
m.p. (HCl) 195 – 200°	
Ultraviolet	λ_{max} 224 nm, ε 17,300
Infrared (HCCl₃)	2100 cm⁻¹ (acetylene)
	1670 cm⁻¹ (diene)
Optical rotation (HCl)	$[\alpha]_D^{25} = -125.9°$
(c, 1.1, C₂H₅OH)	
Allodihydrohistrionicotoxin	
m.p. (HCl) 247 – 250°	
Optical rotation (HCl)	$[\alpha]_D^{25} = -43.4°$
(c 1.2, C₂H₅OH)	

Table 11 *(continued)*

Tetrahydrohistrionicotoxin	
Ultraviolet (C$_2$H$_5$OH)	λ_{max} 228 nm, ε 3900
Infrared (HCCl$_3$)	1670 cm^{-1} (diene)
Isotetrahydrohistrionicotoxin	
Ultraviolet (C$_2$H$_5$OH)	λ_{max} 228 nm, ε 19,200
Infrared (HCCl$_3$)	1950 cm^{-1} (allene)
	1665 cm^{-1} (diene)
Octahydrohistrionicotoxin	
Natural	
m.p. (HBr) 180 – 181°	*(253)*
Ultraviolet (C$_2$H$_5$OH): end absorption	
Synthetic racemate	*(109)*
m.p. (HCl) 151 – 154°	
Perhydrohistrionicotoxin	
"Natural"	
m.p. (HCl) 184 – 186°	
Optical rotation (HCl)	$[\alpha]_D^{25} = -34.6, -36.2°$
(c 1.0, C$_2$H$_5$OH, CHCl$_3$)	
Synthetic racemate	
m.p. (HCl) 159 – 161°	*(28)*
Synthetic (2S)	
Optical rotation (HCl)	$[\alpha]_D^{25} = -34.5, -36.0°$
(c 1.0, C$_2$H$_5$OH, CHCl$_3$)	
Synthetic (2R)	
Optical rotation (HCl)	$[\alpha]_D^{25} = +35.8°$
(c 1.0, CHCl$_3$)	

B. Syntheses

A synthesis of the azaspiro-[5.5]-undecan-8-ol ring system characteristic of the histrionicotoxins was first reported in 1975 in 30% overall yield by GOSSINGER *et al.* (Scheme XXI) (*116*). The carbon-13 magnetic resonance spectral peaks for azaspiro-[5.5]-undecan-8-ol were as follows: 67.5 (C-8), 52.0 (C-6), 40.8 (C-2), 40.0, 38.5, 37.4, 33.7, 27.5, 20.0, 15.9. The base peak in the mass spectra was at m/z 126. No major ion at m/z 96 was included in the mass spectral tabulation. Infrared and proton magnetic resonance data were reported. The synthetic approach was not successful for the synthesis of 7-n-butylazaspiro-[5.5]-undecan-8-ol, a desamyl-analog of perhydro-histrionicotoxin. This compound has now been synthesized by TAKAHASHI *et al.* (*246*).

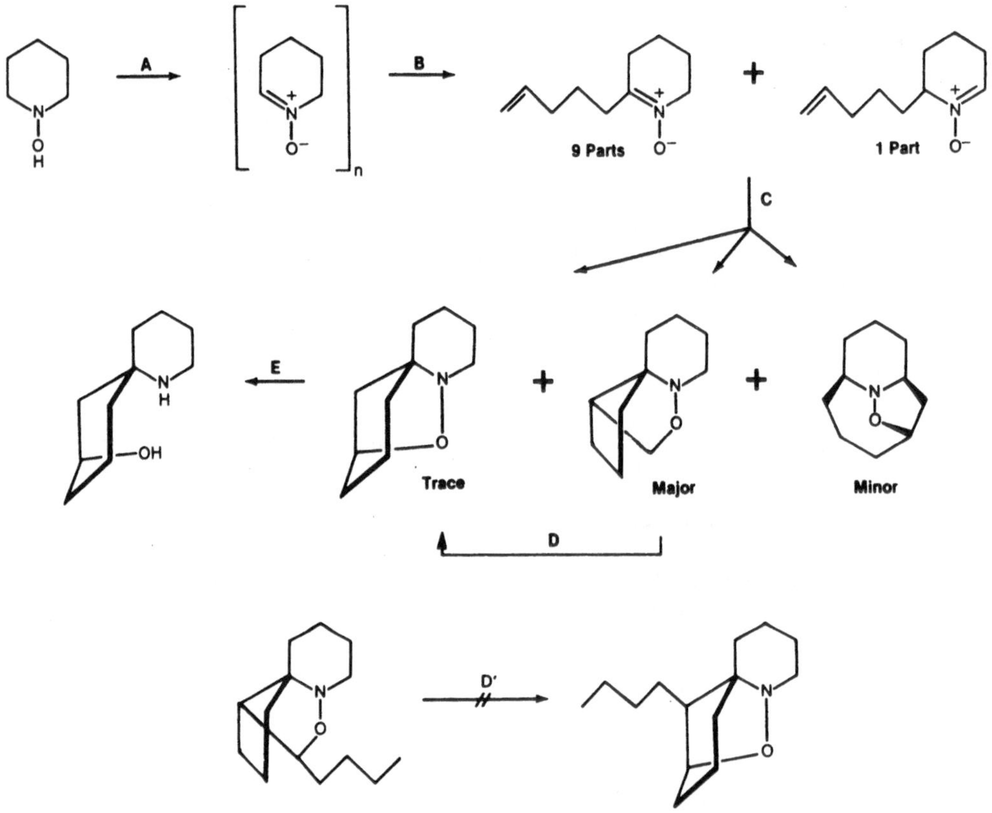

Scheme XXI. Synthesis of azaspiro-[5.5]-undecan-8-ol (*115*). The route proved unsuccessful for synthesis of 7-butylazaspiro-[5.5]-undecan-8-ol (see reaction **D′**). **A** i) HgO. **B** i) 5-Pentenyl MgBr, ii) HgO. **C** i) 110°. **D** i) 190°. **E** i) Raney nickel, H_2

A number of laboratories have conducted studies on the synthesis of perhydro-(dodecahydro)-histrionicotoxin. A spiroketolactam intermediate has figured prominently in such efforts. Various synthetic routes to this compound are presented in Scheme XXII. A number of syntheses of perhydrohistrionicotoxin and other analogs of histrionicotoxin have now been reported: These are summarized in Schemes XXIII—XXX.

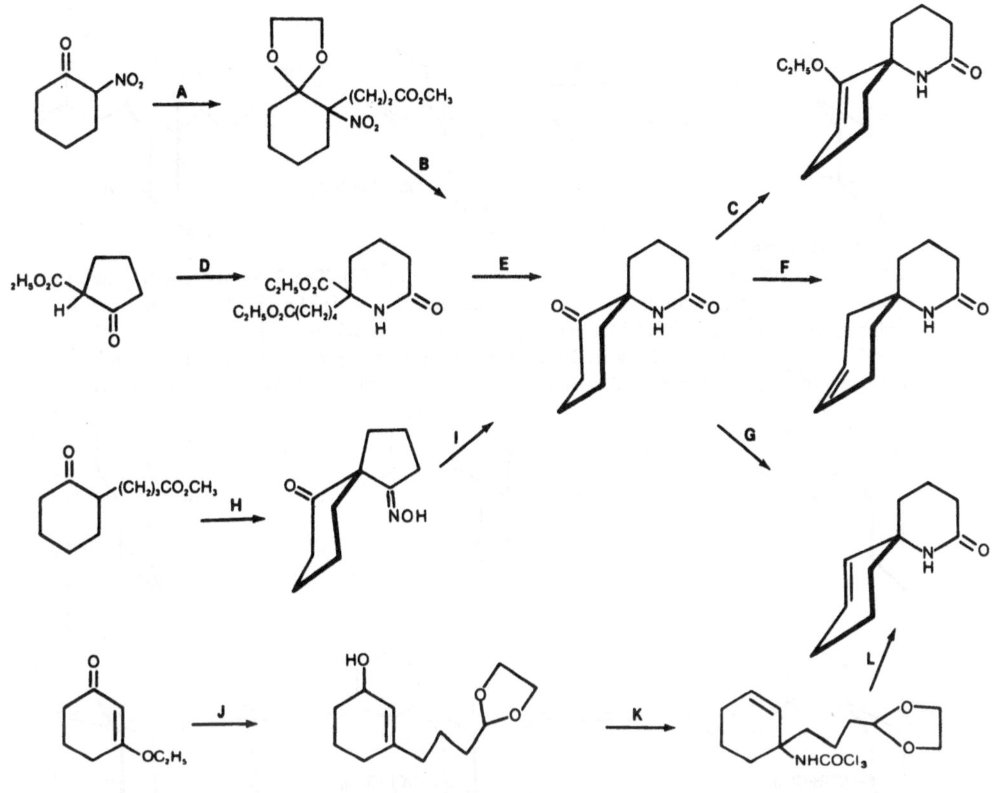

Scheme XXII. Synthesis of precursors for the elaboration of histrionicotoxins (reaction **ABC**, *28*; **DEFG**, *166*; **HI**, *44*; **JKL**, *205*). For further references to syntheses related to the azaspiro-[5.5]-undecane system see (*147*). **A** i) HOCH₂CH₂OH, H⁺, ii) CH₂=CHCO₂CH₃, iii) NaOH, iv) SOCl₂, v) CH₂N₂, vi) a. (C₂H₅)₃N, b. AgBF₄, CH₃OH. **B** i) Raney nickel, H₂, ii) F₃CCO₂H. **C** i) (C₂H₅O)₃CH, H⁺, ii) Heat. **D** i) a. KOH, b. Br(CH₂)₄CO₂C₂H₅, ii) HN₃. **E** i) NaH, ii) NaOH, heat. **F** i) Tosyl hydrazide, ii) Heat, hexamethylphosphoric triamide, NaH. **G** i) Tosyl hydrazide, ii) BuLi. **H** i) Polyphosphoric acid, ii) HOCH₂CH₂OH, H⁺, iii) NH₂OH, iv) H⁺. **I** i) Polyphosphoric acid. **J** i) (CH₂O)₂CH(CH₂)₃MgCl, ii) LiAlH₄. **K** i) NaH, ii) Cl₃CCN, iii) Heat. **L** i) H₂C₂O₄, ii) AgO, iii) OH⁻, iv) H⁺, CH₃OH

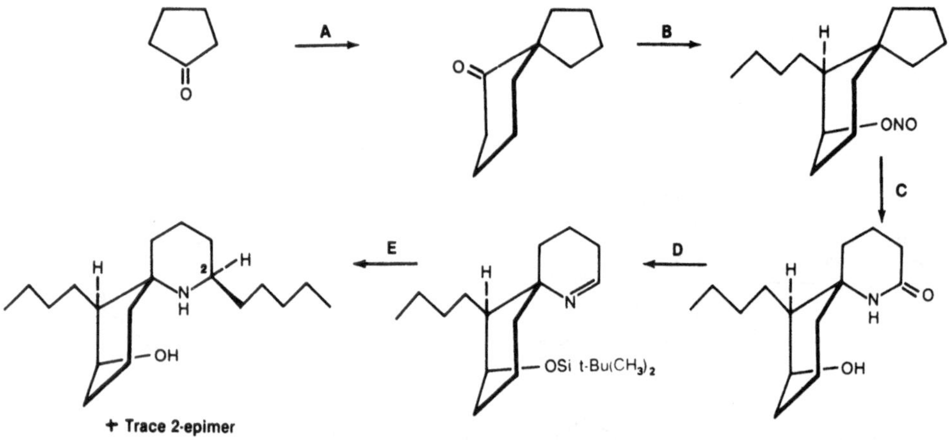

Scheme XXIII. Synthesis of dl-perhydrohistrionicotoxim (*72*). The 2-epimer was the major product if the free alcohol was used in reaction **E**. **A** i) Mg, TiCl₄ (Formation of cyclopentanone pinacol), ii) H₂SO₄, heat. **B** i) n-ButylLi, ii) SOCl₂, pyridine, iii) a. Diborane, b. H₂O₂, OH⁻, iv) ClNO. **C** i) hv, ii) Tosyl chloride, pyridine. **D** i) LiAlH₄, ii) (CH₃)₂t-BuSiCl, iii) N-Bromosuccinimide, iv) t-AmylOK. **E** i) n-AmylLi, ii) Tetrabutylammonium fluoride

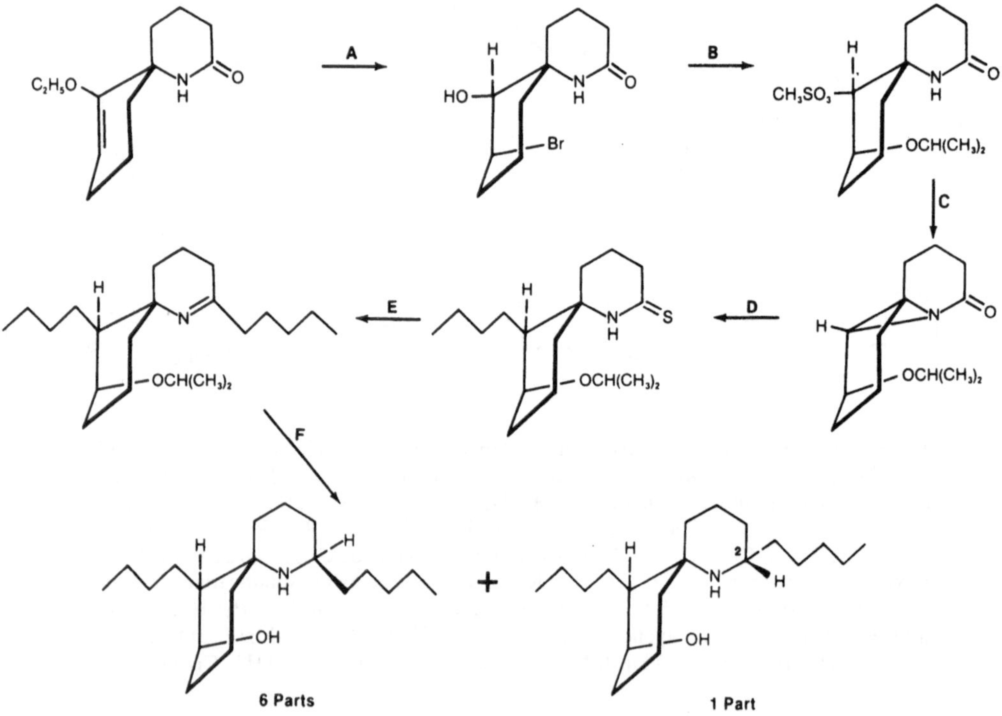

Scheme XXIV. Synthesis of dl-perhydrohistrionicotoxin (*28*). For starting material see Scheme XXII. Reduction in tetrahydrofuran or with NaBH₄ in methanol for reaction **F**. (ii) gave mainly the 2-epimer. **A** i) Br₂, ii) NaBH₄. **B** i) i-PrONa, ii) CH₃SO₂Cl, pyridine. **C** i) NaH. **D** i) (n-Bu)₂CuLi, ii) P₂S₅. **E** i) Triethyloxonium tetrafluoroborate, ii) n-PentylLi, Al(i-Bu)₂H. **F** i) BBr₃, ii) AlH₃, cyclohexane

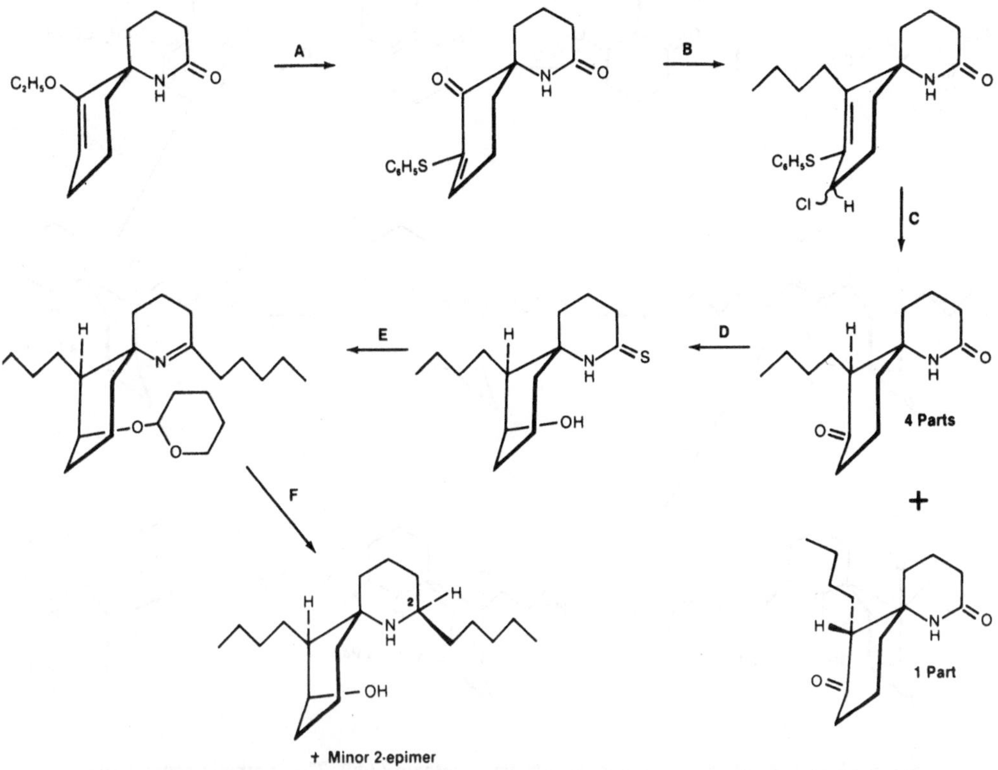

Scheme XXV. Synthesis of dl-perhydrohistrionicotoxin (28, 109). For starting material see Scheme XXII. **A** i) Phenylsulfenyl chloride. **B** i) n-ButylMgCl, ii) SOCl₂. **C** i) Zn, HCl, ii) HBr, iii) CH₃ONa (Isomerization of products). **D** i) Li or Ca, NH₃, ii) Ac₂O, pyridine, iii) P₂S₅, iv) OH⁻. **E** i) Dihydropyran, H⁺, ii) Triethyloxonium tetrafluoroborate, iii) n-PentylLi, Al(i-Bu)₂H. **F** i) H⁺, ii) AlH₃, cyclohexane

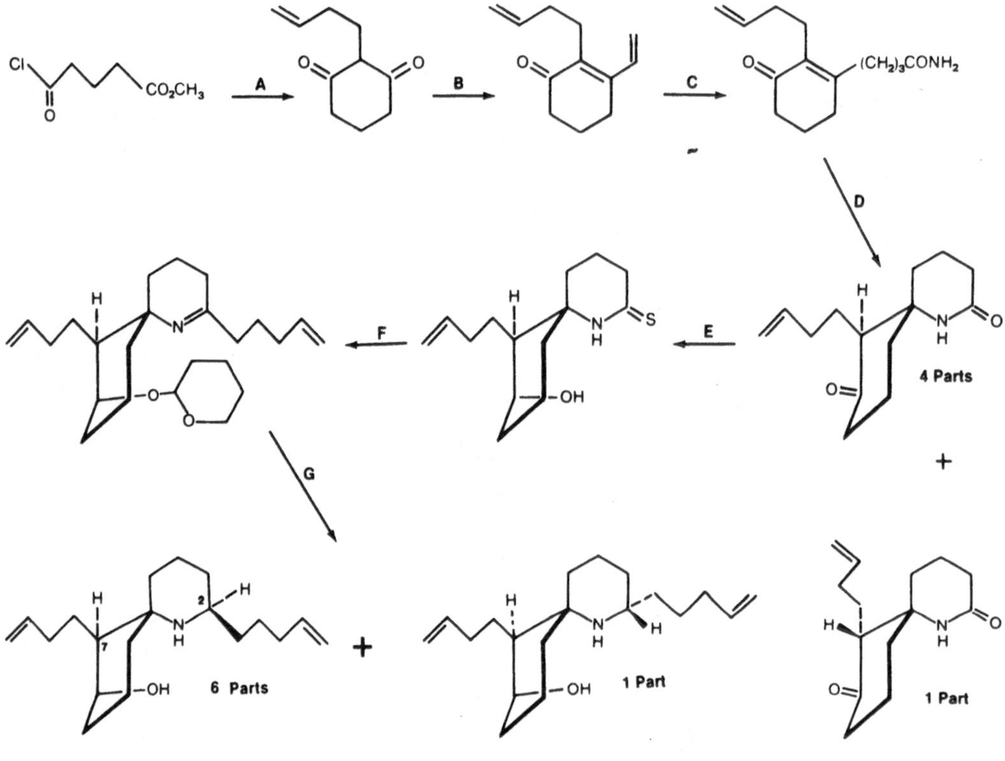

Scheme XXVI. Synthesis of dl-octahydrohistrionicotoxin (*109*). The synthetic route proved satisfactory for synthesis of perhydrohistrionicotoxin (dodecahydro) and for decahydro analogs containing either 7-butenyl-2-pentyl or 7-butyl-2-pentenyl side chains. **A** i) (4-Pentenyl)$_2$Cd, ii) t-BuOK. **B** i) H$^+$, C$_2$H$_5$OH, ii) CH$_2$=CHMgBr. **C** i) C$_2$H$_5$O$_2$CCH$_2$CONH$_2$, CH$_3$ONa, ii) NaOH, iii) 100°. **D** i) (C$_2$H$_5$O)$_3$CH, H$^+$, ii) CH$_3$ONa (Isomerization of products). **E** i) Li or Ca, NH$_3$, ii) Ac$_2$O, pyridine, iii) P$_2$S$_5$, iv) OH$^-$. **F** i) Dihydropyran, H$^+$, ii) Triethyloxonium tetrafluoroborate, iii) 4-PentenylLi, Al(i-Bu)$_2$H. **G** i) H$^+$, ii) AlH$_3$, cyclohexane

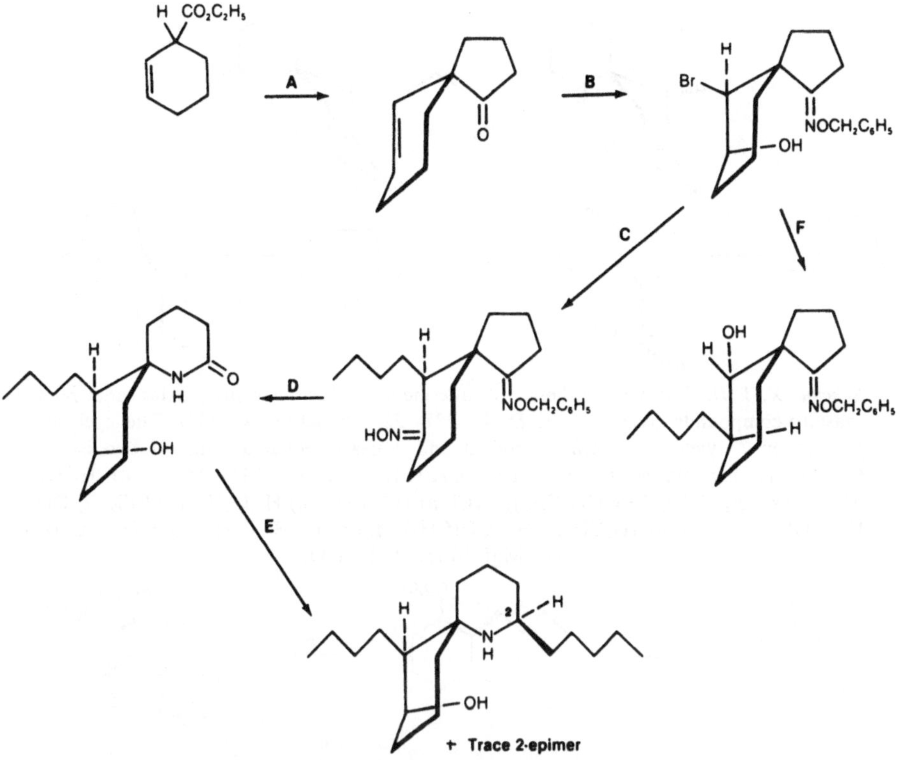

Scheme XXVII. Synthesis of dl-perhydrohistrionicotoxin (*73, 74*). **A** i) Br(CH₂)₄CO₂C₂H₅, ii) NaH, iii) H₂SO₄ – H₂O, heat. **B** i) NH₂OH, ii) C₆H₅CH₂Br, iii) N-Bromosuccinimide, H₂O. **C** i) CrO₃, ii) NH₂OH, iii) l-ButynylLi, iv) Pd/C, H₂. **D** i) TiCl₃, ii) Pd/C, H₂, iii) Na, NH₃, iv) Tosyl chloride, pyridine. **E** i) LiAlH₄, ii) (CH₃)₃t-BuSiCl, iii) N-Bromosuccinimide, iv) t-AmylOK, v) n-AmylLi, vi) Tetrabutylammonium fluoride. **F** i) n-PrONa, ii) Butyl-pentynylCuLi, hexamethylphosphoric triamide

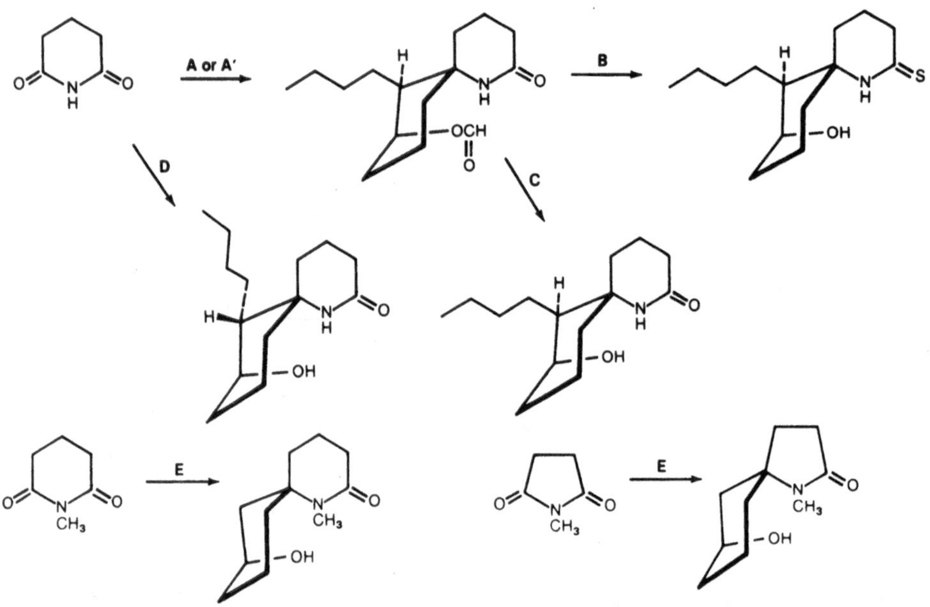

Scheme XXVIII. Further syntheses of intermediates for histrionicotoxins and related azaspiro-compounds (reactions **A, D, E,** *225—227*; Reaction **A′,** *101*). The cyclization of reaction **A′** yields the desired product and a nearly equal amount of two isomeric 6,5-azaspiro formate esters. **A** i) trans-$CH_3(CH_2)_3CH=CH(CH_2)_3MgBr$, ii) HCO_2H. **A′** i) trans-$CH_3(CH_2)_3CH=CH(CH_2)_3MgCl$, ii) H^+, heat, iii) HCO_2H. **B** i) P_2S_5, ii) OH^-. **C** i) OH^-. **D** i) *cis*-$CH_3(CH_2)_3CH=CH(CH_2)_3MgBr$, ii) HCO_2H, iii) OH^-. **E** i) 4-PentenylMgI, ii) HCO_2H, iii) OH^-

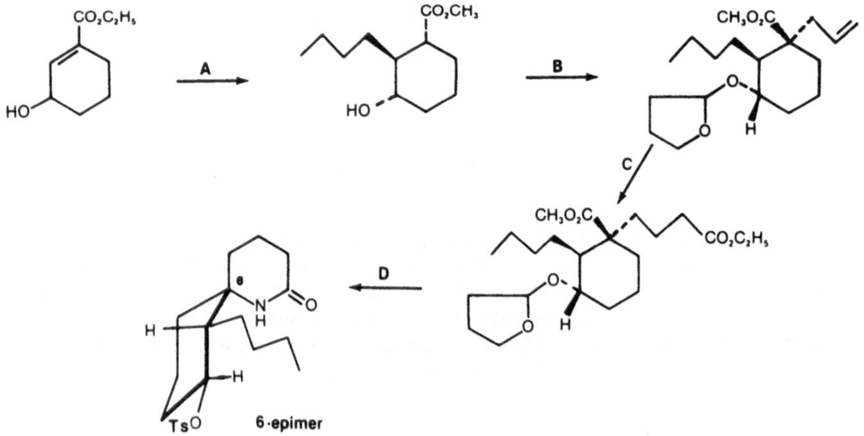

Scheme XXIX. Synthesis of a 6-epimer of the spirohydroxylactam intermediate for perhydro-histrionicotoxins (*141*). Reaction **B** was found to result in an inversion. The structure of the final lactam was confirmed by X-ray analysis. An alternative route afforded the spirohydroxyl-actam precursor of dl-perhydrohistrionicotoxin. **A** i) $(CH_3)_2$t-BuSiCl, ii) n-ButylCu, $AlCl_3$, iii) OH^-, iv) Ac_2O, v) NaOBr, vi) CH_2N_2. **B** i) Dihydrofuran, H^+, ii) $CH_2=CHCH_2Br$, base. **C** i) OsO_4, ii) triethylphosphonoacetate, iii) PtO_2, H_2. **D** i) KH, ii) OH^-, heat, iii) NH_2OH, iv) Tosyl chloride, pyridine

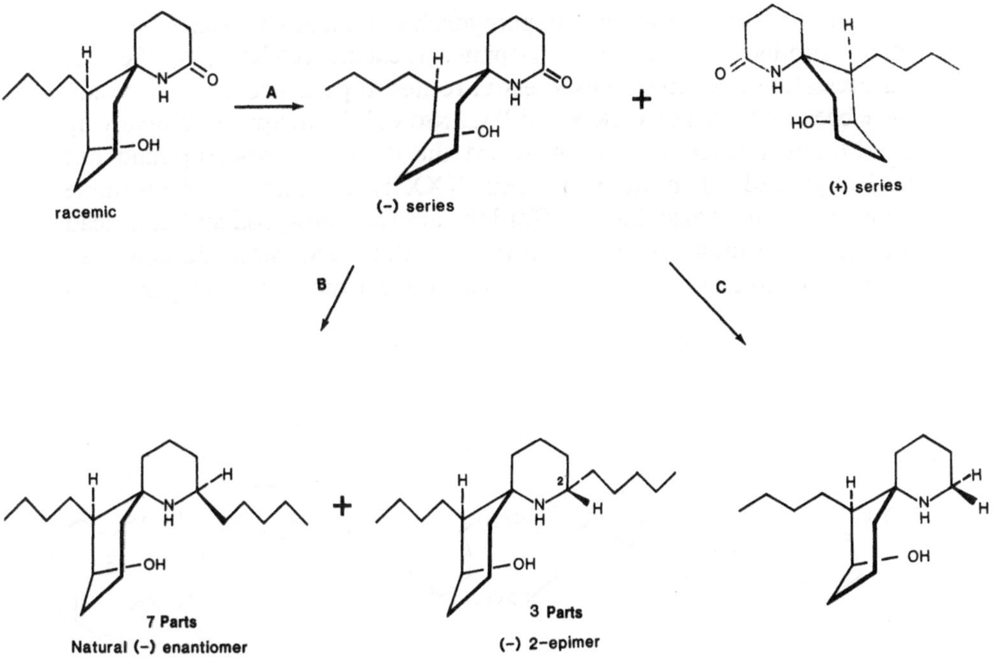

Scheme XXX. Synthesis of (+) and (−) perhydrohistrionicotoxins and (+) and (−) 2-desamylperhydrohistrionicotoxins (246). **A** i) a. S-(+)-α-Methylbenzylisocyanate, b. separation of diastereomeric esters, ii) NaOC$_2$H$_5$. **B** i) Ac$_2$O, pyridine, ii) P$_2$S$_5$, iii) NaOCH$_3$, iv) CH$_3$I, v) n-PentylMgBr, vi) AlH$_3$, cyclohexane. **C** i) Ac$_2$O, pyridine, ii) P$_2$S$_5$, iii) Raney nickel. Reaction **B** in the (+) series afforded the unnatural (+) enantiomer and its 2-epimer. Reaction **C** in the (+) series afforded the corresponding desamyl (+) enantiomer

"Natural" *l*-perhydrohistrionicotoxin has not been detected as yet in extracts from dendrobatid frogs, but instead has been prepared for biological studies by reduction of histrionicotoxin or dihydroisohistrionicotoxin. Reduction of a variety of histrionicotoxins with hydrogen and either palladium on charcoal or platinum oxide had yielded this common perhydro-derivative (86, 253). However, in one case with palladium on charcoal, histrionicotoxin yielded a partially reduced mixture of tetrahydro- and octahydro-derivatives which were isolated as a co-crystallized hydrochloride salt (253). Reduction of histrionicotoxin with hydrogen and Lindlar's palladium catalyst yielded tetrahydrohistrionicotoxin, hexahydrohistrionicotoxins, and a dihydrohistrionicotoxin in which the addition of hydrogen had occurred at the acetylenic terminus of the five carbon side chain (253). This dihydrohistrionicotoxin (**285 E**) has now been isolated as a trace constituent from extracts of *Dendrobates histrionicus* (255).

One synthetic approach to perhydrohistrionicotoxin yielded as the major product the racemic 2,7-epimer (Scheme XXXI) (75). Several unsuccessful or relatively unsuccessful routes to perhydrohistrionicotoxin have been discussed by COREY and BALANSON (70). An approach involving the Mannich reaction did not afford the desired spiro-compound but instead yielded a diene amine (Scheme XXXII). The methylated eneamine on acid treatment also did not afford the desired spiro-product but instead another diene amine. A route involving an ultimate Michael addition of an amine function (Scheme XXXIII, reaction F') did not yield perhydro-

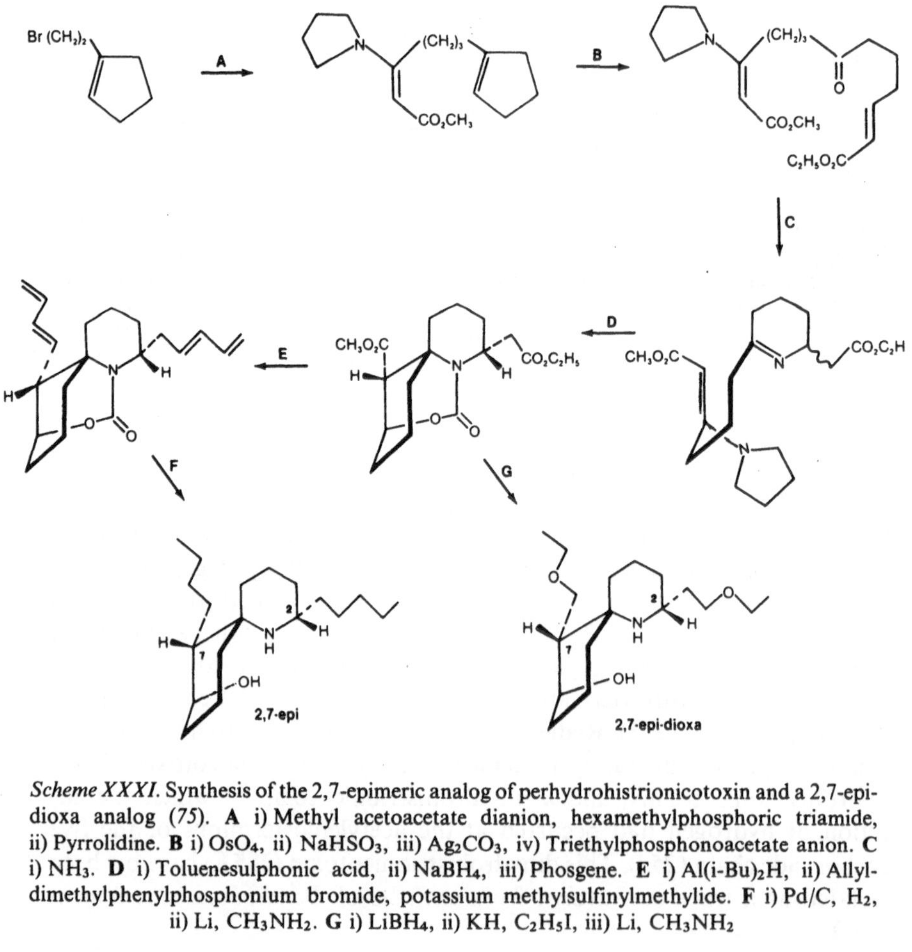

Scheme XXXI. Synthesis of the 2,7-epimeric analog of perhydrohistrionicotoxin and a 2,7-epi-dioxa analog (75). **A** i) Methyl acetoacetate dianion, hexamethylphosphoric triamide, ii) Pyrrolidine. **B** i) OsO₄, ii) NaHSO₃, iii) Ag₂CO₃, iv) Triethylphosphonoacetate anion. **C** i) NH₃. **D** i) Toluenesulphonic acid, ii) NaBH₄, iii) Phosgene. **E** i) Al(i-Bu)₂H, ii) Allyl-dimethylphenylphosphonium bromide, potassium methylsulfinylmethylide. **F** i) Pd/C, H₂, ii) Li, CH₃NH₂. **G** i) LiBH₄, ii) KH, C₂H₅I, iii) Li, CH₃NH₂

histrionicotoxin under a variety of acidic and basic conditions. Interestingly when a mesylate precursor was treated with ammonia for 18 hrs at 60° and "the crude reaction mixture reduced with sodium borohydride in methanol", a low yield (5%) of perhydrohistrionicotoxin was obtained (Scheme XXXIV). It was proposed that ammonia might first add to the enone and followed by displacement of the mesylate in an intramolecular reaction. Photolytic cyclization of the azide intermediate to yield the aziridine was not successful (Scheme XXXIII, reaction D). Instead an imine was obtained. Cyclization to the desired aziridine did occur on heating the azide undoubtedly *via* an intermediate triazoline (reaction F). The aziridine could be reduced in two steps to perhydrohistrionicotoxin and isomers. An attempt to prepare an analogous desbutylaziridine was unsuccessful (Scheme XXXV). Thermal treatment of the intermediate azide yielded instead the enone (reaction A). Reduction of the azide with Lindlar's catalysts, however, afforded a mixture of azaspiro compounds (reaction B).

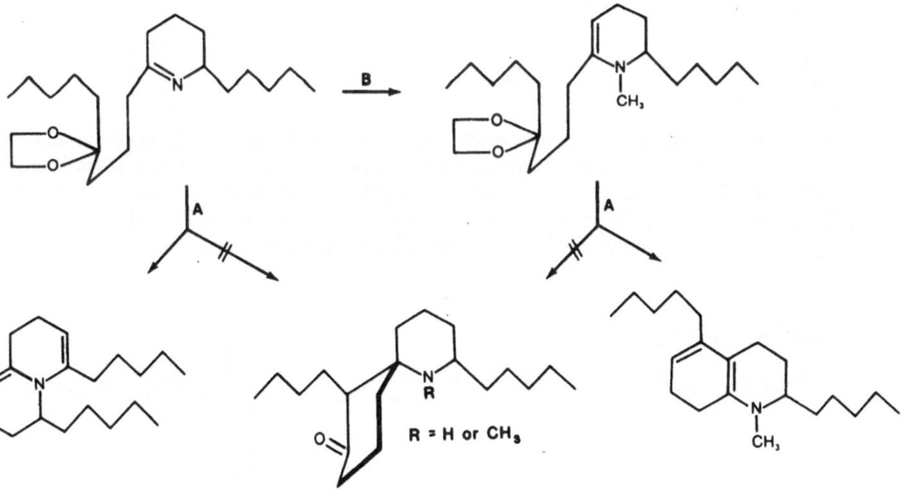

Scheme XXXII. Cyclizations leading to hexahydroquinolines rather than the desired azaspiro-[5.5]-undecan-8-ones (*70*). **A** i) H⁺. **B** i) Methylfluorosulphonate, ii) NaHCO₃

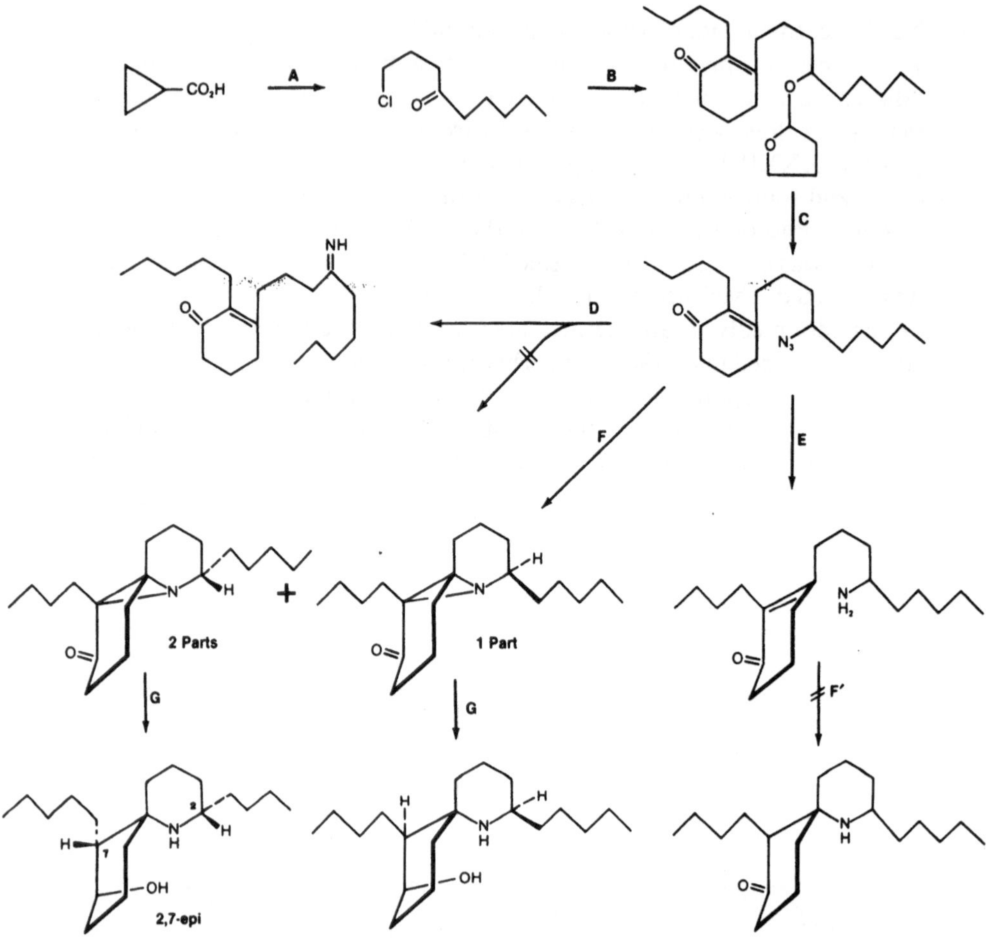

Scheme XXXIII. Further routes to dl-perhydrohistrionicotoxin and its 2,7-epimeric analog
(*70*). **A** i) n-PentylLi, ii) HCl. **B** i) LiAlH₄, ii) Dihydrofuran, iii) a. Li, b. 2-n-Butylcyclohexan-
1.3-diene methyl enol ether. **C** i) AcOH, ii) CH₃SO₂Cl, (C₂H₅)₃N, iii) LiN₃. **D** i) hv. **E**
i) Lindlar's Pd, H₂. **F** i) 136°. **F′** i) Acid or base and heat. **G** i) Li, NH₃, ii) NABH₄· Reaction **G**
undoubtedly affords both 7-butyl stereoisomers

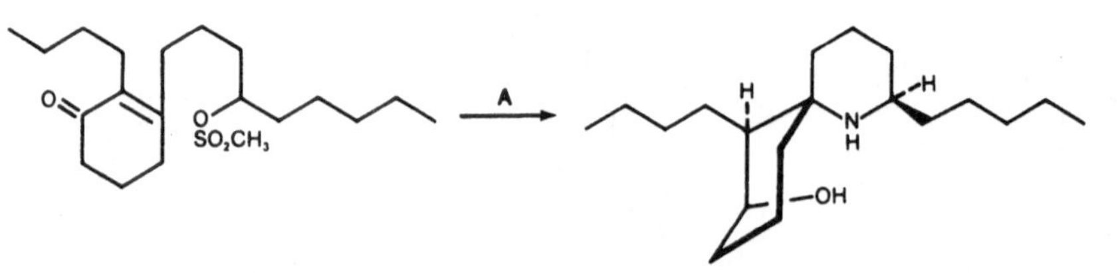

Scheme XXXIV. Synthesis of dl-perhydrohistrionicotoxin (*70*). For the source of the mesylate
see Scheme XXXIII (reaction C, ii).The yield of dl-perhydrohistrionicotoxin in reaction **A** was
about 5%. **A** i) NH₃, ii) NaBH₄

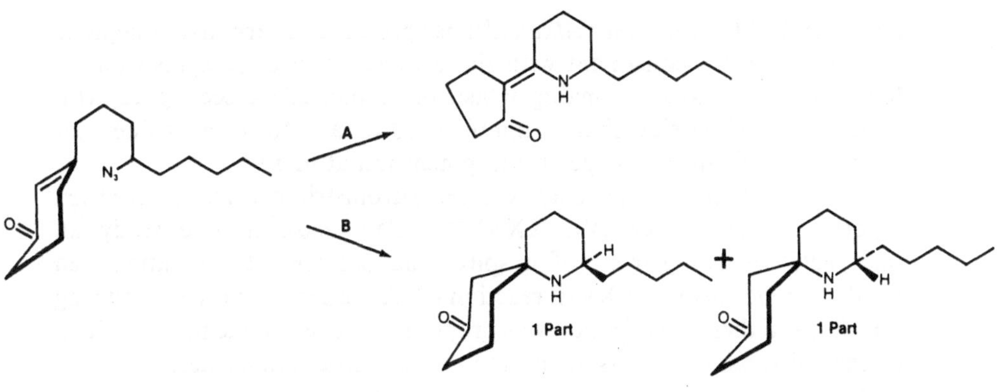

Scheme XXXV. Synthesis of potential precursors of a desamyl analog of perhydrohistrionico-toxin (*70*). Synthesis of the azide was similar to that depicted in Scheme XXXIII (reactions **A**, **B**, **C**). **A** i) 136°. **B** i) Lindlar's Pd, H$_2$

The synthesis of both enantiomers of perhydrohistrionicotoxin and of 2-desamylperhydrohistrionicotoxin was recently reported by TAKAHASHI *et al.* (Scheme XXX) (*246*). A spirohydroxylactam (cf. Scheme XXVIII) was converted to diastereomeric esters with an optically active isocyanate. After separation of the esters, optically active hydroxylactams were regenerated and converted to (−) and (+) perhydrohistrionicotoxins and (−) and (+) 2-desamylperhydrohistrionicotoxins.

N-Methylation of perhydrohistrionicotoxin has been investigated. Perhydrohistrionicotoxin was readily converted to the mono-N-methyl derivative with methyl iodide in acetonitrile or methanol (*253*). The mass spectrum of N-methylperhydrohistrionicotoxin was as follows: m/z 309 (45), 292 (18), 266 (24), 252 (14), 238 (100), 194 (42), 181 (48), 114 (63), 110 (47). The formation of the methiodide of perhydrohistrionicotoxin also was reported (*253*), but this conversion has not been reproduced under a variety of conditions with methyl iodide (unpublished results). It appears likely that conversion to the methiodide proceeds with great difficulty because: of hydrogen bonding from the alcohol function to the nitrogen (T. TAKAHASHI and A. BROSSI, personal communication).

Perhydrohistrionicotoxin has been converted to the deoxy-derivative in a sequence involving reaction with thionyl chloride in pyridine to yield a 7,8-dehydro-derivative followed by reduction with hydrogen and platinum oxide (*86, 255*). The mass spectrum of deoxyperhydrohistrionicotoxin was as follows: m/z 279 (28), 250 (12), 237 (16), 236 (71), 222 (10), 209 (15), 208 (39), 194 (12), 181 (40), 180 (83), 167 (100), 152 (16), 138 (8), 124 (9), 123 (10), 110 (23), 96 (80). Thus, the fragment ion at m/z 96 may be typical of both histrionicotoxins and deoxyhistrionicotoxins. Deoxyhistrionico-toxins may actually represent another class of dendrobatid alkaloids. The compound **219A** (see *Other Alkaloids*) has proton and carbon-13 magnetic resonance spectra consonant with those expected of a deoxyhistrionico-toxin, namely the 8-deoxy analog of histrionicotoxin **235A** (see Fig. 13) with 2-allyl and 7-vinyl side chains (unpublished data). However, it does not afford a significant mass spectral fragment ion at m/z 96.

Other unsuccessful approaches to perhydrohistrionicotoxins have been reported (Schemes XXXVI, XXXVII). (*259, 260*). In one study an unexpected rearrangement of a spiro ring occurred during attempted ketalization (Scheme XXXVII, reaction C). A sequence involving first ring opening and then recyclization was proposed. The rearrangement product, a hexahydroquinoline, was reduced to a decahydroquinoline derivative and acetylated.

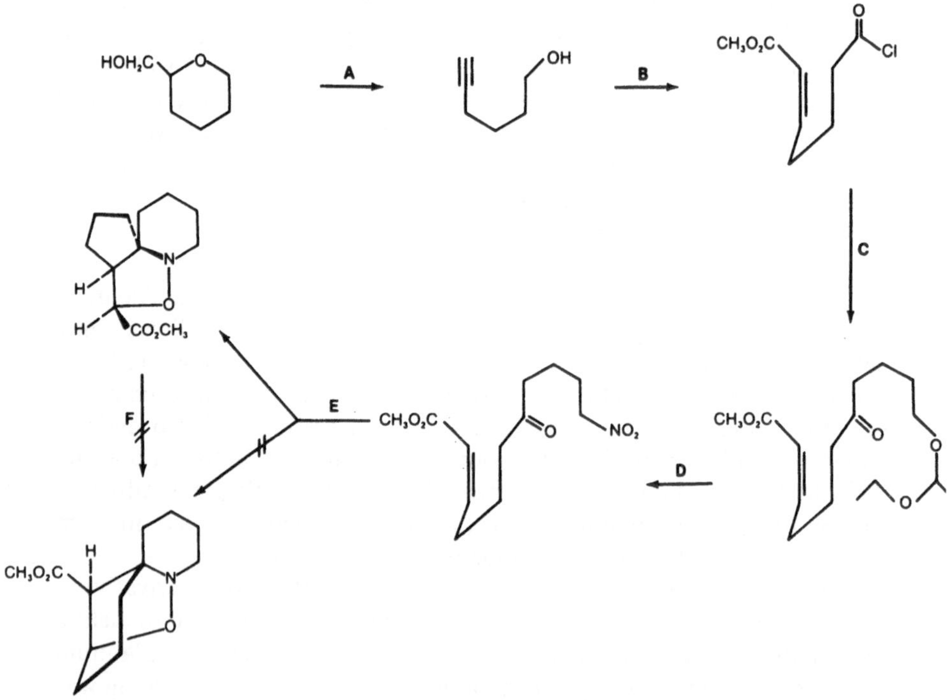

Scheme XXXVI. Cyclization leading to an azaspiro-[4.5]-decane system rather than the desired azaspiro-[5.5]-undecane (*259*). **A** i) SOCl₂, pyridine, ii) NaNH₂, NH₃. **B** i) Dihydropyran, ii) n-ButylLi, methyl chloroformate, iii) Toluenesulphonic acid, iv) Lindlar's Pd, H₂, v) CrO₃, vi) Oxalyl chloride. **C** i) CuI, hexamethylphosphoric triamide, ethyl 4-lithiobutylacetaldehyde acetal. **D** i) HCl, ii) CH₃SO₂Cl, (C₂H₅)₃N, iii) LiBr, iv) NaNO₂. **E** i) Zn, NH₄Cl. **F** i) Heat

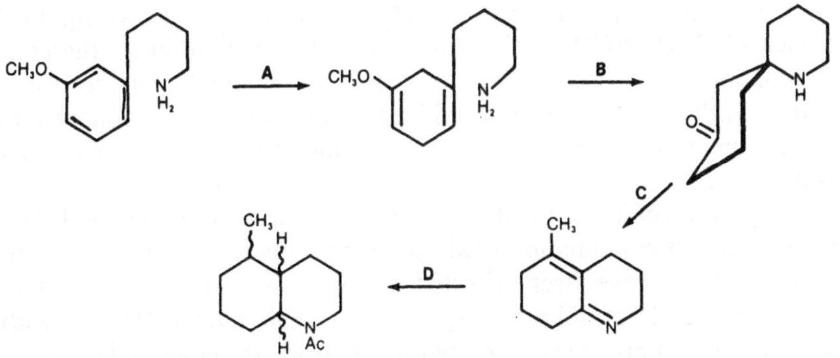

Scheme XXXVII. Synthesis of azaspiro-[5.5]-undecan-8-one and isomerization to a hexahydroquinoline (*260*). **A** i) Li, NH₃. **B** i) HCl. **C** i) HOCH₂CH₂OH, H⁺, heat. **D** i) Pd/C, H₂, ii) Ac₂O

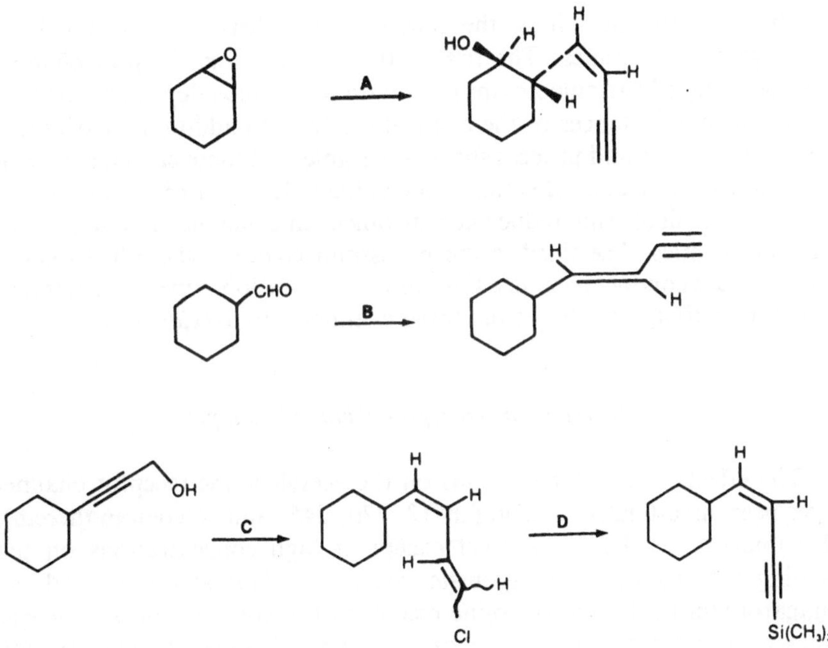

Scheme XXXVIII. Elaboration of eneyne side chains (reaction **A**, *130*; reaction **B, C, D**, *71*). **A** i) HC≡CLi, H₂NCH₂CH₂NH₂, ii) HC≡CSi(CH₃)₃, CuCl, H₂N(CH₂)₄NH₂, iii) Pd/BaSO₄, quinoline, H₂, iv) Tetrabutylammonium fluoride. **B** i) Trimethylsilylpropargylinetriphenyl-phosporane, ii) Tetrabutylammonium fluoride. **C** i) Pd/CaCO₃, H₂, ii) MnO₂, iii) Chloro-methylenetriphenylphosphorane. **D** i) CH₃Li, ii) (CH₃)₃SiCl

18*

A synthetic precursor, namely a tricarbonyl compound, $RCH_2CO(CH_2)_3CO(CH_2)_3COCH_2R$, which might serve in syntheses of both pumiliotoxin-C and histrionicotoxin class alkaloids, has been proposed (275). Initial cyclization could lead to a cyclohexenone, followed by reductive amination and then Michael addition to yield a histrionicotoxin precursor.

The synthesis of histrionicotoxin itself as yet has not been accomplished. Histrionicotoxin contains terminal cis-enyne units in its side chains and synthetic efforts have been directed towards the elaboration of such moieties in simple cyclohexane systems (Scheme XXXVIII) (71, 130). Recently, these approaches were used to elaborate the enyne side chain of gephyrotoxin (107).

C. Biological Activity

1. Action and Mechanisms

Histrionicotoxins inhibit the function of at least three channels in electrogenic membranes: The first is the acetylcholine receptor channel complex, where histrionicotoxins in a time and stimulus-dependent manner reduce the conductances of the channel (7, 170). In addition, histrionicotoxins in a more rapid process shorten the time that these channels remain open. The second locus of action is the voltage-dependent sodium channel where histrionicotoxins reduce conductances in a manner reminiscent of local anesthetics. The third is the potassium channel where histrionicotoxins reduce conductances in a time and concentration-dependent manner. Structure activity profiles at the three channels differ *(vide infra)*.

2. Acetylcholine-receptor Channel Complex

The effects of histrionicotoxins on the acetylcholine receptor channel have been investigated thoroughly (7, 170, 245; and references therein). Histrionicotoxins have no effect except at high concentrations on the binding of nicotinic antagonists such as *d*-tubocurarine and α-bungarotoxin to the acetylcholine receptor (88). Histrionicotoxin instead appears to interact with a site on the open channel, thereby causing blockade of the channel. Certainly the time- and stimulus-dependence of the blockade by histrionicotoxin provides evidence suggestive of an interaction with a site on the open channel. However, a more rapid effect of histrionicotoxins on the acetylcholine receptor channel complex is to shorten the decay time, thus appearing to increase the rate of inactivation of the channel. It is felt that this effect may reflect binding of histrionicotoxin to a site on the closed form of the channel (*31*). The dual effects of histrionicotoxin on the acetylcholine receptor channel complex are reversible on washing. Indeed, the stimulus-dependent blockade of receptor-elicited increases of conductances is reversed in the continued presence of the alkaloid. However, the blockade quickly reinstates itself during a second series of stimuli. "Histrionicotoxin" blocked the carbamylcholine-elicited increase in conductance with a purified acetylcholine receptor complex reconstituted into vesicule preparations (*277*). Thus, such a purified and reconstituted preparation still retained a functional histrionicotoxin-site. Although the compound used was stated to be histrionicotoxin, it appears more likely that it was dl-perhydrohistrionicotoxin.

Certain evidence suggests that histrionicotoxin in a stimulus-dependent manner converts the receptor-channel complex to a state similar if not identical with the so-called desensitized state. Certainly the presence of low concentrations of histrionicotoxins increases the nicotinic agonist-elicited phenomenon known as desensitization (*48*). In addition, the presence of histrionicotoxin, like classical desensitization, increases the apparent affinity of nicotinic agonists for the receptors (*48, 154, 155*). In one study, in electroplax membranes, synthetic dl-perhydrohistrionicotoxin was reported to have no significant effect on desensitization processes (*98*). Further studies will be needed to fully define the mechanism of action of histrionicotoxins on acetylcholine receptor channel complexes and to determine whether these alkaloids also interact with other chemosensitive channels.

Histrionicotoxin has similar effects in neuromuscular preparations from *Rana pipiens* and from the frog *Dendrobates histrionicus,* which produces the alkaloid (*10*). Muscle contractions are potentiated and muscle action potentials prolonged in both species, but blockade of neuromuscular transmission requires higher concentrations in the *Dendrobates* than in the *Rana*. Thus, whatever the site of action of histrionicotoxin it is retained in the frog that produces the alkaloid. It should be pointed out that histrionicotoxin has very low toxicity and that the "toxin" designation is indeed a misnomer. Thus, the dendrobatid frog probably has no reason to completely protect itself from the action of this class of alkaloids.

Histrionicotoxin noncompetitively inhibits nicotine-stimulated secretion of catecholamines from adrenal medulla cells (*164*), suggesting similar interactions of the alkaloid with the nicotinic acetylcholine receptor channel complex in both adrenal medulla and in muscle.

Isodihydrohistrionicotoxin at 4 µM has little or no effect on muscarinic responses in ileum or atrium (*183*). Histrionicotoxins are, however, noncompetitive muscarinic antagonists in neural cell lines (*49*), suggestive of interaction at a nonreceptor site, perhaps at a channel component.

Perhydrohistrionicotoxin is a very weak antagonist of binding of quinuclidinyl benzilate to muscarinic receptors in rat brain membranes (30).

Histrionicotoxins antagonize both acetylcholine and glutamate-elicited excitation of central neurons (155), and have been cited as antagonizing glutamate-responses in invertebrate muscles (245). Histrionicotoxin also has a depressant effect on spontaneous activity of cortical and spinal neurons (114, 155). Perhydrohistrionicotoxin at very low concentrations (< 0.1 μM) blocks endplate currents elicited by iontophoretic acetylcholine in rat neuromuscular preparations, while having no effect on spontaneous miniature endplate currents or endplate currents evoked by nerve stimulation (6, 14). Further studies will be required to clarify the reason for the remarkable potency of perhydrohistrionicotoxin versus responses to iontophoretically applied acetylcholine.

Recently, structure activity relationships have been reported for various histrionicotoxins (245). The onset of the time and stimulus-dependent blockade of evoked conductances by the alkaloids increases three-fold in a series including histrionicotoxin, and dihydro, tetrahydro, octahydro and perhydro-derivatives. Thus, potency in terms of onset of action apparently decreases with increased saturation in the side chains. A related synthetic compound, namely azaspiro-[5.5]-undecan-8-ol, which has no side chains has very low activity. The 2-desamyl analog of perhydrohistrionicotoxin is less active than perhydrohistrionicotoxin itself (246). In the *Torpedo* electroplax, histrionicotoxin is slightly more potent than isodihydrohistrionicotoxin in blocking carbamylcholine-elicited depolarization (155). Quantitative assessment of potencies of histrionicotoxins at acetylcholine receptor channel-complexes, at sodium channels *(vide infra)* and at potassium channels *(vide infra)* is difficult since the effects are dependent not only on concentration but also on time and rate of stimulation. The potencies of four histrionicotoxins with respect to reducing endplate currents are as follows with the isotetrahydrohistrionicotoxin most active: Isotetrahydrohistrionicotoxin > histrionicotoxin ≈ perhydrohistrionicotoxin > octahydrohistrionicotoxin. Histrionicotoxins increase the rate of closing of channels but there are insufficient data to provide a good rank order for potency. Isotetrahydrohistrionicotoxin appears, however, more potent than octahydrohistrionicotoxin, histrionicotoxin and perhydrohistrionicotoxin (7). The azaspiro-[5.5]-undecan-8-ol is very weak in reducing endplate currents and appears to *decrease* the rate of closing of channels (245). The latter effect might, however, have been due to high concentrations of ethanol, the solvent in which the azaspiro-compound was added.

It was stated (245)that the synthetic dl-perhydrohistrionicotoxin was *not* significantly less potent in reducing endplate currents and increasing rate of closure of channels than optically active l-perhydrohistrionicotoxin obtained by reduction of histrionicotoxin or isodihydrohistrionicotoxin. The chemical literature (28, 72) also contains statements suggesting that dl-perhydrohistrionicotoxin had biological activity "identical" with that of l-perhydrohistrionicotoxin derived from histrionicotoxin or dihydroiso-histrionicotoxin. Further studies were however, required to determine whether the d- or l-enantiomers of perhydrohistrionicotoxin did have equivalent potences in biological systems. Such studies have now been initiated with d- and l-enantiomers of perhydrohistrionicotoxin. The natural l-enantiomer appears to have a potency identical with or even slightly less than the unnatural d-enantiomer (246). The two enantiomeric 2-desamylperhydrohistrionicotoxins are also equipotent in blocking indirect evoked muscle twitch. The 2-desamyl analogs did appear to have lower activity than perhydrohistrionicotoxin.

The 2,7-epimer of perhydrohistrionicotoxin has been synthesized (75). While no biological data were available for this racemic compound, the corresponding "dioxa"-2,7 epimer (see Scheme XXXI, reaction G) was stated to have "ca one fourth the biological activity of the naturally derived perhydrohistrionicotoxin". It was stated that the activity was ascertained "using murine nerve/diaphragm preparation by Dr. E. X. ALBUQUERQUE and associates". A synthetic 2-pentenyl-7-butyl analog of histrionicotoxin caused a time-dependent inhibition of responses to iontophoretic acetyl-choline in neuromuscular preparations (27).

All studies of the various histrionicotoxins indicate that the nature of the side chains and even their configuration on the spiro-ring system is not critical to biological activity. Furthermore, the lack of stereoselectivity for d- and l-perhydrohistrionicotoxin suggests a two-point binding of the alkaloids to the histrionicotoxin-site on the acetylcholine receptor-channel complex.

3. Binding of Radioactive Perhydrohistrionicotoxin

A radioactive perhydro-derivative of histrionicotoxin ($[^3H]H_{12}$-HTX; (10, 92) has proved a very useful tool for investigation of the binding sites for histrionicotoxin in membranes and microsacs from Torpedo electroplax preparations (see Table 12 for references). Its use has not been extended to muscle preparations where the density of nicotinic receptor channel complexes are many fold lower than in Torpedo electroplax. Indeed, no specific binding could be detected in the electroplax of the Egyptian electric fish (Malapterurus electricus) where the density of nicotinic receptors also appear very low (97). In membranes from Torpedo electroplax $[^3H]H_{12}$-HTX binds to a site with an affinity constant (K_D) of about 0.4 µM. The density of the binding sites for $[^3H]H_{12}$-HTX is apparently about twice that of the acetylcholine-binding sites in electroplax membranes. The rate of binding of [³H]H$_{12}$-HTX is greatly accelerated by the presence of nicotinic agonists although the apparent binding constant and density of sites appear unaffected (*31, 93*). Nicotinic antagonists have no effect on binding of [³H]H$_{12}$-HTX, providing evidence that the histrionicotoxin-binding site is not at the receptor.

The binding sites for [³H]$_{12}$-HTX were first thought to be mainly associated with a 43,000 mol. wt. subunit of the acetylcholine-receptor channel complex (*241, 242*). This subunit was subsequently found to be non-essential to binding of [³H]H$_{12}$-HTX or to function of the receptor-channel complex (*100, 195*). Recent studies indicate that [³H]H$_{12}$-HTX binds to a 66,000 mol. wt. subunit (*204, 273*). The receptor subunit has a mol. wt. of 40,000.

Binding of [³H]H$_{12}$-HTX to electroplax membranes is antagonized by a number of compounds including local anesthetics, amantadine, phencyclidine, quinacrine, antidepressants, and tetraethylammonium ions (see Table 12 for references). On the basis of electrophysiological studies in neuromuscular preparations, such compounds appear to have actions similar to those of the histrionicotoxins. As yet, the histrionicotoxins are the most potent ligands for this site. The potency of four histrionicotoxins versus binding of [³H]H$_{12}$-HTX is probably in the following order: perhydrohistrionicotoxin > isotetrahydrohistrionicotoxin > octahydrohistrionicotoxin > histrionicotoxin (*89*). This rank ordering does not correspond to the rank order with respect to reduction of indirect elicted twitch or of endplate currents in frog neuromuscular preparations (*vide supra*). The azaspiro-[5.5]-undecan-8-ol is a very weak antagonist of binding of [³H]H$_{12}$-HTX. Perhydrohistrionicotoxin antagonizes the binding of a radioactive local anesthetic, meproadifen (*167*), and of radioactive phencyclidine (*95*) to electroplax membranes.

4. Sodium Channels

The histrionicotoxins show only weak interactions with voltage-dependent sodium channels (*245*). Histrionicotoxins do reduce the rate of rise of action potentials in muscle. The profile of activity for seven histrionicotoxins in this regard is as follows: isotetrahydrohistrionicotoxin > tetrahydrohistrionicotoxin > perhydrohistrionicotoxin > isodihydrohistrionicotoxin > neodihydrohistrionicotoxin > octahydrohistrionicotoxin > histrionicotoxin. Thus histrionicotoxin appeared the least potent alkaloid of this class with respect to classical local anesthetic activity. In frog sciatic nerve, histrionicotoxin at 500 μM causes only a slight blockade of action potentials (*114*).

References, pp. 326—340

5. Potassium Channels

Histrionicotoxins at low concentrations in a time and stimulus-dependent manner block conductances through the potassium channels responsible for termination of action potentials in nerve and muscle (245). Delayed rectification, a reliable indicator of increased potassium conductances, is blocked by histrionicotoxin in frog muscles. The antagonism of potassium channels by histrionicotoxins results in a marked prolongation of action potentials. The profile of activity for eight histrionicotoxins with respect to prolongation of muscle action potentials is as follows: tetrahydrohistrionicotoxin > histrionicotoxin > neodihydrohistrionicotoxin > isodihydrohistrionicotoxin ≅ octahydrohistrionicotoxin > isotetrahydrohistrionicotoxin > perhydrohistrionicotoxin > N-methylperhydrohistrionicotoxin. Thus, the perhydrohistrionicotoxins are the least potent with respect to blockade of potassium channels. Presynaptic prolongation of action potentials by histrionicotoxins will result in greater release of neurotransmitter. Postsynaptically, the resultant enhanced amount of released acetylcholine and the histrionicotoxin-induced prolongation of muscle action potentials probably accounts for the initial potentiation of muscle contractures seen during stimulation in the presence of histrionicotoxins. Soon however, the time and stimulus-dependent blockade of acetylcholine-receptor complexes ensues and nerve stimulation no longer elicits a response. However, direct stimulation of muscle still causes an enhanced contraction due probably to the prolonged action potentials in muscle. The potentiative effect of histrionicotoxin on directly elicited muscle contractions is reversible after washing, but this reversal appears to occur on a slower time course than the reversal of the blockade of the acetylcholine receptor channel by histrionicotoxin. The relative potency of histrionicotoxins with respect to potentiation of muscle contracture has not been quantitated. Perhydrohistrionicotoxin which is very weak with respect to prolongation of action potentials (vide supra) causes only a marginal potentiation of muscle contracture (245).

6. Summary

The various studies on pharmacological activity of histrionicotoxins are summarized in Table 12. It should be repeated that this class of alkaloids are not really "toxins" since even at quite high dosages in mammals, toxicity is not elicited (80, 81). There are several general and specific reviews related to the pharmacology and mechanism of action of histrionicotoxins (7, 9, 89, 90, 125).

Table 12. *Effects of Histrionicotoxins on Biological Systems*

System	Species	Comments	References
Whole animal	Mouse	Relatively nontoxic, $LD_{50} \gg 5$ mg/kg	(80, 81)
Neuromuscular preparation (striated)	Frog, rat, mouse	Blockade neuromuscular transmission and either stimulus or iontophoretically evoked end-plate currents; Enhancement of rate of closing of endplate channels; Prolongation of action potentials; Potentiation of directly evoked muscle contraction. Slight local anesthetic activity at sodium channels. Minimal direct interaction with the acetylcholine receptor.	(6, 7, 9, 10, 14–16, 27, 88, 91, 93, 114, 155, 170, 177, 183, 245, 246, 267)
Muscle cells	Chick	Apparent enhancement of nicotinic agonist-induced desensitization	(48)
Neuronal cells	Mouse neuroblastoma, mouse neuroblastoma-glioma hybrid	Noncompetitive muscarinic antagonist	(49)
Cardiac preparations	Guinea pig atria	No effect on spontaneous activity or on carbamyl-choline responses	(183)
Smooth muscle	Guinea pig ileum	No effect on tension or on acetylcholine-elicited contracture	(183)
Adrenal medulla cells	Bovine	Noncompetitive inhibition of nicotine-elicited release of catecholamines	(164)
Brain	Cat	Depression spontaneous activity cortical and spinal neurons; Blockade excitatory responses to acetylcholine and glutamate	(114, 154)
	Rat synaptosomes	No effect sodium conductances	(179)
	Rat membranes	Virtually no effect on binding of muscarinic antagonist	(30)

Table 12 *(continued)*

System	Species	Comments	References
Electroplax	*Torpedo* membranes, microsacs	Antagonizes binding certain local anesthetics and other agents. Blockade of carbamylcholine-elicited ion flux and depolarization. Increases affinity of agonist for receptor. Binding [^{3}H]H$_{12}$-HTX: Antagonism by local anesthetics and other agents; acceleration by cholinergic agonists	*(2, 3, 9, 10, 13, 20, 21, 29, 30, 31, 37, 38, 88 – 100, 154, 167, 195, 204, 224, 241, 242, 250, 257, 258, 273, 276, 277)*
	Malapterurus	Binding [^{3}H]H$_{12}$-HTX not detectable	*(97)*

V. Gephyrotoxins

A. Structures

In addition to a variety of histrionicotoxins, extracts from the frog *Dendrobates histrionicus* yielded a tricyclic compound with a empirical formula of $C_{19}H_{29}NO$ (*253*). Analysis of the proton magnetic resonance spectrum revealed the presence of a *cis*-CH$_2$CH = CHC ≡ CH side chain like that in histrionicotoxin. However, the mass spectrum of this compound referred to initially as HTX-D was distinctively different than that of the histrionicotoxins. The base peak resulted from the loss of what proved to be a − CH$_2$CH$_2$OH moiety. This moiety formed O-acetyl and O-*p*-bromobenzoyl derivatives (*86*). Hydrogenation of the compound led to a hexahydro-derivative now containing a saturated five carbon side chain. X-ray analysis of a crystal of the hydrobromide salt revealed structure as that of a novel tricyclic alkaloid (*86*). The absolute configuration was assigned as [1S,3aS,5aS,6S(Z),9aR,10R] - dodecahydro - 6 - (2 - penten - 4 - yl) - pyrrolo - [1,2 - a] - 1 - quinoline - 1 - ethanol (Fig. 14). This absolute configuration has been recently questioned based on a synthesis of this enantiomer from L-glutamate (*107*). The synthetic compound was dextrorotatory while the natural compound was levorotatory. The synthesis is depicted in Scheme XLIII (see p. 289) for further details). No satisfactory explanation of the contradictory x-ray and chemical data is apparent. The compound isolated from *Dendrobates histrionicus* and initially referred to as HTX-D was renamed gephyrotoxin (Greek: gephyra meaning bridge) referring in a literal sense to the bridge formed presumably by addition of the nitrogen function to one side chain of a precursor 2,6-disubstituted piperidine (*vide supra*, also *80*). A minor congener, dihydrogephyrotoxin was also isolated and differed only in the presence of a *cis*-$CH_2CH = CH - CH = CH_2$ side chain. Hydrogenation of these gephyrotoxins yielded perhydrogephyrotoxin. Properties of these two alkaloids are reported below in the usual format (see p. 236).

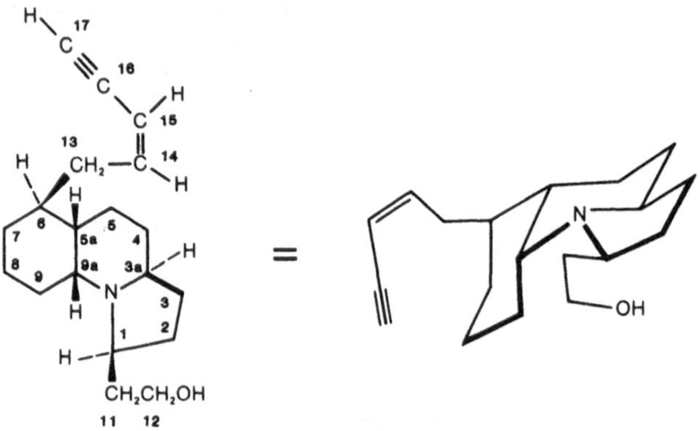

Fig. 14. Structure of gephyrotoxin. The absolute configuration is based on x-ray analysis of hydrobromide and is depicted on the right (*86*). The natural compound is the l-enantiomer (*107*). It is, therefore, inexplicable that synthesis from L-pyroglutamic acid (*107*) of the enantiomeric configuration revealed by x-ray analysis should yield the d-enantiomer. FUJIMOTO and KISHI (*107*) proposed that the true configuration of natural gephyrotoxin must be opposite to that revealed by x-ray analysis. The x-ray analysis was done on gephyrotoxin isolated from the same population of frogs as the material shown by optical rotation measurements to be the levorotatory enantiomer. However, the two samples of gephyrotoxin were from extracts obtained in different years

Gephyrotoxin Class

(Perhydrobenzoindolizidines = dodecahydropyrrolo-[1,2-a]-quinolines)

287C. Gephyrotoxin, $C_{19}H_{29}NO$, 0.20, 218°, m/z 287(5), 286(3), 242(100), 222(45), 122(14). H_6-derivative, m/z 293(5), 292(3), 250(16), 248(100), 222(32). O-acetyl derivative, m/z 329(3), 264(45), 242(100).

289B. Dihydrogephyrotoxin, $C_{19}H_{31}NO$, 0.25, 217°, m/z 289(4), 288(3), 244(100), 222(49), H_4-derivative.

The proton magnetic resonance spectral assignments for gephyrotoxin are reported in Fig. 15. Detailed mass spectra are tabulated for gephyrotoxin, dihydrogephyrotoxin and perhydrogephyrotoxin in Table 13. Other properties of the gephyrotoxins including carbon-13 magnetic resonance peaks are documented in Table 14. Distribution of the two tricyclic gephyrotoxins in dendrobatid frogs is reported in Table 22. The tricyclic gephyrotoxins have a very limited distribution occuring only in certain populations of *Dendrobates histrionicus*.

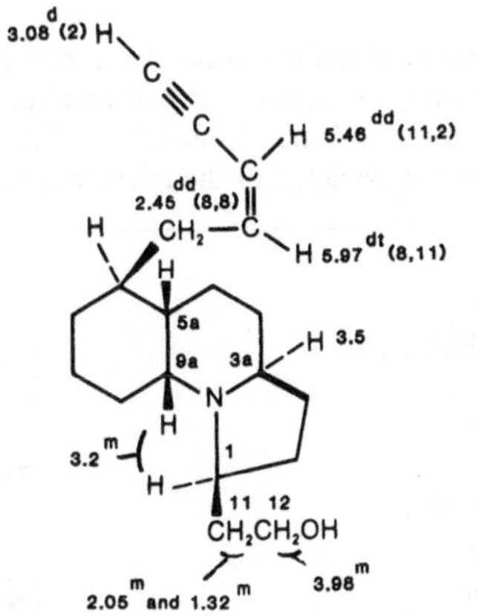

3.08 d (2) H

C

C

H 5.46 dd (11,2)

C

2.45 dd (8,8) CH₂—C

H 5.97 dt (8,11)

H

H

5a

9a　3a

H 3.5

N

H

3.2 m

H

1

11　12

CH₂CH₂OH

2.05 m and 1.32 m　　3.96 m

Fig. 15. Proton magnetic resonance spectral assignments for gephyrotoxin (86). The spectrum (CDCl₃, 100 MHz) of natural gephyrotoxin is depicted in this reference. A spectrum (C₆D₆, 100 MHz) of synthetic dl-gephyrotoxin is available in the microfilm supplement of (108). The spectrum (CDCl₃, 250 MHz) of dl-perhydrogephyrotoxin is available in the microfilm supplement of (211). Chemical shifts for protons in perhydrogephyrotoxin are as follows: 3a-H 3.5m; 5a-H 2.55m; 1-H and 9a-H 3.3m; 11-H (1H) 2.05m; 12-(2H) 4.00d (data from 207, 211)

Table 13. *Mass Spectra of Tricyclic Gephyrotoxins* (86, 253)
Major loss of $-CH_2CH_2OH$

	Gephyro-toxin	Dihydro-gephyrotoxin	Perhydro-gephyrotoxin
$C_{19}H_{35}NO$			293 (5)
$C_{19}H_{34}NO$			292 (3)
$C_{19}H_{31}NO$		289 (4)	
$C_{19}H_{30}NO$		288 (3)	
$C_{19}H_{29}NO$	287 (12)		
$C_{19}H_{28}NO$	286 (11)		
$C_{19}H_{28}NO$	270 (2)		
$C_{17}H_{32}N$			250 (16)
$C_{17}H_{31}N$			249 (15)
$C_{17}H_{30}N$			248 (100)
$C_{17}H_{27}N$		245 (21)	
$C_{17}H_{26}N$		244 (100)	
$C_{17}H_{25}N$	243 (21)		
$C_{17}H_{24}N$	242 (100)		
$C_{14}H_{24}NO$	222 (50)	222 (49)	222 (32)
$C_{14}H_{22}N$		204 (9)	
$C_{14}H_{20}N$		202 (10)	
$C_{12}H_{18}NO$	192 (7)		192 (3)
$C_{12}H_{16}NO$	190 (4)	190 (3)	
$C_{12}H_{18}N$	176 (5)	176 (4)	176 (3)
$C_{10}H_{14}N$	148 (5)	148 (7)	148 (4)
$C_8H_{12}N$	122 (21)	122 (20)	122 (9)
$C_5H_{11}N$	85 (27)		
C_5H_9N	83 (40)		

Table 14. *Other Physical and Chemical Properties of Tricyclic Gephyrotoxins*
Data on gephyrotoxin from (*86*); data on optical rotations from (*107*); data on perhydrogephyrotoxin, including infrared data, from (*207*). The infrared spectrum of perhydrogephyrotoxin is depicted in the microfilm supplement of (*108*)

Gephyrotoxin	
m.p. 231 – 232°	
Ultraviolet (C₂H₅OH)	λ_{max} 225 nm, ε 8400
Infrared (HCCl₃)	2120 cm^{-1} (acetylene)
Synthetic d-enantiomer	
Optical rotation	$[\alpha]_D^{25}$ +50.0°
(c 1.0, C₂H₅OH)	
Natural l-enantiomer	
Optical rotation	$[\alpha]_D^{25}$ −51.5°
(c 1.0, C₂H₅OH)	
Perhydrogephyrotoxin	
Synthetic	
Infrared	
Proton magnetic resonance	4.0^{d,t} (11, 2), 3.7^m, 3.2^m
(CDCl₃, 90 MHz)	
Carbon-13 magnetic resonance	60.1, 56.8, 56.3, 51.1, 41.8,
(C₆D₆)	40.5, 33.4 (two carbons), 32.9,
	32.0, 31.6, 28.6, 27.2, 27.1,
	25.9, 23.5, 21.1, 17.3, 14.8

No further tricyclic gephyrotoxins have been identified in extracts from dendrobatid frogs. However, three simpler bicyclic indolizidine alkaloids have been included in the gephyrotoxin class (*80*). These indolizidines, like the tricyclic gephyrotoxins, contain in a literal sense a bridge formed by the addition of the nitrogen to one side chain of a 2,6-disubstituted piperidine. The proposed structures were tentative in nature and were based on a number of considerations. First the compounds were saturated bicyclic alkaloids. They did not form N-acetyl derivatives. For each alkaloid, two fragments dominated the mass spectrum. These corresponded to loss of either a four carbon or a three carbon side chain. In the simplest member of this group, gephyrotoxin **223 AB** (C₁₅H₂₉N), the side chains were −C₃H₇ and −C₄H₉, while gephyrotoxin **239 AB** (C₁₅H₂₉NO) contained a hydroxyl group in the 3-carbon side chain and gephyrotoxin **239 CD** (C₁₅H₂₉NO) contained a hydroxy group in the 4-carbon side chain (*80*). On the basis of this information and biosynthetic considerations, the three compounds were proposed to be 3,5-disubstituted indolizidines with the four carbon substituent at the 3-position and the five carbon substituent at the 5-position (see Fig. 16). At the time this proposal was made the structures of all of the simpler dendrobatid alkaloids, *i.e.* those of pumiliotoxin C, the histrionicotoxins, and the tricyclic gephyrotoxins, were consonant with a biosynthetic origin from 2,6-disubstituted piperidines with side chains having either lengths of three, five, seven or nine carbons (see Fig. 27). Therefore, it was thought that the three bicyclic gephyrotoxins might have arisen from a 2,6-disubstituted piperidine with a five carbon side chain and a seven carbon side chain. The nitrogen atom was then to have formed the gephyrotoxin "bridge" by addition to the seven carbon side chain.

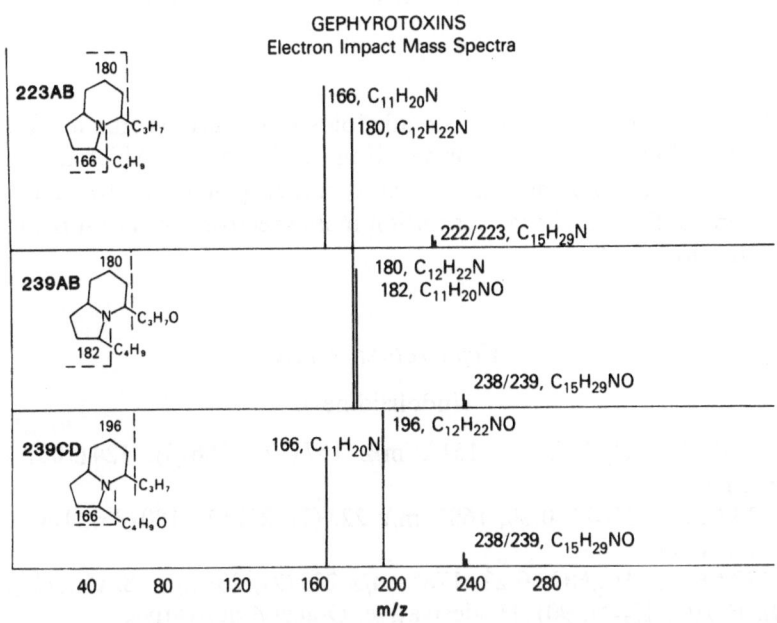

Fig. 16. Mass spectra and proposed tentative structures for gephyrotoxins **223 AB**, **239 AB**, and **239 CD**. The positions of the hydroxyl group in the side chains of **239 AB** and **239 CD** are unknown. The structure of **223 AB** has been confirmed and is shown in Fig. 17

The tentative structure for gephyrotoxin **223 AB** has now been confirmed by comparison with synthetic compounds *(vide infra)*. Gephyrotoxin **223 AB** proved to be identical by gas chromatographic and mass spectral analysis with (5E,9E)-3-butyl-5-propylindolizidine (Fig. 17) *(244)*. The other three possible diastereomers of this structure differed in gas chromatographic *(244, vide infra)* and mass spectral (Table 15) properties from gephyrotoxin **223 AB**. A fourth compound has been recently included in the bicyclic gephyrotoxin class (MYERS and DALY, unpublished results). It would appear, based on a mass spectral properties and biosynthetic considerations, that this compound **167 B** is 5-n-propylindolizidine.

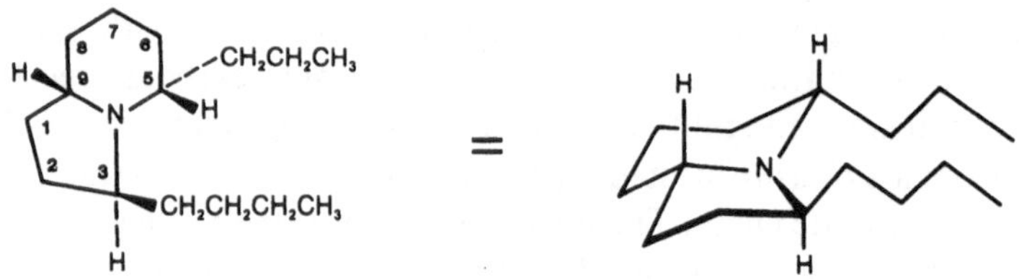

Fig. 17. Structure of gephyrotoxin **223AB**. The absolute configuration is unknown, but is arbitrarily depicted so that the absolute configuration at C-5 corresponds to C-2 of pumiliotoxin C which also has a n-propyl substituent

Properties of the four bicyclic alkaloids in the gephyrotoxin class are reported below in the usual format. Empirical formulas which are based only on analogy and chemical and chromatographic properties and have not been confirmed by high resolution mass spectrometry are surrounded by quotation marks.

Gephyrotoxin Class

(Indolizidines)

167B. "$C_{11}H_{21}N$", −, 151°, m/z 167(12), 166(5), 124(100). H_0-derivative.

223AB. $C_{15}H_{29}N$, 0.30, 160°, m/z 223(1), 222(2), 180(85), 166(100). H_0-derivative.

239AB. $C_{15}H_{29}NO$, 0.22, 178°, m/z 239(2), 238(3), 182($C_{11}H_{20}NO$, 100), 180($C_{12}H_{22}N$, 90). H_0-derivative. O-acetyl derivative.

239CD. $C_{15}H_{29}NO$, 0.16, 179°, m/z 239(4), 238(3), 196($C_{12}H_{22}NO$, 100), 166($C_{11}H_{20}N$, 60). H_0-derivative. O-acetyl derivative.

Properties of synthetic gephyrotoxin **223AB** are provided in Table 16. Gephyrotoxins **239AB** and **239CD** (Fig. 16) presumably have the same stereochemistry as gephyrotoxin **223AB** (see Fig. 17). This seems reasonable. However, the stereochemistry of the simpler dendrobatid alkaloids has not been completely predictable (see discussion of possible biosynthetic pathways, p. 320. Gephyrotoxin **223AB** and its bicyclic hydroxy-congeners **239AB** and **239CD** have not as yet been isolated in sufficient quantities for crystallization, optical rotations or nuclear magnetic resonance spectra. Distribution of these bicyclic (indolizidine) alkaloids in dendrobatid frogs is reported in Table 22. Gephyrotoxins **223AB, 239AB,** and **239CD** occur in only two sibling species, *Dendrobates histrionicus* and *Dendrobates occultator*, while **167B** has only been detected in an undescribed species from Panama.

B. Syntheses

The gephyrotoxins, both of the tricyclic type such as gephyrotoxin itself
— and of the bicyclic type — such as gephyrotoxin **223 AB** — have been the
subject of recent synthetic efforts. A stereoselective synthesis of perhydrogephyrotoxin has been designed, taking advantage in a key step of a 2-
azonia-(3,3)-sigmatropic rearrangement (Scheme XXXIX) (*207*). This
synthesis provided racemic perhydrogephyrotoxin in a 3% yield
(Scheme XL), while a more recent synthesis using a dienamido Diels-Alder
strategy provided perhydrogephyrotoxin in an overall yield of 15%
(Scheme XLI) (*211*). The l-epimer was also obtained as a minor by-product
in the latter synthesis. Another synthesis has recently been described and led
to both perhydrogephyrotoxin and gephyrotoxin itself (Scheme XLII)
(*108*). Perhydrogephyrotoxin was obtained in an overall yield of 9% and
gephyrotoxin in 5%. Optically active gephyrotoxin was also synthesized by
this route using L-pyroglutamic acid (*107*) in order to arrive at the absolute
configuration reported on the basis of an x-ray analysis of the hydrobromide salt of natural gephyrotoxin (*86*). The product (Scheme XLIII)
which was expected to be identical with natural gephyrotoxin was, however,
the d-enantiomer, while the natural compound was the l-enantiomer (see
Table 14). The absolute configuration of the synthetic gephyrotoxin, based
on L-glutamate as a starting material, appears beyond question. Similarly,
the x-ray analysis of a crystal of natural gephyrotoxin HBr presents no
ambiguities and clearly gives the same absolute configuration (I. KARLE,
personal communication). Further studies will be needed to resolve this
apparently irreconcilable set of results. It should be noted that x-ray
crystallography and optical rotations were with samples of gephyrotoxin
isolated from the same population of frogs but in different years. For the
present, the absolute configuration of gephyrotoxin based on x-ray analysis
is presented in Fig. 14 as representing the natural compound.

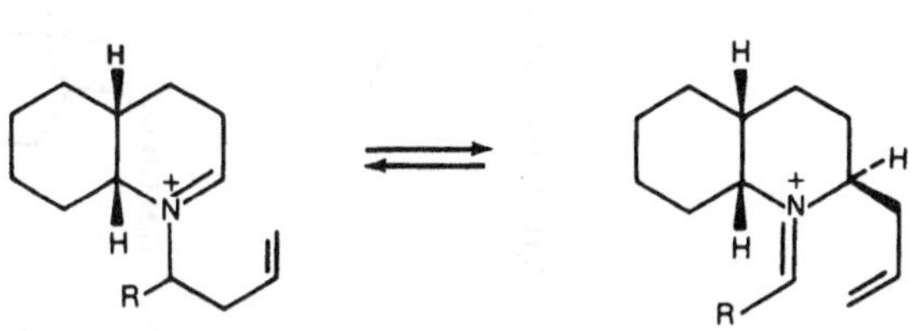

Scheme XXXIX. Sigmatropic rearrangements (*207*)

References, pp. 326—340

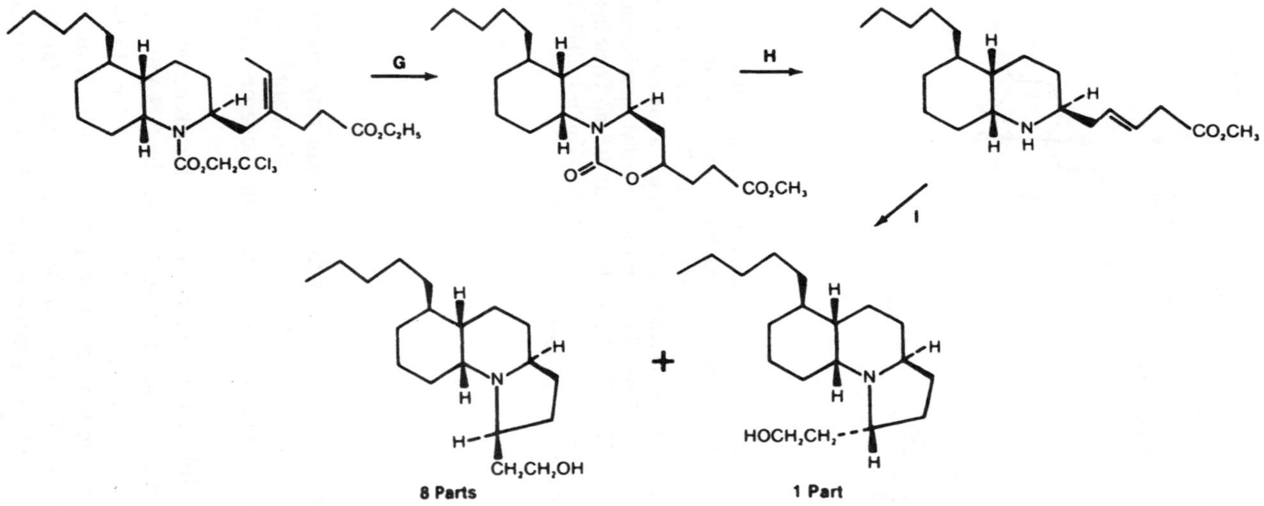

Scheme XL. Synthesis of dl-perhydrogephyrotoxin (207). The last step of reaction B yields a 1 : 1 mixture of stereoisomers. Reaction C yields a 3 : 2 mixture of stereoisomers. A i) 110°. B i) Formylmethylenetriphenylphosphorane, ii) Pyridinium toluenesulphonate, CH_3OH, iii) Pd/C, H_2, iv) 2-Methoxy-2-methylbutenal, v) $NaBH_4$. C i) H^+, 80°. D i) $C_6H_5CH_2Br$, ii) NaOH. E i) $NaBH_4$, ii) $CH_3C(OC_2H_5)_3$, H^+. F i) C_2H_5SH, BF_3, $(C_2H_5)_2O$, ii) $C_6H_5CH_2Br$, ii) NaOH. E i) $NaBH_4$, ii) $CH_3C(OC_2H_5)_3$, H^+. F i) C_2H_5SH, BF_3, $(C_2H_5)_2O$, ii) $ClCO_2CH_2CCl_3$. G i) O_3, ii) $NaBH_4$, iii) LiOH, iv) NaH, v) CH_2N_2. H i) a. $LiN(i-Pr)_2$, b. $NaIO_4$, ii) Zn, AcOH. I i) CH_3ONa, ii) $LiAlH_4$

19*

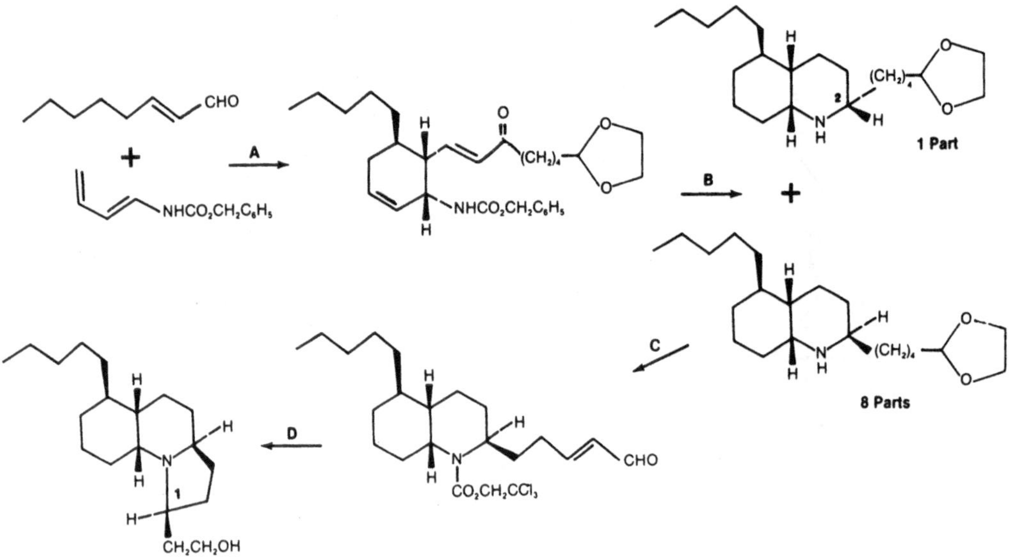

Scheme XLI. Synthesis of dl-perhydrogephyrotoxin (211).Reaction **B** (iii) with LiAlH₄ affords 1 part of the 2-epimer and 8 parts of the desired isomer, while with NaBH₄ the major product is the 2-epimer. Enol silylation is effected with the triflate in reaction **C** (iii). Reaction **D** is best conducted with the dimethylacetal which is formed in reaction **D** (i). This reaction yield some of the 1-epimer. **A** i) Heat, ii) Lithium dimethyl 2-oxo-7-(ethylenedioxy)-heptyl phosphonate. **B** i) Pd/C, CF₃CO₂H, H₂, ii) NaOH, iii) LiAlH₄. **C** i) ClCO₂CH₂CCl₃, ii) HCl, iii) Trimethylsilyl triflate, iv) Pb(OAc)₂. **D** i) Pyridinium toluenesulphonate, CH₃OH, ii) Zn/PbO, iii) HCl, iv) CH₃ONa, v) NaBH₄

Initial studies on yet another route into the gephyrotoxin ring system were reported in 1980 (Scheme XLIV) (123). This route was then applied to a successful *de facto* synthesis of perhydrogephyrotoxin (Scheme XLV) (124).

The first of the four possible enantiomeric pairs of stereoisomers corresponding to the structure proposed for gephyrotoxin **223AB** was synthesized by MACDONALD in 1980 (Scheme XLVI) (175). It might be noted that the final catalytic reduction (reaction D) was anomalous in yielding only one isomer (racemic 5E,9E) in a 38% yield. The catalytic reduction was also anomalous in a related synthesis of the pharaoh ant trail pheromone, again yielding only one of the expected two isomeric 3-butyl-5-methylindolizidines (175). Initial gas chromatographic comparison of the

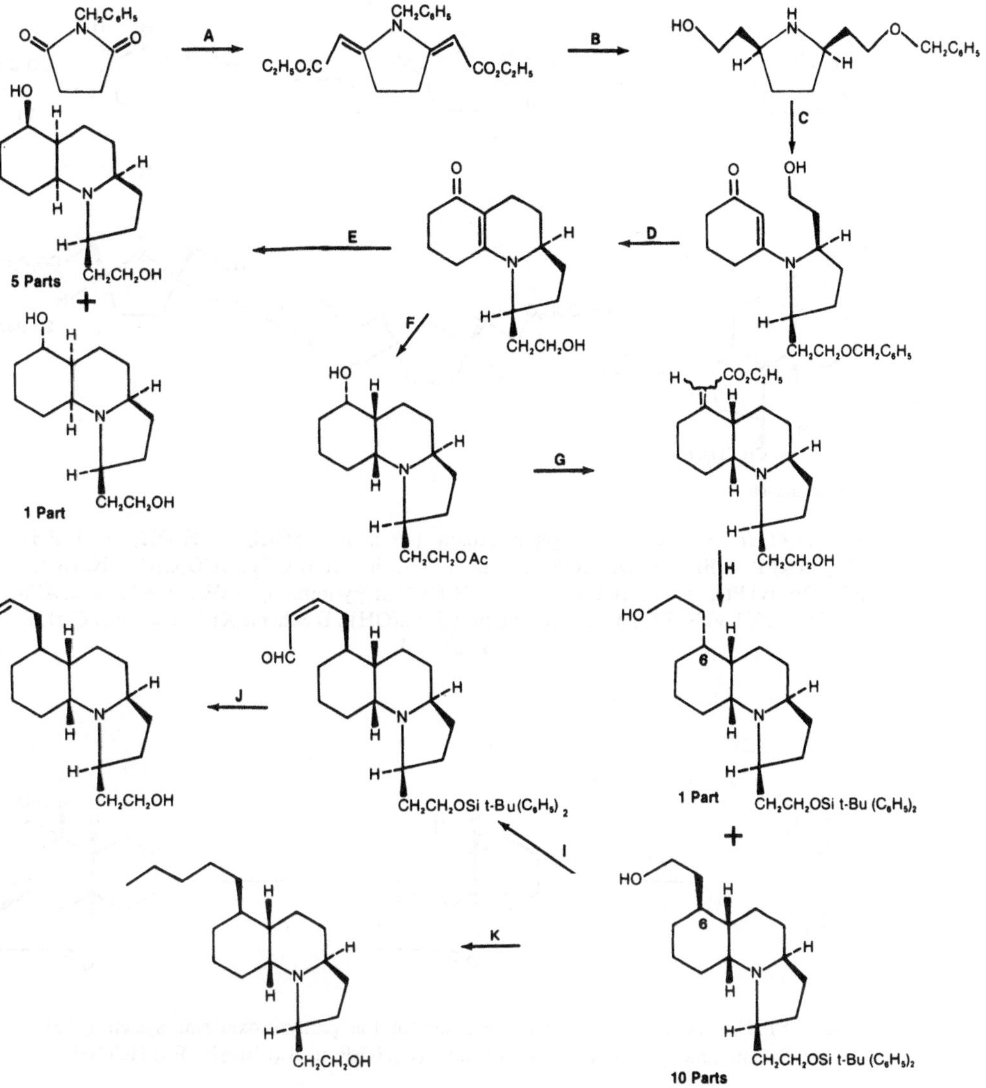

Scheme XLII. Synthesis of dl-gephyrotoxin and dl-perhydrogephyrotoxin (*108*). The major isomer from reaction **B** (i) was *cis*. An x-ray analysis confirmed the relative stereochemistry of an aminoalcohol obtained from reaction **F** (i). Reaction **H** with (i) Li, NH₃ , ii) (C₆H₅)₂t-BuSiCl, iii) LiAlH₄ afforded 1 part of the desired product and 1 part of the 6-epimer. **A** i) C₂H₅OC≡CMgCl, ii) HCl, iii) C₂H₅OC≡CMgCl, iv) HCl. **B** i) Pd/C, H₂, ii) C₆H₅OCOCl, iii) LiBH₄, iv) KH, v) C₆H₅CH₂Br, vi) Ba(OH)₂. **C** i) Cyclohexa-1,3-dione, pyridinium toluenesulphonate. **D** i) CH₃SO₂Cl, ii) LiBr, iii) Pd/C, H₂. **E** i) Pd/C, H₂. **F** i) Pd/Al₂O₃, H₂, ii) Ac₂O. **G** i) Oxalyl chloride, CH₃SOCH₃, (C₂H₅)₃N, ii) C₂H₅O₂CC CMgCl, iii) CH₃MgCl. **H** i) (C₆H₅)₂t-BuSiCl, ii) Rh/Al₂O₃, H₂, iii) LiAlH₄. **I** i) Pyridinium chlorochromate, ii) C₂H₅OCH=CHP(C₆H₅)₃, C₂H₅ONa, iii) Toluenesulphonic acid. **J** i) Chloromethyltriphenylphosphonium chloride, n-butylLi, ii) CH₃Li, (CH₃)₃SiCl, iii) Tetrabutylammonium fluoride. **K** i) Pyridinium chlorochromate, ii) CH₃CH₂CH=P(C₆H₅)₃, iii) Rh/Al₂O₃, H₂, iv) Tetrabutylammonium fluoride

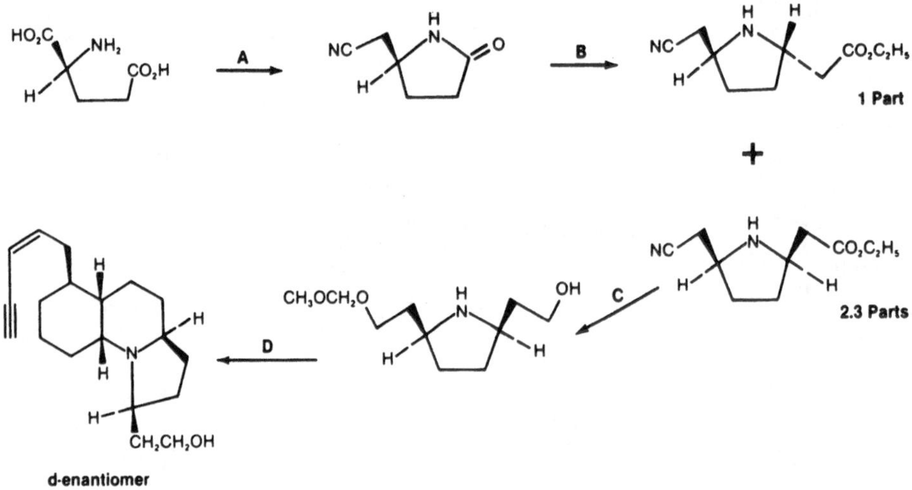

Scheme XLIII. Synthesis of d-gephyrotoxin (*107*). **A** i) SOCl₂, C₂H₅OH, ii) LiBH₄, iii) (C₆H₅)₃P, CBr₄, iv) NaCN/Al₂O₃. **B** i) P₂S₅, ii) CH₃COCHBrCO₂C₂H₅, NaHCO₃, iii) KOH, iv) Pt/C, HClO₄, CH₃OH. **C** i) C₆H₅OCOCl, pyridine, ii) LiBH₄, iii) KH, iv) a. Al(i-Bu)₂H, b. HCl, v) NaBH₄, vi) CH₃OCH₂Br, vii) Ba(OH)₂. **D** Scheme XLII, reactions **C, D, E, F, H, I, J.**

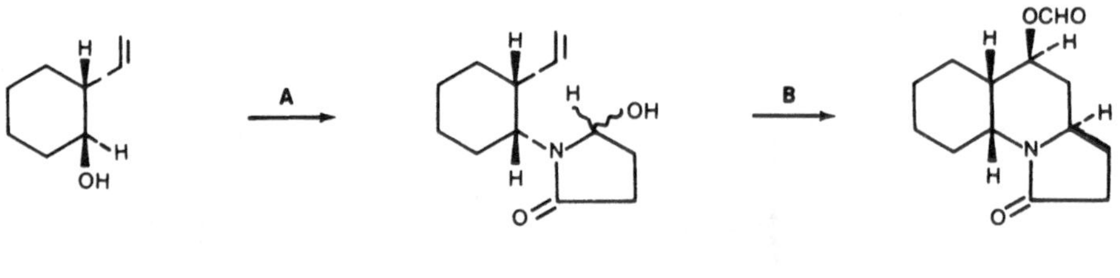

Scheme XLIV. Model synthesis of a precursor for the gephyrotoxin ring system (*123*). **A** i) Succinimide, diethyl azodicarboxylate, (C₆H₅)₃P, ii) Al(i-Bu)₂H. **B** i) HCO₂H

synthetic 5E,9E-3-butyl-5-propylindolizidine with natural **223AB** suggested that the two compounds were separable. Subsequent studies could not substantiate this separation and the stereoisomer synthesized by MACDONALD was shown to be identical with natural gephyrotoxin **223AB** (*244*) (Table 15, 16). The remaining three stereoisomers had been synthesized as shown in Schemes XLVII (D. J. HART and Y.-M. TSAI, personal communication) and XLVIII (*244*). The four stereoisomers emerged from a 10% SP-100 column at 90° in the following order i) 5Z,9Z, ii) 5E,9Z; iii) 5Z,9E; iv) 5E,9E with retention times of 17.4, 19.9, 22.7, 24.2 min. The mass

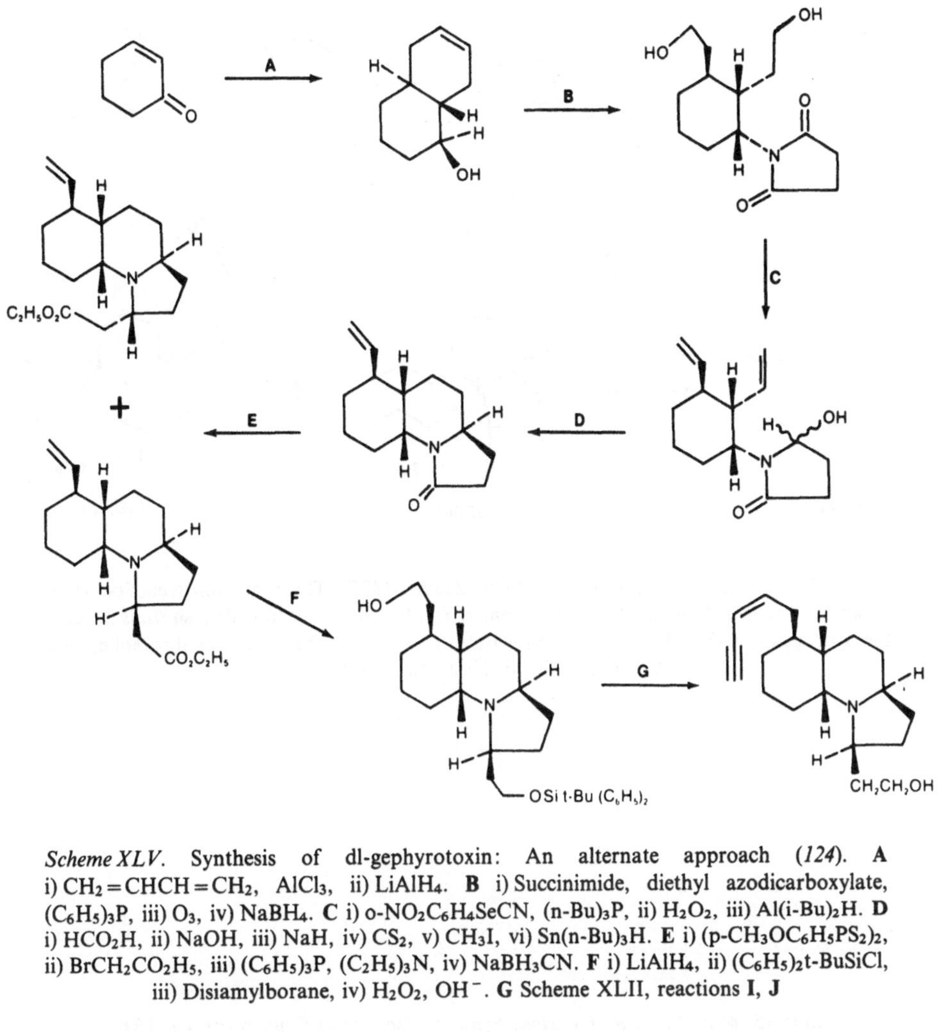

Scheme XLV. Synthesis of dl-gephyrotoxin: An alternate approach (*124*). **A** i) CH₂=CHCH=CH₂, AlCl₃, ii) LiAlH₄. **B** i) Succinimide, diethyl azodicarboxylate, (C₆H₅)₃P, iii) O₃, iv) NaBH₄. **C** i) o-NO₂C₆H₄SeCN, (n-Bu)₃P, ii) H₂O₂, iii) Al(i-Bu)₂H. **D** i) HCO₂H, ii) NaOH, iii) NaH, iv) CS₂, v) CH₃I, vi) Sn(n-Bu)₃H. **E** i) (p-CH₃OC₆H₅PS₂)₂, ii) BrCH₂CO₂H₅, iii) (C₆H₅)₃P, (C₂H₅)₃N, iv) NaBH₃CN. **F** i) LiAlH₄, ii) (C₆H₅)₂t-BuSiCl, iii) Disiamylborane, iv) H₂O₂, OH⁻. **G** Scheme XLII, reactions **I, J**

spectrum and retention time of the last isomer was identical with natural gephyrotoxin **223 AB**. The same set of four stereoisomers was prepared in an iso-series win which the butyl and propyl side chains are interchanged (Scheme XLIX). This set of four stereoisomers of the iso-series cochromatographed isomer for isomer with those of the other series but the mass spectra of the 5E,9E-iso-compound distinguished it from gephyrotoxin **223 AB** (Table 15). The carbon-13 magnetic resonance peaks for carbons 3, 5 and 9 have been assigned for 3-butyl-5-methylindolizidine alkaloids related to pharaoh ant trail pheromone (*243*). The assignments for the ant and frog alkaloids are given in Table 17.

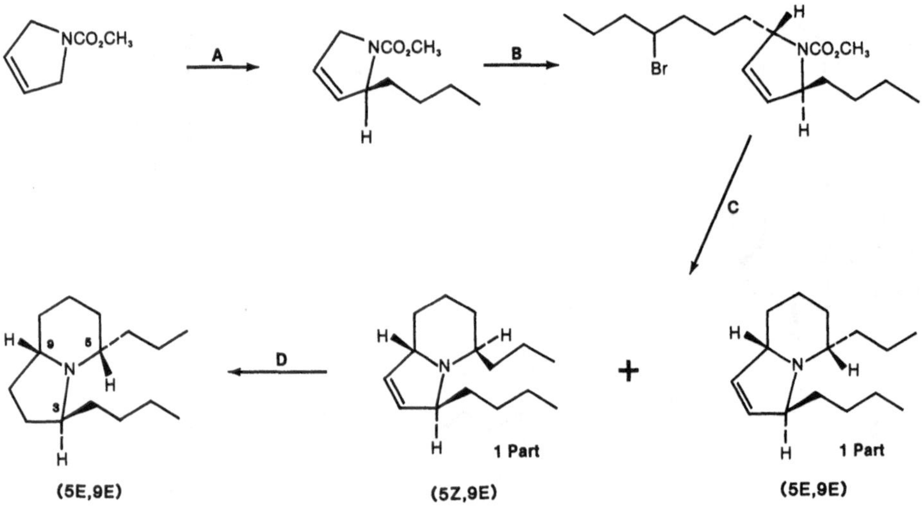

Scheme XLVI. Synthesis of dl-gephyrotoxin **223AB** (*175*). The reduction (reaction **D**) is anomalous in affording onyl the 5E,9E-isomer. The E and Z refer to hydrogen *trans* or *cis* to the hydrogen at C-3. **A** i) LiN(i-Pr)₂, ii) n-ButylBr. **B** i) LiN(i-Pr)₂, ii) 1,4-Dibromoheptane. **C** i) (CH₃)₃SiI, ii) Na₂CO₃. **D** i) PtO₂, H₂

Table 15. *Mass Spectra of Various Synthetic Isomers of Gephyrotoxin* **223AB**
The peak at m/z 180 corresponds to loss of the n-propyl substituent from C-3 and the peak at m/z 166 to loss of the n-butyl substituent from C-5 in the 3-butyl-5-propyl series (see Fig. 17). The Z and E refer to configuration of hydrogens (*cis* or *trans*) with respect to the hydrogen at C-3. All intensities relative to base peak set equal to 100 (*244*)

Ion (m/z)	Intensity of Ion 3-Butyl-5-propyl series				5-Butyl-3-propyl isomer
	5Z, 9Z	5E, 9Z	5Z, 9E	5E, 9E[a]	5E, 9E
223	<2	<2	<2	<2	<2
180	28	100	100	90	70
166	100	75	80	100	100

[a] Gephyrotoxin **223AB**.

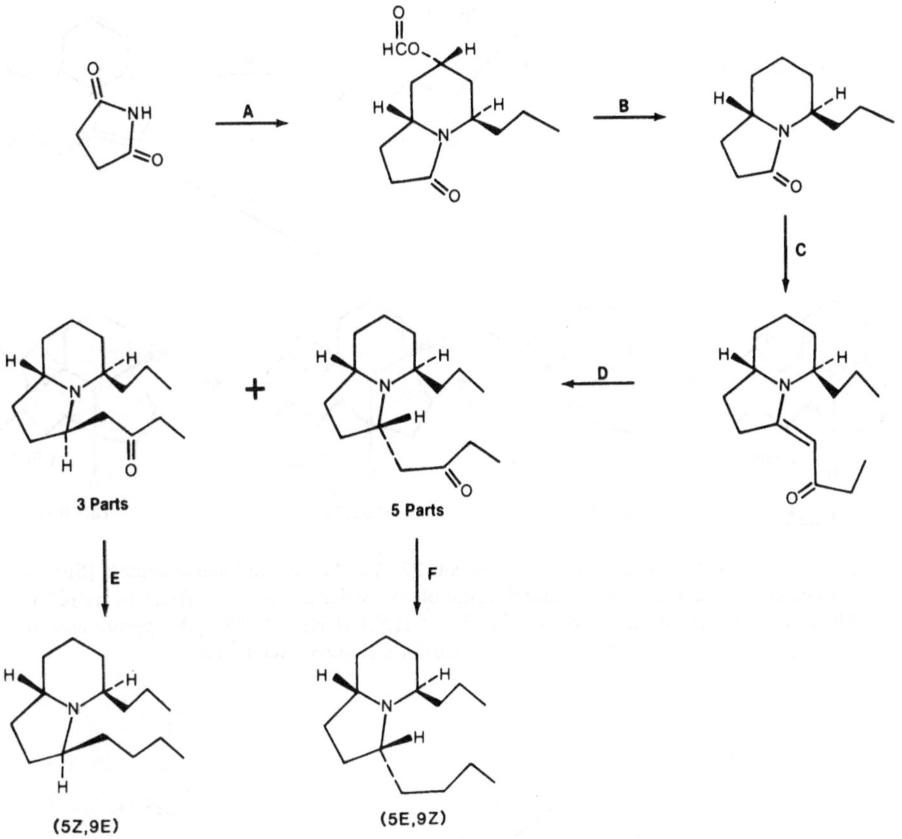

Scheme XLVII. Synthesis of stereoisomers of gephyrotoxin **223AB** (D. H. HART and Y.-M. TSAI, unpublished, cited in *244*). For E and Z terminology see Scheme XLVI. **A** i) 1-Hepten-4-ol, diethyl azodicarboxylate, $(C_6H_5)_3P$, ii) $Al(i-Bu)_2H$, iii) HCO_2H. **B** i) NaOH, ii) NaH, iii) CS_2, iv) CH_3I, v) $Sn(n-Bu)_3H$. **C** i) P_2S_5, ii) $BrCH_2COCH_2CH_3$, iii) $(C_6H_5)_3P$, $(C_2H_5)_3N$. **D** i) $NaBH_3CN$, H^+, ii) CrO_3. **E** i) $HS(CH_2)_2SH$, HCl, ii) Li, $C_2H_5NH_2$. **F** i) $LiAlH_4$, ii) NaH, iii) CS_2, iv) CH_3I, v) $Sn(n-Bu)_3H$

Table 16. *Physical and Spectral Properties of Synthetic dl-Gephyrotoxin* **223AB** (*175*)

Gephyrotoxin **223AB**	
Infrared	1190 cm^{-1}
Proton magnetic resonance	2.48^m (3H)
(CDCl$_3$)	1.78^m (6H)
	1.36^{br} (14H)
	0.96^m (6H)
Carbon-13 magnetic resonance	58.5, 57.7, 55.9, 34.7, 30.9, 29.6,
(CDCl$_3$)	28.7, 28.3, 25.4, 24.3, 23.6, 21.9,
	17.9, 13.4, 13.0

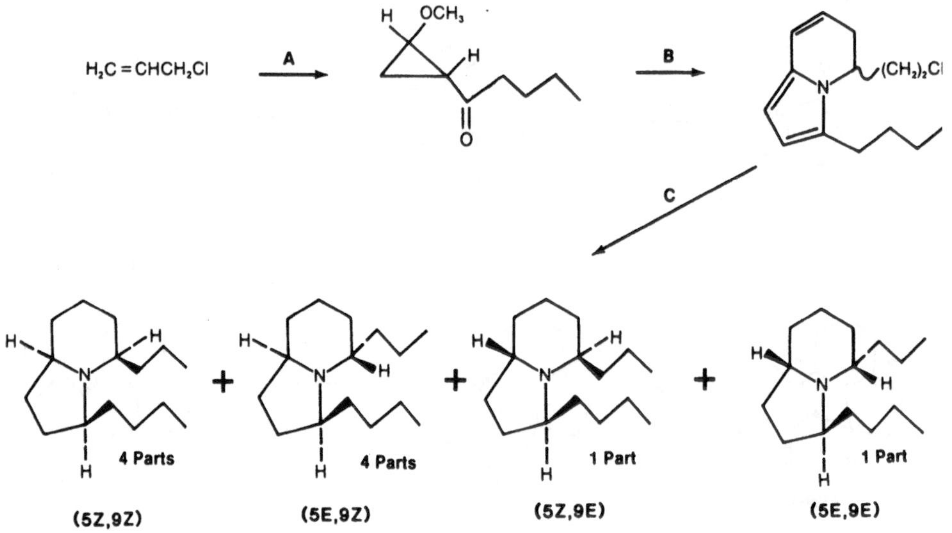

Scheme XLVIII. Synthesis of dl-gephyrotoxin **223AB** (5E,9E) and stereoisomers (SPANDE, unpublished, cited in *244*). For E and Z terminology see Scheme XLVI. **A** i) CH$_3$(CH$_2$)$_3$COCl, AlCl$_3$, ii) CH$_3$OH, iii) Base. **B** i) CH$_3$(CH$_2$)$_2$CHNH$_2$(CH$_2$)$_2$CH(OCH$_3$)$_2$, pyridinium tri-fluoroacetate. **C** i) Rh/Al$_2$O$_3$, C$_2$H$_5$OH, AcOH, H$_2$

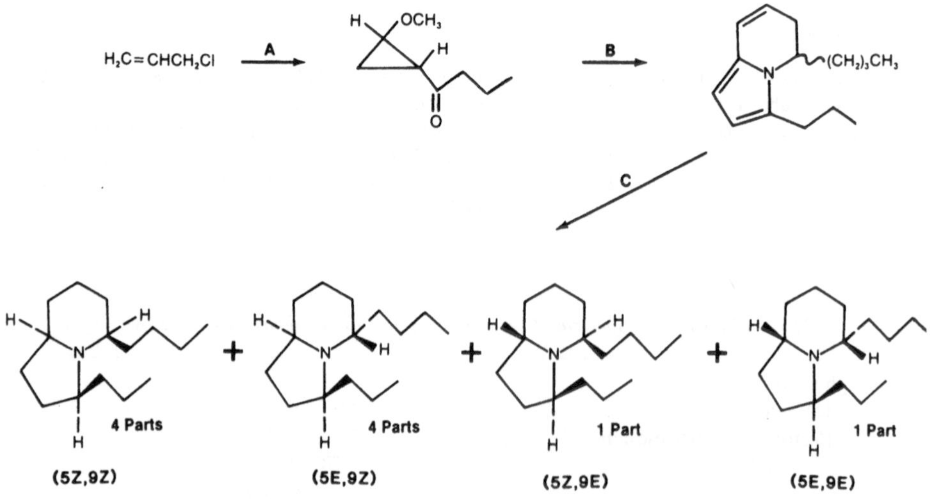

Scheme XLIX. Synthesis of an *iso*-series of compounds related to gephyrotoxin **223AB** (SPANDE, unpublished, cited in *244*). For E and Z terminology see Scheme XLVI. **A** i) CH$_3$(CH$_2$)$_2$COCl, AlCl$_3$, ii) CH$_3$OH, iii) Base. **B** i) CH$_3$(CH$_2$)$_2$CHNH$_2$(CH$_2$)$_3$CH(OCH$_3$)$_2$, pyridinium trifluoroacetate. **C** i) Rh/Al$_2$O$_3$, C$_2$H$_5$OH, AcOH, H$_2$

References, pp. 326—340

Table 17. *Carbon-13 Magnetic Resonance Spectral Assignments for Bicyclic Gephyrotoxin Isomers (3-butyl-5-propylindolizidines) and for Ant Trail Pheromone Isomers (3-butyl-5-methylindolizidines)*
(for *Z* and *E* designations see Table 15) (*243, 244*)

R = Methyl or Propyl	Chemical Shift of Proton[a]		
	C-3	C-5	C-9
Gephyrotoxin isomers			
3-Butyl-5-propyl			
5Z,9Z	65.3	62.3	67.7
5E,9Z	56.4	52.7	58.4
5Z,9E	54.8	52.3	58.5
5E,9E[b]	57.7	55.9	58.5
Ant pheromone isomers			
3-Butyl-5-methyl			
5Z,9Z[c]	61.8	59.7	67.6
5E,9Z	55.4	47.2	58.7
5Z,9E	55.0	48.4	59.4
5E,9E	56.6	51.8	58.9

[a] Solvent $CDCl_3$; [b] Synthetic gephyrotoxin **223AB**; [c] Ant trail pheromone.

C. Biological Activity

The biological activity of gephyrotoxin(s) will require further investigation (see Table 18). In a preliminary study (*183*) gephyrotoxin at 5 μM reduced acetylcholine but not histamine-elicited contractures of guinea pig ileum, and at 10 μM caused a reduction in the response to acetylcholine in guinea pig atrium. Thus, gephyrotoxin appeared to act as a relatively weak muscarinic antagonist. Gephyrotoxin *p*-bromobenzoate was also a weak antagonist of acetylcholine in ileum. At 10 μM gephyrotoxin had virtually no effect on direct or indirect elicited twitch in rat phrenic nerve diaphragm. It should be noted that like the histrionicotoxins, gephyrotoxin is in reality not a "toxin", showing little toxicity in mammals at relatively high doses (Table 18) (*80*). Gephyrotoxin **239 CD** at 4 mg/kg in white mice caused long-lasting locomotor difficulties and prostration (*80*).

Table 18. *Effects of Gephyrotoxin on Biological Systems*

System	Species	Comments	Reference
Whole animal	Mouse	Relatively nontoxic $LD_{50} > 10$ mg/kg	(80)
Neuromuscular preparation (striated)	Rat	No effect at $10 \mu M$	(183)
Cardiac preparation	Guinea pig atrium	Blockade muscarinic responses	(183)
Smooth muscle	Guinea pig ileum	Blockade muscarinic response	(183)

VI. Pumiliotoxin-A Class

A. Structures

After the batrachotoxins, the next examples of dendrobatid alkaloids to be isolated were the Panamanian poison frog *Dendrobates pumilio* (77). These were pumiliotoxin A and B. Approximately 1.5 mg of each compound was isolated from extracts of 20 frog skins. The molecular formulae were $C_{19}H_{33}NO_2$ and $C_{19}H_{33}NO_3$, respectively, and the nuclear magnetic resonance spectra were thought to be reminiscent of those expected of steroidal alkaloids. The compounds formed O-methyl ethers with methanolic hydrogen chloride. Initial studies suggested the formation of an N-acetyl derivative with, in addition, either one or two O-acetyl groups for pumiliotoxin A and B, respectively. Subsequently, additional amounts of pumiliotoxin A and B were isolated from *Dendrobates pumilio*, e.g., from 250 skins, 17 mg of pumiliotoxin A, 20 mg of pumiliotoxin B, and 16 mg of pumiliotoxin C were isolated (see ref. 84). Further investigation of these samples indicated that pumiliotoxin A and B were bicyclic alkaloids with two double bonds, only one of which underwent facile hydrogenation (80). In addition the formation of neutral N-acetyl derivatives could not be confirmed (unpublished results). Analysis of nuclear magnetic spectra and mass spectra indicated that pumiliotoxin A and B differed only in the terminal portion of a side chain; $-CH = CCH_3CHOHCH_2CH_3$ in pumiliotoxin A and $-CH = CCH_3CHOHCHOHCH_3$ in pumiliotoxin B. The allylic hydroxyl group of the side chain was undoubtedly the source of the frustrating instability of pumiliotoxin A and B under acid conditions. This instability thwarted repeated efforts to prepare a crystalline salt or derivative. Because of this, ten years after their initial isolation, the structures of pumiliotoxin A and B were still undeciphered.

During this period, structures of other dendrobatid alkaloids such as pumiliotoxin C, the various histrionicotoxins and gephyrotoxin were determined. In addition alkaloid profiles had been delineated in many species of dendrobatid frogs. Thin-layer and gas chromatographic characteristics, mass spectra and chemical properties with respect to hydrogenation and acetylation had been determined for over one hundred alkaloids (*80, 190*; MYERS and DALY, unpublished results). Some twenty-three of these alkaloids had appeared to be related in structure to pumiliotoxin A and B based on mass spectra similarities; namely a prominent ion of C_4H_8N (m/z 70) accompanied usually by either a prominent ion of $C_{10}H_{16}NO$ (m/z 166) or $C_{10}H_{16}NO_2$ (m/z 182). Pumiliotoxin A and B and eight other alkaloids exhibited the $C_{10}N_{16}NO$ ion, while eleven alkaloids exhibited the $C_{10}H_{16}NO_2$ ion. The eleven alkaloids which exhibited the $C_{10}H_{16}NO_2$ ion (m/z 182) where deemed an allo-series. The remaining four alkaloids showed the C_4H_8N ion but the other most intense fragment ion was at either 84, 110, 138, 150, or 168. All of these compounds were assigned tentatively to a pumiliotoxin-A class (*80*) but in the present review only the alkaloids exhibiting a major peak at 166 or 182 have been presented as members of the pumiliotoxin-A or allopumiliotoxin class. The remaining alkaloids are documented under *Other Alkaloids* until further characterization is forthcoming. Two further alkaloids considered as possible members of the pumiliotoxin-A class (*190*) do not have a *major* fragment ion of C_4H_8N and these alkaloids, **251F**, and **269B** and a related congener **247** are also documented under *Other Alkaloids*.

It had been hoped that during screening of various extracts a source of a simpler alkaloid related to pumiliotoxin A and B would be discovered and that such a simpler alkaloid would provide the key to this class of alkaloids. One such simpler alkaloid was pumiliotoxin **251D**. It was certainly less complex than pumiliotoxin A and B, with an empirical formula of $C_{16}H_{29}NO$ and only one double bond. The alkaloid **251D** occurred in a number of species, including an Ecuadorian poison frog, *Dendrobates tricolor*. It did not, however, appear to be a major alkaloid in extracts from some eight specimens of this frog collected in 1974. At the time the remarkable volatility of pumiliotoxin **251D** was not appreciated; indeed as the result of further work **251D** turned out to be a major component of the alkaloids from *Dendrobates tricolor*. Further extracts of this frog were obtained not as a source of **251D**, but because another alkaloid present in trace amounts exhibited remarkable pharmacological properties. This alkaloid, whose molecular formula is still uncertain (see *Other Alkaloids*), was a potent analgesic. Extracts from skins of some 750 *Dendrobates* *tricolor* provided much less than 1 mg of the analgesic alkaloid, but by serendipity did provide some 21 mg of pumiliotoxin **251 D**, the long sought simpler analog of pumiliotoxin A and B (*84*).

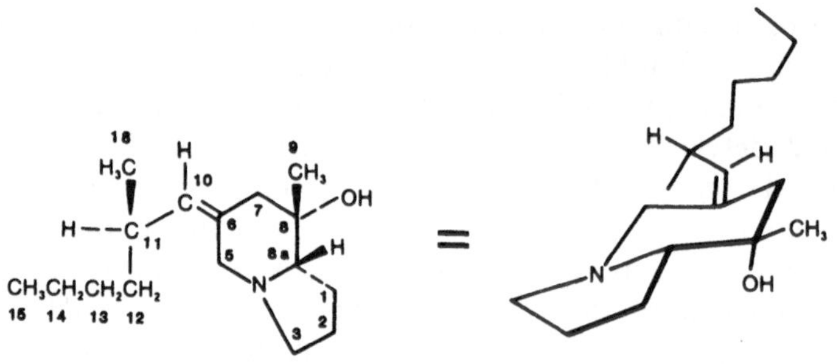

Fig. 18. Structure of pumiliotoxin **251 D**. Configuration depicted at the right is based on x-ray analysis of the HCl salt (*84*)

X-ray analysis of a crystal of the hydrochloride of **251 D** showed that the structure and absolute configuration of pumiliotoxin **251 D** was 8 - hydroxy - 8 - methyl - 6 - (2′ - methylhexylidene) - 1 - azabicyclo - [4.3.0]- nonane (Fig. 18) (*84*). Unlike the other dendrobatid alkaloids, this compound could not be simply derived by biosynthetic ring closure of a postulated precursor 2,6-disubstituted piperidine. It contained the bridge characteristic of the gephyrotoxins and was thus an indolizidine like the bicyclic gephyrotoxin **223 AB** (*vide supra*). The anomalous substituent, however, was at what would have corresponded to C-5 of a precursor 2-substituted piperidine. The proton and carbon-13 magnetic resonance spectra of pumiliotoxin **251 D** are presented in Figs. 19 and 20 and the mass spectrum in Table 19. Further physical and spectral properties of pumiliotoxin **251 D** are presented in the listing of various pumiliotoxins of this class (p. 308) and in Table 20. The hydroxyl group in pumiliotoxin **251 D** is very unreactive and provides acetyl and trimethylsilyl derivatives only with great difficulty (*84*). The cryptic nature of the hydroxyl group probably contributes to the volatility of pumiliotoxin **251 D** and to relatively high R_f values (0.52) on silica gel thin layer chromatography ($CHCl_3 : CH_3OH$, 9 : 1). Hydrogenation of **251 D** yields a dihydro-derivative with a more reactive hydroxyl group and an R_f value of 0.26. Acetyl and trimethylsilyl derivatives of dihydropumiliotoxin **251 D** are readily formed. Reduction of the double bond from the less hindered aspect should yield a dihydro-derivative with the 8-methyl axial and the 6-hexyl substituent and the 8-hydroxy now equatorial.

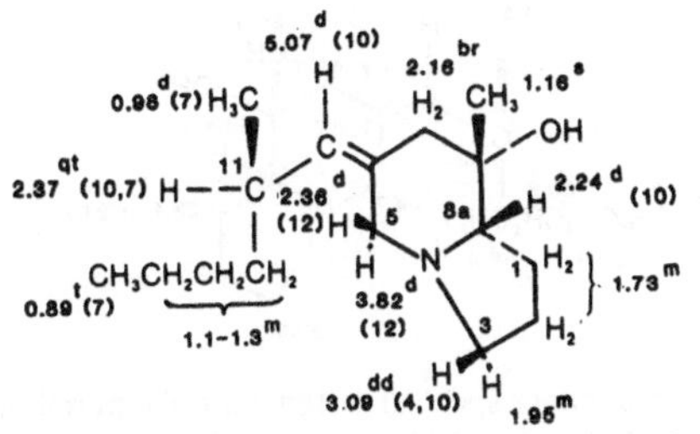

5.07 ^{d} (10)

0.98 ^{d} (7) H₃C   H   2.16 ^{br} 1.16 ^{s}
                          H₂ CH₃
                    C   —OH
^{qt} 11
2.37 (10,7) H——C ^{d} 2.36   8a   H 2.24 ^{d} (10)
              (12) H   5
      CH₃CH₂CH₂CH₂   H   N   H₂
0.89 ^{t} (7)   3.82 ^{d} } 1.73 ^{m}
       1.1–1.3 ^{m} (12)   3   H₂
              dd   H H
      3.09 (4,10)   1.95 ^{m}

Fig. 19. Proton magnetic resonance spectral assignments for pumiliotoxin **251D** (free base, CDCl₃, 84). Certain resonance peaks (H-5αβ; H-3β; H-1″; H-8α) have been reassigned based on subsequent decoupling experiments with pumiliotoxin B (256). Resonance peaks for the HCl salt (CD₃OD) are as follows: H-5α 3.37; H-5β 4.36; H-10 5.31; and H-11 2.41. The spectrum (CDCl₃, 220 MHz) of the free base is depicted in DALY et al. (84)

Table 19. *Mass Spectra of Pumiliotoxin* **251D**, *Pumiliotoxin A and Pumiliotoxin B* (77, 80, 84) Only diagnostically useful peaks are reported. Major fragmentation pathway for these compounds, their dihydro-derivatives and the allo-series are discussed in (80, 84). The mass spectrum of pumiliotoxin A is depicted in (77)

Fragment	PTX **251D**	PTX A	PTX B
$C_{19}H_{33}NO_3$			323 (10)
$C_{19}H_{32}NO_3$			322 (2)
$C_{19}H_{33}NO_2$		307 (5)	
$C_{19}H_{32}NO_2$		306 (1)	306 (5)
$C_{19}H_{31}NO$		290 (3)	290 (2)
$C_{17}H_{28}NO_2$		278 (4)	278 (12)
$C_{17}H_{28}NO$		262 (2)	262 (1)
$C_{17}H_{26}NO$		260 (1)	260 (2)
$C_{16}H_{29}NO$	251 (20)		
$C_{16}H_{28}NO$	250 (5)		
$C_{15}H_{26}NO$	236 (8)		
$C_{16}H_{28}N$	234 (5)		
$C_{16}H_{27}N$	233 (5)		
$C_{15}H_{24}N$	218 (4)		
$C_{14}H_{26}N$	208 (13)		
$C_{13}H_{20}NO$	206 (9)	206 (10)	206 (15)
$C_{12}H_{20}NO$	194 (20)	194 (16)	194 (26)
$C_{12}H_{19}NO$	193 (5)	193 (5)	193 (22)
$C_{12}H_{18}N$	176 (10)	176 (10)	176 (15)
$C_{10}H_{16}NO$	166 (85)	166 (85)	166 (75)
$C_{10}H_{16}N$	150 (18)		
$C_{10}H_{14}N$	148 (6)		
$C_9H_{16}N$	138 (5)	138 (5)	
$C_6H_{12}NO$	114 (5)	114 (6)	114 (4)
$C_6H_{10}NO$	112 (15)	112 (10)	112 (8)
$C_5H_{10}N$	85 (6)	84 (10)	84 (8)
C_4H_8N	70 (100)	70 (100)	70 (100)

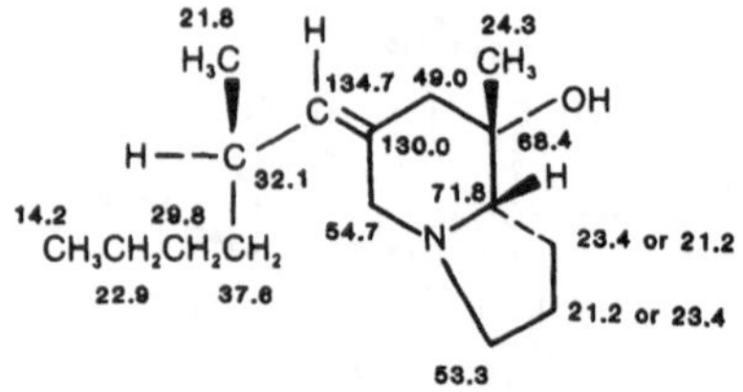

Fig. 20. Carbon-13 magnetic resonance spectral assignments for pumiliotoxin **251 D** (free base, CDCl₃). Frequencies of the 6,10-dihydro-derivative were stated to be not remarkably different with, of course, the exception of C-6 and C-10 (*84*)

Table 20. *Properties of Synthetic d-Pumiliotoxin* **251 D** (*206*)

m.p. 205 – 206°	
Optical rotation	
HCl	$[\alpha]_D^{25} + 31.4°$
(c 0.62, CH₃OH)	
Free base (contains small amount of the C-11 epimer)	
(c 1.6, CHCl₃)	$[\alpha]_D^{25} - 3.1°$

Once the structure of pumiliotoxin **251 D** was known, the structures of a number of the other alkaloids of the pumiliotoxin-A class could be derived from mass and nuclear magnetic resonance spectra. The typical mass spectra of this class of compounds with major ions at C_4H_8N (m/z 70) and either $C_{10}H_{16}NO$ (m/z 166) or $C_{10}H_{16}NO_2$ (m/z 182, allo-compounds) (*80*) would appear to derive from the indolizidine portion of the molecule. Thus, the allopumiliotoxins must contain an additional hydroxyl group in the indolizidine system. The hydroxyl is not at the 9-(methyl)-position since this would result in a major loss of CH_2OH, which is not observed in allo-compounds. The additional hydroxyl group undergoes facile acylation and, hence is a secondary hydroxyl. The tertiary hydroxyl at the 8-position of pumiliotoxin **251 D** is highly hindered, as mentioned previously, and forms O-acetyl and O-trimethylsilyl-derivatives only with difficulty. Analysis of the mass spectra of perhydro derivatives in the allopumiliotoxin-A series suggests that the additional hydroxyl of the indolizidine ring in this series is at the 7-position adjacent to the other ring hydroxyl. Nuclear magnetic resonance spectra of two isomeric allopumiliotoxins, minor alkaloids from *Dendrobates auratus,* confirms the position of the additional hydroxyl at position 7 and indicates that both 7-hydroxy epimers occur in this alkaloid series (*128, 256, vide infra*).

Analysis of the nuclear magnetic resonance spectra had now provided the long sought key to the structures of pumiliotoxin A and B. Comparison

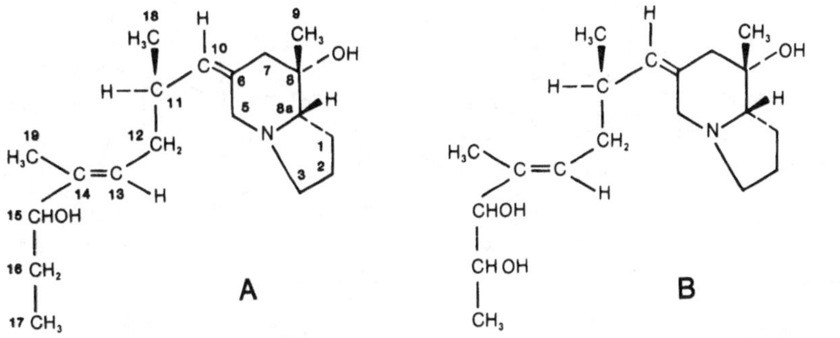

Fig. 21. Structures of **A** pumiliotoxin A and **B** pumiliotoxin B. The configuration at the 13,14-double bond has been shown to be E (*256*)

of their carbon-13 spectra with that of pumiliotoxin **251D** allowed identification of all the carbons of the indolizidine portion and of initial carbons of the side chain, namely $=CHCHCH_3-$. The remaining seven carbon moiety was readily formulated as shown in Fig. 21, although some question remained as to the configuration of the double bond. It was initially tentatively assigned the less stable Z-configuration, but subsequent studies based on nuclear magnetic resonance analysis of the aldehyde formed on periodate oxidation of the glycol in pumiliotoxin B suggested that the compounds contained the 13,14-double bond in the more stable E-configuration (*256*). Nuclear-Overhauser effects on the C-15 proton magnetic resonance peak of pumiliotoxin B itself substantiated this assignment. Interpretation of the carbon-13 and proton magnetic resonance spectra of pumiliotoxin A, B and **251D** had not been successful prior to x-ray analysis of a crystal of **251D** in large part due to the misinterpretation of the C-8a resonance; since the C-8a signal is found at 71.8 ppm it was incorrectly assumed that C-8a had an oxygen substituent. Once this misassignemt was recognized, interpretation of spectra of pumiliotoxin A, B, and **251D** was relatively straightforward. Proton and carbon-13 magnetic resonance spectral assignments for pumiliotoxin A and B are documented in Figs. 22 and 23. Further physical and spectral properties of pumiliotoxin A and B are given in the listing of pumiliotoxins of this class (p. 308). Detailed mass spectral data are presented in Table 19. The absolute configurations of the hydroxyls at carbon-15 and 16 of pumiliotoxin A or B are unknown. Recent results suggest the *threo* configuration for pumiliotoxin B based on comparison of nuclear magnetic resonance spectra with model compounds (T. TOKUYAMA and M. UEMURA, unpublished data).

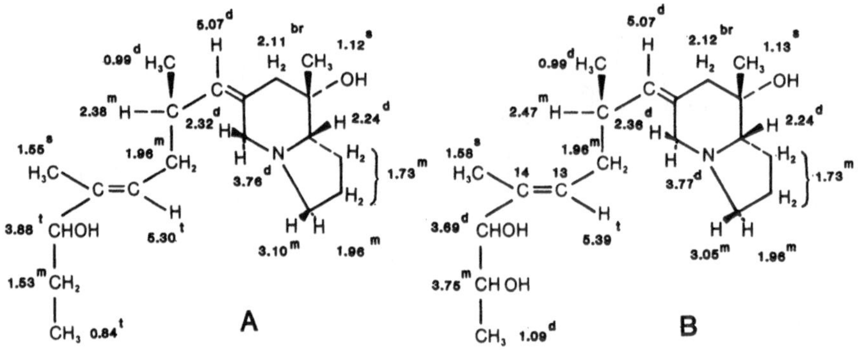

Fig. 22. Proton magnetic resonance assignment for **A** pumiliotoxin A and **B** pumiliotoxin B (free base, CDCl₃) (84). Proton magnetic resonance spectra (CDCl₃, 220 MHz) are depicted and discussed in (84). Certain resonance peaks (H-5αβ, H-1, H-8a) have been reassigned based on subsequent decoupling experiments with pumiliotoxin B (256). The lack of coupling between the olefinic proton at C-13 and the C-19 methyl protons even at 360 MHz suggested the Z-configuration for the substituents on the 13,14 double bond (84) but subsequent data did not support this interpretation (256)

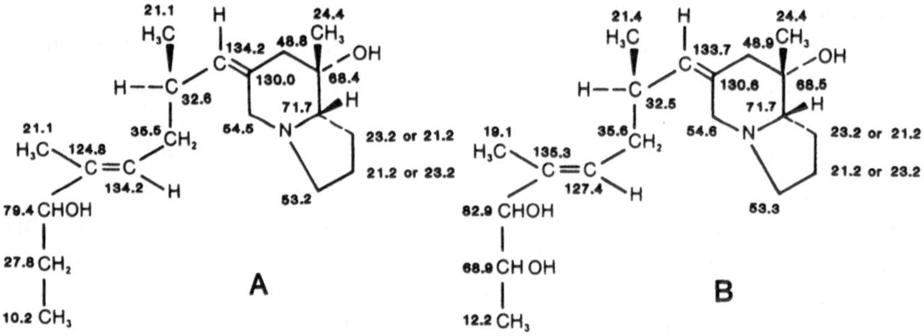

Fig. 23. Carbon-13 magnetic resonance assignments for **A** pumiliotoxin A and **B** pumiliotoxin B (free base, CDCl₃) (84)

Boronate and dimethylsilanate derivatives form very readily from pumiliotoxin B. Indeed, during gas chromatography on OV-1 columns, a portion of pumiliotoxin B was often converted to a dimethylsilanate, thus leading to the incorrect postulate that three dendrobatid alkaloids **379, 381** and **395** were present (80). All of these were subsequently shown to be artefacts representing dimethylsilanates formed from pumiliotoxin B, a dihydropumiliotoxin B and an alkaloid now designated pumiliotoxin **339,** Most of the compounds of the pumiliotoxin-A and allopumiliotoxin-A class have not been isolated in sufficient quantity or purity to allow nuclear magnetic resonance spectroscopy. However, based on mass spectra of the alkaloids and their perhydro-derivatives and on biosynthetic conside-rations, structures can be proposed for some of these compounds (Fig. 24). Two isomeric allopumiliotoxins were isolated from *Dendrobates auratus* in sufficient quantities for nuclear magnetic resonance spectroscopy (*128, 256*). Tentative structures of these two 7-hydroxy-isomers are presented in Fig. 25.

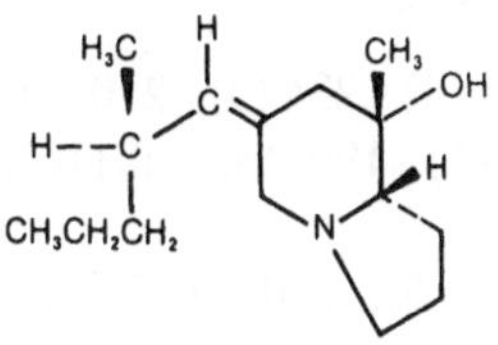

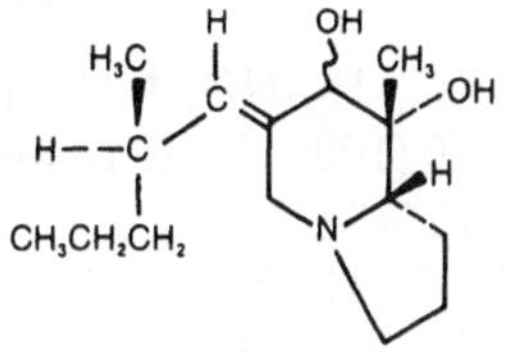

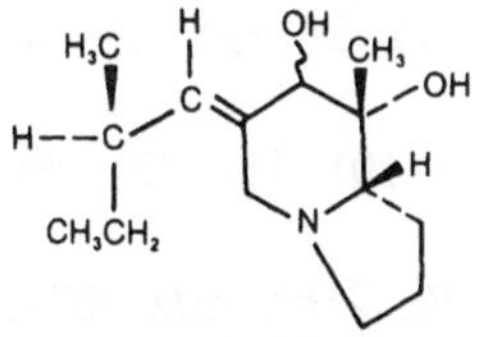

Fig. 24. Tentative structures for alkaloids of the pumiliotoxin A (**237A**) and allopumilio-toxin A (**237B, 253, 267A**) class (see pp. 308, 309 for further properties of these alkaloids)

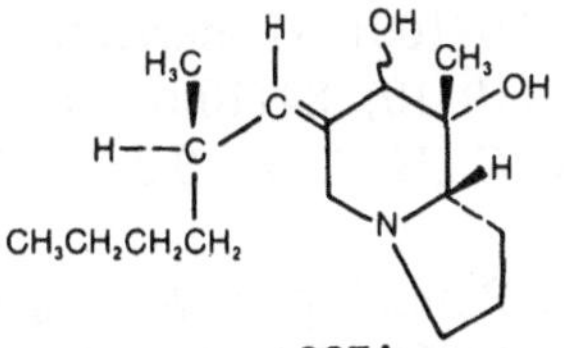

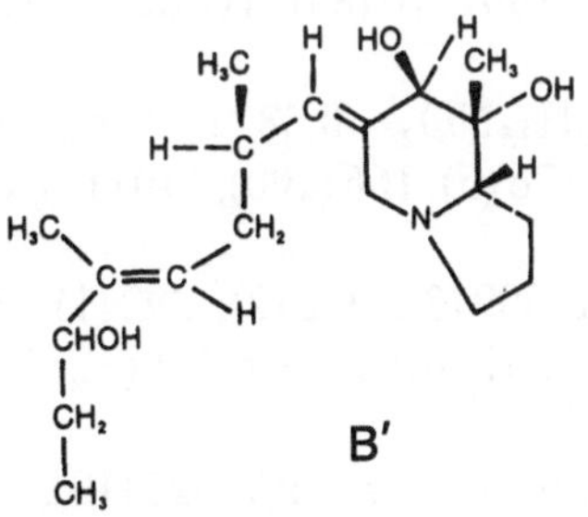

Fig. 25. Structures for the 7-hydroxy isomers of allopumiliotoxin **323B** (formerly referred to as allopumiliotoxin B) (*80*). Isomer **B′** appears to have a 7-equatorial hydrogen (proton magnetic resonance peak 3.67⁵ ppm) and isomer **B″** a 7-axial hydrogen (3.73⁵). The assignments are based on proton and carbon-13 magnetic resonance data from (*128, 256*)

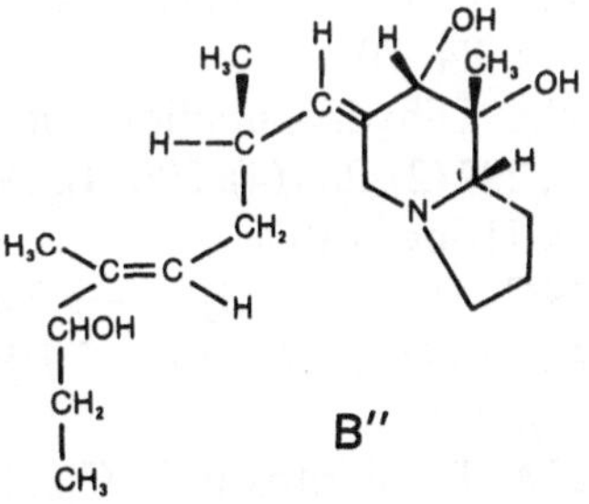

Properties of pumiliotoxin A, B and **251 D** and other alkaloids of the pumiliotoxin-A class are reported below in the usual format; empirical formulas which are based only on analogy and chemical and chromatographic properties and which have not been confirmed by high resolution mass spectrometry are in quotation marks. Certain trivial names employed earlier (*80*) are no longer used now that the structures of these alkaloids are better understood.

Pumiliotoxin A-Class

237 A. $C_{15}H_{27}NO$, 0.58, 167°, m/z 237 (4), 236 (3), 220 (3), 194 (12), 166 (40), 70 (100). H_2-derivative, m/z 239 (5), 196 (5), 168 (10), 110 (30), 84 (100), 70 (45).

251 D. $C_{16}H_{29}NO$, 0.52, 172°, m/z 251 (5), 250 (3), 236 (2), 234 (1), 194 (5), 176 (3), 166 (78), 70 (100). H_2-derivative, m/z 253 (18), 110 (70), 84 (100), 70 (55).

281 A. "$C_{17}H_{31}NO_2$", −, 205°, m/z 281 (4), 280 (2), 264 (2), 194 (12), 166 (72), 70 (100). H_2-derivative, m/z 283 (1), 282 (2), 266 (4), 208 (40), 138 (10), 110 (10), 84 (100), 70 (85).

297 B. $C_{18}H_{35}NO_2$, 0.35, 222°, m/z 297 (10), 166 (92), 70 (100). H_4-derivative.

307 A. Pumiliotoxin A, $C_{19}H_{33}NO_2$, 0.36, 216°, m/z 307 (5), 290 (3), 278 (4), 206 (10), 194 (16), 176 (10), 166 (85), 70 (100). H_2-derivative, m/z 309 (3), 308 (2), 210 (8), 166 (30), 110 (45), 84 (100), 70 (40). H_4-derivative, m/z 311 (3), 110 (32), 84 (100), 70 (55). O-acetyl derivative.

307 B. Isopumiliotoxin A, $C_{19}H_{33}NO_2$, 0.36, 211°, m/z 307 (12), 306 (4), 290 (2), 194 (24), 193 (45), 166 (100), 70 (56). H_2-derivative, m/z 309 (1), 280 (3), 208 (10), 138 (25), 110 (60), 84 (100), 70 (45). H_4-derivative, m/z 311 (1), 110 (25), 84 (100), 70 (30).

307 D. "$C_{18}H_{31}NO_3$", −, 234°, m/z 307 (8), 306 (5), 292 (3), 290 (5), 264 (4), 262 (11), 206 (11), 194 (18), 166 (100), 70 (85). H_4-derivative, m/z 311, 110, 84, 70.

309 A. Dihydropumiliotoxin A, "$C_{19}H_{35}NO_2$", 0.38, 218°, m/z 309 (9), 308 (3), 292 (2), 280 (4), 206 (4), 194 (15), 176 (5), 166 (100), 110 (10), 84 (20), 70 (51). H_2-derivative.

309 C. "$C_{19}H_{35}NO_2$", −, 210°, m/z 309 (3), 308 (2), 292 (1), 280 (4), 266 (4), 194 (15), 166 (100), 70 (90). H_2-derivative, m/z 311 (3), 110 (25), 84 (100), 70 (30).

323 A. Pumiliotoxin B, $C_{19}H_{33}NO_3$, 0.17, 230°, m/z 323 (10), 306 (5), 290 (2), 278 (12), 260 (2), 206 (15), 194 (26), 193 (22), 176 (15), 166 (75), 70 (100). H_2-derivative, m/z 325 (5), 166 (25), 110 (24), 84 (100), 70 (38). H_4-derivative, m/z 327 (3), 326 (2), 312 (1), 310 (2), 282 (14), 264 (12), 110 (30), 84 (100), 70 (43). O-acetyl derivatives.

Allopumiliotoxin A-Class

237B. "$C_{14}H_{23}NO_2$", 0.58, 168°, m/z 237 (1), 236 (2), 182 (60), 114 (30), 112 (25), 70 (100).

253A. $C_{15}H_{27}NO_2$, 0.30, 179°, m/z 253 (4), 252 (1), 236 (22), 210 ($C_{12}H_{20}NO_2$, 3) 208 (2), 192 (5), 182 ($C_{10}H_{16}NO_2$, 16), 114 ($C_6H_{12}NO$, 27), 112 ($C_6H_{10}NO$, 26), 70 (C_4H_8N, 100). H_2-derivative, m/z 255 (3), 238 (8), 110 (50), 84 (100), 70 (65).

267A. $C_{16}H_{29}NO_2$, 0.31, 186°, m/z 267 (15), 250 (33), 182 (18), 114 (52), 112 (44), 70 (100). H_2-derivative, m/z 269 (6), 252 (18), 110 (50), 84 (100), 70 (60). O-acetyl derivative.

297A. $C_{17}H_{31}NO_3$, 0.13, 225°, m/z 297 (3), 296 (4), 280 (9), 236 (2), 210 (3), 194 (4), 193 (3), 182 (21), 114 (27), 112 (16), 70 (100). H_2-derivative, m/z 299 (4), 282 (12), 256 (6), 224 (4), 110 (50), 84 (100), 70 (75).

307C. $C_{19}H_{33}NO_2$, 0.39, 214°, m/z 307 (9), 290 (11), 182 (62), 70 (100). H_4-derivative, m/z 311 (3), 294 (3), 268 (5), 236 (2), 110 (30), 84 (100), 70 (35).

323B. $C_{19}H_{33}NO_3$, 0.20, 228°, m/z 323 (5), 306 (10), 210 (4), 209 (3), 182 (50), 114 (20), 70 (100). H_4-derivative.

325. "$C_{19}H_{35}NO_3$", 0.20, 232°, m/z 325 (3), 308 (12), 280 (3), 210 (2), 182 (56), 114 (16), 112 (9), 70 (100). H_2-derivative.

339. $C_{19}H_{33}NO_4$, 0.10, 243°, m/z 339 (3), 322 (3), 294 (4), 276 (5), 222 (4), 209 (15), 208 (11), 192 (14), 182 (45), 114 (25), 70 (100). H_4-derivative, m/z 343 (3), 342 (2), 324 (3), 138 (10), 110 (25), 84 (100), 70 (65).

341A. $C_{19}H_{35}NO_4$, 0.48, 222°, m/z 341 (4), 324 (3), 323 (1), 306 (1), 298 (3), 266 (4), 254 (7), 210 (5), 182 (10), 114 (10), 112 (40), 84 (22), 70 (100). H_2-derivative, m/z 343 (1), 342 (2), 328 (2), 266 (10), 138 (100), 110 (5), 84 (15), 70 (15).

341B. "$C_{19}H_{35}NO_4$", −, 223°, m/z 341 (1), 324 (4), 182 (60, 114 (20), 112 (20), 70 (100).

357. $C_{19}H_{35}NO_5$, 0.30, 240°, m/z 357 (3), 340 (8), 324 (8), 272 (4), 182 (20), 138 (10), 110 (20), 84 (20), 70 (100).

Three alkaloids from the frog *Dendrobates bombetes* were previously included in the pumiliotoxin-A class based primarily on the presence of a significant fragment ion of C_4H_8N (m/z 70) (*190*). However all three lack a major ion at m/e 166 or 182 and instead show anomalous base peaks with odd masses in their mass spectra. Alkaloid **251F** ($C_{16}H_{29}NO$) shows a base peak of $C_7H_{13}N$ (m/z 111) and a higher analog **265B** ($C_{17}H_{31}NO$) shows a base peak at $C_8H_{15}N$ (m/z 125). The third compound, **247**, is a trace constituent in *Dendrobates bombetes*. It shows a base peak at m/z 109 and a major fragment ion at m/z 70 and probably is an unsaturated analog of **251F**. In addition, some four other alkaloids had been tentatively placed in the pumiliotoxin-A class based primarily on a major ion of C_4H_8N (m/z 70)

(80). None exhibit a major ion at m/z 166 or 182, but instead show other ions of even mass ranging from m/z 84 to m/z 168. These alkaloids **251E, 281B, 341A, 351** and the above mentioned alkaloids **247, 251F, 265B,** are for the present included under *Other Alkaloids* until further characterization is possible.

The pumiliotoxin-A class of alkaloids with a major fragment ion at m/z 166 is widely distributed in frogs of the genus *Dendrobates* occurring in 14 of the 20 species in Table 22. The allopumiliotoxin-A subclass with a major fragment ion at m/z 182 occurs in 16 of the 20 species in Table 22.

B. Syntheses

A synthesis of optically active pumiliotoxin **251D,** identical in physical and spectral properties with the natural compound has been realized using L-proline as starting material (Scheme L) (*206*). The overall yield was about 20%. One important comparison has not been carried out, namely optical rotation, due to the present lack of sufficient quantities of the natural compound. The hydrochloride of synthetic pumiliotoxin **251D** was dextrorotatory, while the free base appeared to be weakly levorotatory (Table 20). The chiral precursor for the side chain of pumiliotoxin **251D** was prepared in an overall yield of about 80% (reactions H, I).

A lower homolog of **251D** also was prepared (reactions C, D). The carbon-13 magnetic resonance peaks ($CDCl_3$) of this lower homolog which does not contain the 18-CH_3 of the pumiliotoxin-A class alkaloids are as follows: 131.4^s, 128.0^d, 71.7^d, 68.4^s, 54.6^t, 52.8^t, 48.9^t, 32.1^t, 27.2^t, 24.3^q, 23.3^t, 22.3^t, 21.1^t, 13.9^q. Proton magnetic resonance data were also reported. The compound though lacking the 18-CH_3 group exhibits a major mass spectral fragmentation at m/z 166 indicating that the pathway proposed earlier (*84*) to yield this fragment in the pumiliotoxin-A class alkaloids is incorrect.

C. Biological Activity

1. Potentiation of Muscle Contraction and Calcium Translocation

Pumiliotoxin B, a representative member of the pumiliotoxin-A class of alkaloids potentiates both direct and indirect evoked contractions of striated muscle (*24, 26, 183*). In addition, pumiliotoxin B prolongs muscle contraction *(vide infra)*. The potentiation of muscle contracture appears to

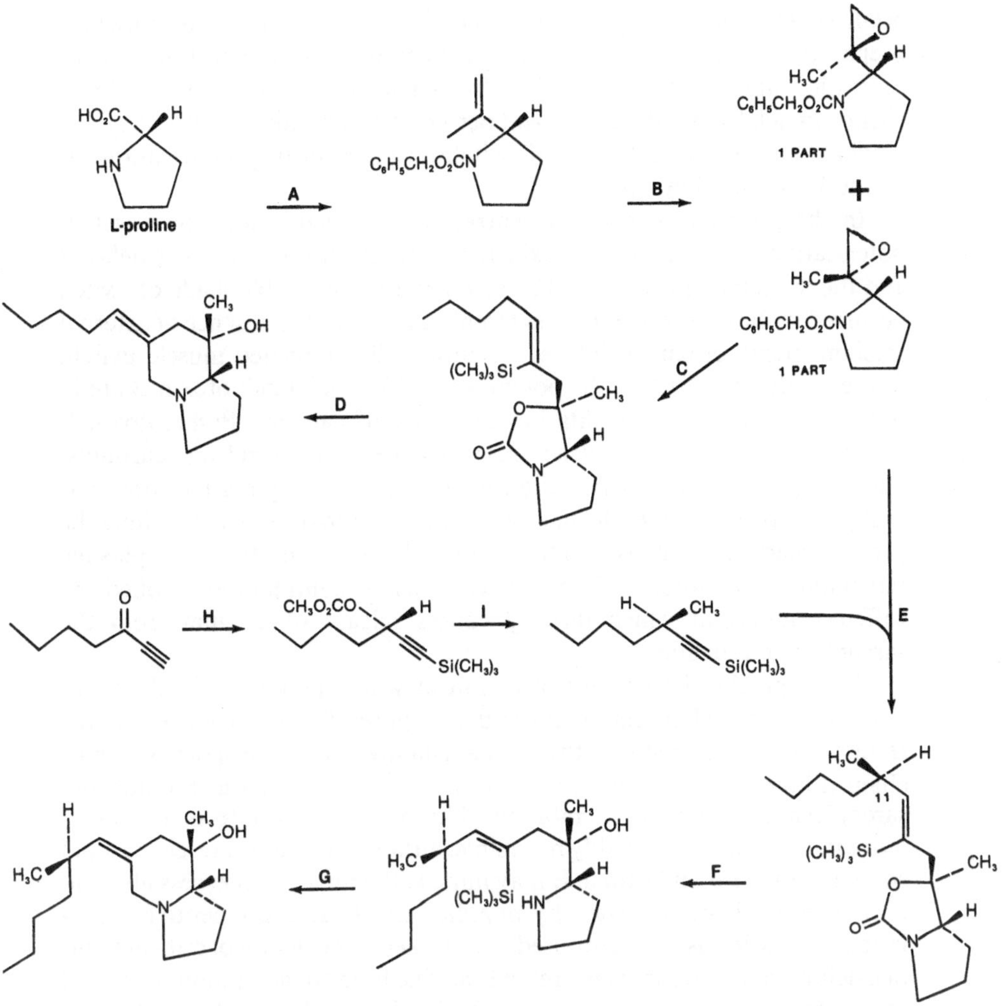

Scheme L. Synthesis of l-pumiliotoxin **251 D** (*206*). Reaction E affords a minor amount of the epimer at C-11 due to the fact that the silyl alkyne was only 82% optically pure. Optically pure silyl alkyne could be isolated by chromatographic separation of a chiral carbamate (BELL, K. H., Ph. D. Thesis, University of California at Irvine, 1981). **A** i) $C_6H_5CO_2Cl$, ii) CH_2N_2, iii) CH_3MgI, iv) $SOCl_2$, pyridine. **B** i) m-Chloroperbenzoic acid. **C** i) $(CH_3)_3SiC \equiv C(CH_2)_3CH_3$, Al(i-Bu)$_2H, CH_3$Li. **D** i) KOH, ii) Paraformaldehyde, iii) H$^+$. **E** i) Al(i-Bu)$_2$H, CH$_3$Li. **F** i) KOH. **G** i) Paraformaldehyde, ii) H$^+$. **H** i) β-3-Pinanyl-9-borabicyclo-[3.3.1]-nonane, ii) (CH$_3$)$_3$SiCl, iii) CH$_3$O$_2$CCl. **I** i) (CH$_3$)$_2$CuMgBr

be linked to a mechanism involving facilitation of calcium influx into the muscle fiber and/or facilitation of release of calcium from the sarcoplasmic reticulum. The effect of pumiliotoxin B on the magnitude of muscle twitch is time and stimulus-dependent. No blockade of neuromuscular transmission occurred with a low concentration of pumiliotoxin B in a rat neuromuscular preparation (*183*), while in a frog neuromuscular preparation a concentration and stimulus-dependent blockade of both direct and indirect elicited twitch followed the transient period of potentiated twitch (*24, 26*). A maximal six-fold potentiation of twitch occurred in frog preparations at about 3 µM pumiliotoxin B.

In the presence of high concentrations of calcium ion, the onset of potentiative effect of pumiliotoxin B on muscle contraction was delayed and the maximal effect was reduced. Furthermore, no blockade of twitch occurred in the presence of high external calcium. In the absence of external calcium, pumiliotoxin B did not potentiate direct elicited muscle twitch. These results suggest that the potentiative effects of pumiliotoxin B are in some way linked to a facilitation of influx of calcium. Pumiliotoxin B restored directly evoked contractures in a glycerol-shocked frog neuromuscular preparation in which excitation-contraction coupling had been virtually disrupted. This result suggests that pumiliotoxin B may facilitate the process leading to evoked release of calcium from the sarcoplasmic reticulum. Pumiliotoxin B enhanced muscle contractures evoked by caffeine, an agent which directly causes release of calcium from the sarcoplasmic reticulum.

Although the site or sites of action at which pumiliotoxin B acts to potentiate striated muscle contracture requires further delineation, the results strongly suggest that the mechanism involves a transport system or channel involved in calcium fluxes across both the plasma membrane and sarcoplasmic reticulum membrane. Pharmacological effects of pumiliotoxin B appear to be readily reversible, although in whole muscle the lipid solubility of the pumiliotoxin B may impede the removal of excess alkaloid (*24*). The blockade of twitch by pumiliotoxin B during repetitive stimulation of muscle has been proposed to be linked to depletion of critical pools of calcium in the sarcoplasmic reticulum, due both to facilitation of evoked release by pumiliotoxin B and to blockade of reuptake of calcium through the alkaloid's inhibitory effects on the calcium-dependent ATPase of sarcoplasmic reticulum *(vide infra)*. That pumiliotoxin B may displace critical internal pools of calcium was suggested in experiments with the calcium chelator EGTA. After treatment of muscle with EGTA and pumiliotoxin B, reintroduction of calcium could restore the potentiative actions of the alkaloid only if pumiliotoxin B were removed and then subsequently reapplied. Thus, no marked potentiation of contractions occurred in media containing EGTA and pumiliotoxin B or during subsequent incubation with media containing calcium and pumiliotoxin B. Marked potentiation of contractions occurred only after washing and incubation with media containing calcium but not pumiliotoxin B, followed by reinstatement of alkaloid.

Pumiliotoxin B had no apparent effect on sodium, potassium or chloride conductances or on resting membrane potential (24). The potentiative effects of pumiliotoxin B were fully manifest in preparations in which acetylcholine receptors were blocked by α-bungarotoxin. Dantrolene, an agent which appears to interfere with release of calcium from the sarcoplasmic reticulum and methoxyverapamil, a purported antagonist of calcium channels, only partially prevented the effects of pumiliotoxin B on muscle contractions.

Structure-activity correlations for pumiliotoxin-A class alkaloids have not been delineated because of the limited supplies available from nature. Pumiliotoxin B is more active in *potentiating* frog muscle contraction than is pumiliotoxin A, which in turn is much more active than pumiliotoxin **251D** (cited in 24). Thus, hydroxyl groups in the side chain appear important to at least one biological activity of this class of alkaloids.

Further evidence for facilitation by pumiliotoxin B of calcium translocation across membranes derives from crayfish skeletal muscle (24). Pumiliotoxin B enhanced calcium-dependent action potentials in crayfish muscle with a half maximal effect at about $25 \mu M$. Pumiliotoxin B also markedly potentiated directly evoked contractions of crayfish muscle.

In guinea pig atrium, pumiliotoxin B $(1.5 - 7.5 \mu M)$ caused a dose-dependent potentiation of spontaneous contractions and increased the rate of contractions (183). The effects were readily reversible and could not be blocked by a β-adrenergic antagonist.

Pumiliotoxin B caused dose-dependent contractions and peristaltic movements in guinea pig ileum segments (183). The effects were blocked by tetrodotoxin suggesting that nerve activity and perhaps facilitation of transmitter release by pumiliotoxin B *(vide infra)* were involved. Neither muscarinic nor histamine-antagonists prevented the effects of pumiliotoxin B.

2. Prolongation of Muscle Contractions and Inhibition of Calcium-Dependent ATPase

Pumiliotoxin B both potentiated and prolonged contractions in frog neuromuscular preparations (24, 26). The mechanism involved in prolongation of contractions appears to be linked to inhibition of the calcium-dependent ATPase of sarcoplasmic reticulum. This enzyme is primarily responsible for reuptake of calcium released into the cytoplasm and thereby termination of muscle contractions. The prolongation of muscle contraction occurred with pumiliotoxin B even in calcium-free medium and was still present even after the potentiative effects on twitch had reversed during repetitive stimulation and been replaced by a marked reduction in twitch.

The prolongation of twitch elicited by pumiliotoxin B resulted in a fusion of muscle twitches during tetanic stimulation.

There was an excellent correlation between the effects of pumiliotoxins on the time course of muscle twitch and their inhibitory effects on calcium-dependent ATPase of sarcoplasmic reticulum (247). Pumiliotoxin **251D** and pumiliotoxin B were potent both with respect to prolongation of muscle twitch and with respect to inhibition of calcium-dependent ATPase, while pumiliotoxin A was relatively inactive in both paradigms. It should be noted that pumiliotoxin B caused a maximal prolongation of muscle twitches at concentrations of 1 – 3 µM or less, while half maximal inhibitory effects on calcium-dependent ATPase *in vitro* required nearly 25 µM. Furthermore, pumiliotoxin B inhibited only about 50% of the calcium-dependent ATPase activity in sarcoplasmic reticulum. The inhibition of the enzyme by pumiliotoxin B was not readily reversible. In contrast to results with pumiliotoxin B, inhibition of the calcium-dependent ATPase by pumiliotoxin **251D** was virtually complete, although again requiring about a 25 µM concentration of the alkaloid for the half maximal inhibitory effect. Other data suggested that pumiliotoxin B is a relatively specific inhibitor for calcium-dependent ATPases, but that pumiliotoxin **251D** may be a general ATPase inhibitor. Thus, the latter alkaloid inhibited both sodium-potassium-dependent-ATPase and magnesium-dependent ATPase of rat brain synaptosomes, while pumiliotoxin B and pumiliotoxin A had little effect on these enzymes. Half maximal inhibition of synaptosomal ATPases required about 30 µM pumiliotoxin **251D**.

Pumiliotoxin B potentiated and prolonged muscle contractions in chick neuromuscular preparations and inhibited chick calcium-dependent ATPase (26). Remarkably, in muscles from a dystrophic strain of chicks, pumiliotoxin B had no effect neither on contractions nor on the calcium-dependent ATPase. Thus, in confirmation of other data, calcium mechanisms appear aberrant in the dystrophic muscles of this strain of chicks.

3. Effects on Nerve

Pumiliotoxin B had calcium-dependent effects on evoked release of acetylcholine from nerve terminals in frog neuromuscular preparations (26). The alkaloid in a dose and stimulus-dependent manner caused repetitive endplate potentials in response to a single stimulation of nerve. This effect on neurotransmitter release was strongly calcium-dependent and probably involved facilitation by pumiliotoxin B of evoked calcium transport through plasma membranes and/or membranes of the endoplasmic reticulum of the nerve terminal.

References, pp. 326—340

4. Summary

Pumiliotoxin B and its congeners appear to have selective effects on nerve and muscle linked to facilitation of evoked calcium translocation across both plasma membranes and internal membranes of calcium storage organelles. In addition, these alkaloids appear to also affect muscle function by inhibiting the reuptake of calcium into storage organelles *via* an inhibition of the calcium-dependent ATPase. Further studies are required to delineate the nature of the sites whereby such alkaloids facilitate calcium translocation. Pumiliotoxin A and B are relatively toxic compounds with a minimum lethal doses of 1.5 to 2.5 mg/kg in mice (*77*). Allopumiliotoxin **267A** appeared less toxic and at 2 mg/kg caused locomotor difficulties but no deaths (*80*). The effects of the pumiliotoxin-A class of alkaloids, primarily those of pumiliotoxin B, on biological systems are summarized in Table 21.

Table 21. *Effects of Alkaloids of the Pumiliotoxin-A Class on Biological Systems*

System	Species	Comments	References
Whole animal	Mouse	Locomotor difficulties, clonic convulsions and death. Minimum lethal dose 2.5 and 1.5 mg/kg for pumiliotoxin A and B respectively	(*77, 80*)
Neuromuscular preparation (striated)	Rat, frog, chick	Potentiation and prolongation of muscle contractions, facilitation of neurotransmitter release	(*24, 26, 183*)
	Dystrophic chick	No effect	(*26*)
Cardiac preparation	Guinea pig atrium	Positive chronotropic and inotropic effect	(*183*)
Smooth muscle	Guinea pig ileum	Elicits rhythmic muscle contractures	(*183*)
Enzymes	Ca^{++}-ATPase	Inhibition	(*24, 26, 247*)
	Na^{+}-K^{+}-ATPase and Mg^{++}-ATPase	Inhibition by pumiliotoxin **251D** but not by pumiliotoxin B	(*247*)
	Acetylcholinesterase	No effect	Cited in (*24*)

VII. Other Alkaloids

A. Structures, Properties, and Occurrence

More than fifty alkaloids from dendrobatid frogs cannot, at present, be *positively* assigned to the classes represented by pumiliotoxin C, by the histrionicotoxins, by the bicyclic or tricyclic gephyrotoxins, or by the pumiliotoxin-A and allopumiliotoxin groups. A number of these compounds were previously assigned to either the pumiliotoxin-C, hydroxypumiliotoxin-C, or pumiliotoxin-A classes (*80*). In the present review they are grouped under *Other Alkaloids* and their properties are reported below. Most are minor or trace compounds, and isolation of sufficient quantities for establishment of definitive structures may prove difficult. Certain of these other alkaloids do occur in significant amounts, but in dendrobatid species which will be difficult to collect in large numbers. Thus, it appears that further progress will occur mainly as a result of advances in spectral techniques. In some instances the identity of such compounds may be established by comparison with synthetic material as was the case for the bicyclic gephyrotoxin **223 AB** (*244, vide supra*). In view of the pharmacological activities exhibited by most classes of alkaloids isolated from dendrobatid frogs, further studies on the as yet unclassified compounds appear warranted. Alkaloid **219 A** at 4 mg/kg in mice caused paralysis of hind limbs, salivation and piloerection (*80*).

The distribution of these "other alkaloids" together with the distribution of pumiliotoxins, histrionicotoxins and gephyrotoxins in dendrobatid frogs is documented in Table 22.

One dendrobatid alkaloid not listed in Table 22 is a potent analgesic (unpublished results). The presence of this compound in extracts from the frog *Dendrobates tricolor, D. espinosai*, and *D. pictus* was suggested by a marked Straub-tail reaction after injection of small amounts of crude alkaloid fraction (*80*). A Straub-tail reaction is typical of the opiate-class of analgesics. With the Straub-tail as a bioassay, small amounts of the active principle were isolated. The substance was a trace alkaloid and appeared to be present in quantities of less than 1 µg per frog skin. The true parent ion of this analgesic alkaloid has not been detected apparently because of facile pyrolysis during attempts to carry out direct probe mass spectral analysis spectrometry. Instead, an apparent parent ion of $C_{10}H_{11}NCl$ has been detected both with direct probe analysis and with combined gas chromatography-mass spectral analysis. The presence of a halogen is surprising and the structure of this trace alkaloid is under investigation.

Recently, two "indole" alkaloids, calycanthine and chimonanthine, and a piperidinyldipyridine alkaloid, noranabasamine (Fig. 26) were isolated as minor compounds in extracts from the Colombian poison-dart frog

References, pp. 326—340

Phyllobates terribilis (*254*). The occurence of alkaloids with such structures in a dendrobatid frog is truly remarkable. Apparently these alkaloids, calycanthine ($[\alpha]_D^{25}$-570°, CH_3OH) chimonanthine ($[\alpha]_D^{25}$ + 280°, CH_3OH) and noranabasamine ($[\alpha]_D^{25}$ − 14.4°, CH_3OH) are optical antipodes of the corresponding compound from the plant kingdom. Calycanthine is an alkaloid derived biosynthetically from tryptamine. Calycanthine, while truly not an indole alkaloid, has been included in this category of alkaloids because of its biosynthetic relationship to other *Calycanthus* alkaloids such as chimonanthine which are indolic in character. Noranabasamine is a member of a family of alkaloids related to nicotine.

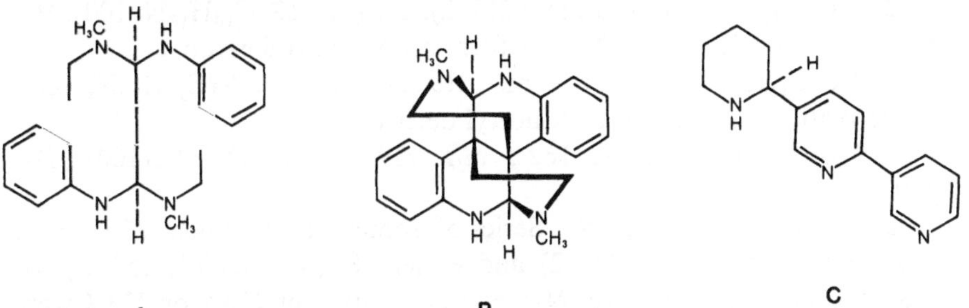

Fig. 26. Structures of **A** chimonanthine, **B** calycanthine, and **C** noranabasamine: Minor alkaloids isolated from skin extracts of the poison-dart frog *Phyllobates terribilis* (*254*). Structures are not intended to depict absolute configurations

Properties of most of the more typical "other alkaloids" from dendrobatid frogs are documented in the following entries which are in the usual format; empirical formulae which are based on analogy and chemical and chromatographic properties and which have not been confirmed by high resolution mass spectrometry are in quotation marks.

Other Alkaloids

167A. "$C_{11}H_{21}N$", −, 151°, m/z 167(1), 166(1), 138(100). H_0-derivative.

181A. "$C_{12}H_{23}N$", −, 152°, m/z 181(2), 180(1), 152(100). H_0-derivative.

181B. "$C_{12}H_{23}N$", −, 153°, m/z 181(2), 180(2), 138(100). H_0-derivative.

185. ?, 0.20, 153°, m/z 185(1), 170(100). H_0-derivative.

195B. "$C_{13}H_{25}N$", 0.23, 156°, m/z 195(2), 194(1), 138(100). H_0-derivative.

195C. "$C_{13}H_{25}N$", −, 153°, m/z 195(65), 152(100). H_0-derivative.

197. ?, $-$, 160°, m/z 197 (1), 180 (100), 126 (35). H_0-derivative.

203. $C_{14}H_{21}N$, 0.33, 158°, m/z 203 (1), 202 (2), 138 (100). H_6-derivative, m/z 209, 138. Did not form an N-acetyl derivative.

205. "$C_{14}H_{23}N$", $-$, 158°, m/z 205 (1), 204 (2), 138 (100). H_4-derivative, m/z 209, 138.

207. "$C_{14}H_{25}N$", $-$, 158°, m/z 207 (1), 206 (1), 138 (100). H_2-derivative, m/z 209, 138.

209 A. "$C_{13}H_{25}NO$", $-$, 162°, m/z 209 (5), 168 (100). H_2-derivative, m/z 211, 168.

209 B. "$C_{13}H_{25}NO$", $-$, 163°, m/z 209 (5), 138 (100). H_2-derivative, m/z 211, 138.

217. $C_{15}H_{23}N$, 0.40, 166°, m/z 217 (2), 216 (3), 152 ($C_{10}H_{18}N$, 100). H_6-derivative, m/z 223, 152. Did not form an N-acetyl derivative.

219 A. $C_{15}H_{25}N$, 0.32, 165°, m/z 219 (1), 218 (2), 178 ($C_{12}H_{20}N$, 100). H_4-derivative, m/z 223, 180. N-acetyl derivative.

219 B. "$C_{15}H_{25}N$", $-$, 162°, m/z 219 (1), 218 (2), 152 (100). H_4-derivative, m/z 223, 152.

223 A, B, C, D. "$C_{15}H_{29}N$". Series of isomeric compounds, 0.28 – 0.32, 158 – 160°, m/z 223 (1), 222 (2) and either 180 (**A**), 166 (**B**), 152 (**C**), or 138 (**D**) (100). H_0-derivative. N-acetyl derivatives of **223 A** or **223 B** were obtained in certain extracts of *D. auratus* and *D. histrionicus*.

223 E. "$C_{14}H_{25}NO$", $-$, 163°, m/z 223 (2), 222 (3), 168 (100). H_2-derivative, m/z 223, 168.

223 F. "$C_{14}H_{27}NO$", $-$, 192°, m/z 253 (1), 138 (100). H_2-derivative. m/z 225, 182.

225. "$C_{14}H_{27}NO$", $-$, 164°, m/z 225 (3), 224 (6), 208 (2), 168 (100), 152 (25). H_0-derivative.

231 A. "$C_{16}H_{25}N$", 0.30, 166°, m/z 231 (2), 230 (1), 166 (100). H_6-derivative, m/z 237, 166. Did not form an N-acetyl derivative.

231 B. "$C_{16}H_{25}N$", 0.30, 166°, m/z 231 (2), 230 (1), 152 (100). H_6-derivative m/z 237, 152.

233. "$C_{16}H_{27}N$", $-$, 167°, m/z 233 (2), 232 (2), 166 (100). H_4-derivative, m/z 237, 166.

235 B. "$C_{16}H_{29}N$", $-$, 166°, m/z 235 (1), 234 (1), 138 (100). H_2-derivative, m/z 237, 138.

237 C. "$C_{16}H_{31}N$", $-$, 168°, m/z 237 (2), 236 (1), 180 (100). H_0-derivative.

237 D. "$C_{16}H_{31}N$", $-$, 163°, m/z 237 (1), 236 (2), 138 (100). H_0-derivative.

237 E. "$C_{15}H_{27}NO$", 0.25, 180°, m/z 237 (1), 236 (3), 208 (70), 152 (100). H_2-derivative, m/z 239, 152.

239 A. "$C_{15}H_{29}NO$", $-$, 178°, m/z 239 (2), 238 (3), 182 (100). H_0-derivative.

239 B. "$C_{15}H_{29}NO$", −, 178°, m/z 239 (2), 238 (3), 180 (100). H_0-derivative.

239 C. "$C_{15}H_{29}NO$", −, 179°, m/z 239 (2), 238 (3), 196 (100). H_0-derivative.

239 E. "$C_{15}H_{29}NO$", −, 176°, m/z 239 (2), 238 (3), 210 (40), 152 (100). H_0-derivative.

239 F. "$C_{15}H_{29}NO$", 0.30, 176°, m/z 239 (1), 168 (100). H_0-derivative. O-acetyl derivative.

239 G. "$C_{15}H_{29}NO$", −, 178°, m/z 239 (1), 238 (3), 138 (100). H_0-derivative.

241. ?, −, 180°, m/z 241 (2), 240 (3), 166 (100), 126 (48).

243. $C_{17}H_{25}N$, 0.36, 182°, m/z 243 (2), 242 (1), 202 ($C_{14}H_{20}N$, 100). H_8-derivative, m/z 251, 208. N-acetyl derivative.

247. "$C_{16}H_{25}NO$", −, 175°, m/z 247 (5), 246 (3), 230 (5), 170 (25), 166 (25), 150 (40), 109 (100), 70 (50). H_4-derivative, m/z 251, 232, 111, 70.

251 A. "$C_{17}H_{33}N$", −, 170°, m/z 251 (2), 208 (6), 152 (100). H_0-derivative.

251 B. "$C_{16}H_{29}NO$", −, 184°, m/z 251 (2), 234 (4), 138 (100). H_2-derivative, m/z 253, 138.

251 C. "$C_{16}H_{29}NO$", −, 190°, m/z 251 (2), 234 (4), 154 (100). H_2-derivative, m/z 253, 154.

251 E. "$C_{15}H_{25}NO_2$", −, 175°, m/z 251 (3), 250 (1), 234 (2), 168 (30), 84 (18), 70 (100).

251 F. $C_{16}H_{29}NO$, 0.25, 184°, m/z 251 (20), 250 (23), 236 (10), 234 (3), 222 ($C_{14}H_{24}NO$, 13), 220 ($C_{15}H_{26}N$, 25), 194 ($C_{12}H_{20}NO$, 30), 166 (10), 164 ($C_{11}H_{18}N$, 14), 152 ($C_{10}H_{18}N$, 32), 150 ($C_{10}H_{16}N$, 16), 112 (40), 111 ($C_7H_{13}N$, 100), 98 ($C_6H_{12}N$, 52), 70 (15). H_0-derivative. O-acetyl derivative.

253 B. "$C_{16}H_{33}NO$", −, 192°, m/z 253 (1), 138 (100). H_0-derivative.

257 A. "$C_{18}H_{25}N$", 0.30, 188°, m/z 257 (1), 256 (2), 216 (100). H_8-derivative, m/z 265, 222.

257 B. ?, 0.35, 192°, m/z 257 (60), 256 (100), 152 (20).

265 A. ?, 0.35, 198°, m/z 265 (50), 264 (100), 222 (58), 180 (72).

265 B. $C_{17}H_{31}NO$, 0.28, 193°, m/z 265 (18), 264 (22), 250 (12), 236 (17), 234 (20), 194 (16), 166 (25), 152 (10), 126 (30), 125 (100), 112 (28), 70 (20). H_0-derivative.

267 B. ?, −, 208°, m/z 267 (7), 266 (4), 250 (1), 170 (100), 152 (4), 112 (13).

269 A,B. $C_{19}H_{27}N$. Probably two isomeric compounds, 0.35, 207°, m/z 269 (4), 268 (12), either 204 (**A**) or 202 (**B**) (100). H_{10}-derivative, m/z 279, 208. N-acetyl derivatives.

275. $C_{19}H_{33}N$, 0.28, 198°, m/z 275 (3), 274 (2), 260 (5), 152 ($C_{10}H_{18}N$, 100). H_4-derivative, m/z 279, 278, 152. Did not form an N-acetyl derivative.

281 B. "$C_{18}H_{35}NO$", $-$, 200°, m/z 281 (4), 264 (12), 208 (25), 206 (20), 150 (65), 98 (5), 96 (20), 70 (100). H_0-derivative.

283 B. $C_{17}H_{33}NO_2$, 0.36, 197°, m/z 283 ($<$1), 282 (1), 254 (2), 212 ($C_{12}H_{22}NO_2$, 40), 152 ($C_{10}H_{18}N$, 23), 140 ($C_9H_{18}N$, 100). H_0-derivative.

283 C. $C_{17}H_{33}NO_2$, 0.40, 195°, m/z 283 ($<$1), 282 (1), 240 (5), 226 ($C_{13}H_{24}NO_2$, 28), 224 ($C_{15}H_{30}N$, 10), 166 ($C_{11}H_{20}N$, 60), 126 ($C_8H_{16}N$, 100). H_0-derivative.

285 D. ?, $-$, 190°, m/z 285 (3), 270 (2), 256 (2), 180 (35), 140 (100).

289 A. "$C_{20}H_{35}N$", $-$, 216°, m/z 289 (2), 287 (2), 274 (3), 152 (100). H_4-derivative, m/z 293, 152.

291 B. "$C_{19}H_{33}NO$", 0.12, 221°, m/z 291 (2), 290 (3), 276 (6), 168 (100). H_4-derivative, m/z 295, 168.

291 C. "$C_{19}H_{33}NO$", 0.20, 220°, m/z 291 (1), 290 (2), 276 (4), 210 (10), 152 (100). H_4-derivative, m/z 295, 152.

295. "$C_{19}H_{37}NO$", 0.09, 224°, m/z 295 (3), 278 (4), 138 (100).

301. "$C_{21}H_{35}N$", $-$, 213°, m/z 301 ($<$1), 260 (100).

309 B. "$C_{20}H_{39}NO$", 0.09, 220°, m/z 309 (1), 152 (100).

351. "$C_{21}H_{37}NO_3$", 0.15, 230°, m/z 351 (6), 350 (2), 336 (4), 152 (38), 138 (65), 70 (100).

As the structures of further dendrobatid alkaloids are elucidated, possible biosynthetic pathways for many of these alkaloids may become clearer. At present, it would appear that a 2,6-disubstituted piperidine is an attractive precursor for pumiliotoxin C, the histrionicotoxins and the gephyrotoxins (Fig. 27). The genesis of alkaloids of the pumiliotoxin-A class cannot result from simple ring closures of a precursor 2,6-disubstituted piperidine. Possible biosynthetic pathways for dendrobatid alkaloids have been discussed (80) in terms of ring closures leading to pumiliotoxin C, the histrionicotoxins and gephyrotoxins similar to those shown in Fig. 27. Pumiliotoxin **251 D** now has been included although its origin must involve a rather different pathway, either i) a migration of a side chain (Fig. 27), ii) formation of a decahydroquinoline through cyclization of one side chain followed by loss of three of the ring carbons and prior or subsequent cyclization of a three carbon side chain to form an indolizidine, or iii) some other unexpected route. It should be noted that pumiliotoxin C, the histrionicotoxins and gephyrotoxin **223 AB** could all be derived from ring closure of a cis-2,6-disubstituted piperidine while gephyrotoxin would derive from a trans-2,6-disubstituted piperidine. Gephyrotoxin is not widely distributed in dendrobatid frogs having been detected in only one species out of twenty, while histrionicotoxins occur in nearly half and pumiliotoxin C in about one fourth of the species (Table 22). Alkaloids related to pumiliotoxin **251 D**, that is the pumiliotoxin-A class and its allo-subclass, occur in about three quarters of the species. Clearly, biological aspects of these unique secondary metabolites of dendrobatid frogs should continue to be fascinating to scientists in future years as will further structure elucidation, synthesis, and pharmacological investigation.

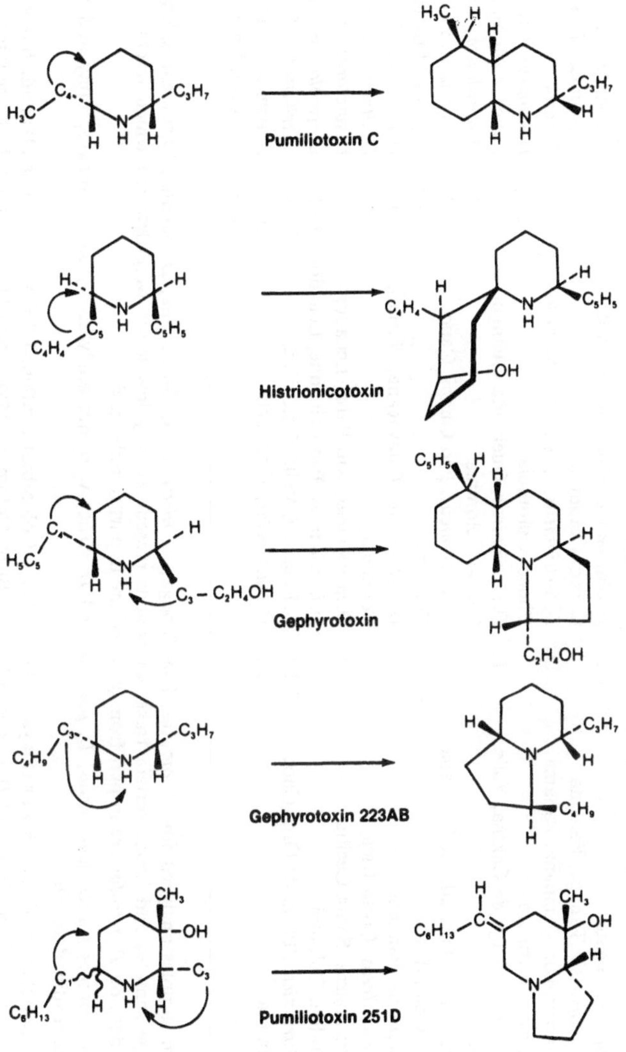

Fig. 27. Possible biosynthetic pathways from a 2,6-disubstituted piperidine to various dendrobatid alkaloids. The absolute configuration suggested for the precursor of gephyrotoxin **223 AB** is arbitrary. The absolute configuration of gephyrotoxin is based on x-ray analysis (*86*)

Table 22. *Distribution of Pumiliotoxin C, Histrionicotoxins (HTX), Gephyrotoxins (GTX), Pumiliotoxin-A (PTX-A) and Allopumiliotoxin-A (Allo-PTX-A) Class and Other Alkaloids in Various Species of Dendrobates (80, 189, 190)*

The references contain detailed information on sample size, collection sites and gas chromatographic traces for the two populations of *Dendrobates bombetes* (see Fig. 12 for gas chromatographic traces). Certain data in the footnotes is from MYERS and DALY (unpublished). Six populations of *Dendrobates femoralis*, a widespread Amazonian species have been analyzed and an alkaloid (**181B**, trace) was tentatively identified in one population (MYERS and DALY, unpublished)

I. *D. abditus*, Ecuador	VIIC. *D. histrionicus*, Quebrada Vicordo, Choco, Colombia	XA. *D. minutus*, Cerro Campana, Panama
IIA. *D. auratus*, Isla Taboga, Panama	D. *D. histrionicus*, Quebrada Docordo, Choco, Colombia	B. *D. minutus*, El Llano-Carti Road, Panama
B. *D. auratus*, Rio Campana, Panama	E. *D. histrionicus*, Quebrada Guangui, Cauca, Colombia	C. *D. minutus*, Quebrada Guangui, Cauca, Colombia
III. *D. azureus*, Surinam	F. *D. histrionicus*, Rio Guapi, Cauca, Colombia	XI. *D. occultator*, Cauca, Colombia
IVA. *D. bombetes*, Lago de Calima, Valle, Colombia	G. *D. histrionicus*, Guayacana, Narino, Colombia	XII. *D. parvulus*, Putamayo, Colombia
B. *D. bombetes*, Quebrada de la Chapa, Valle, Colombia	H. *D. histrionicus*, Rio Baba, Ecuador	XIII. *D. pictus*, Loreto, Peru
V. *D. fulguritis*, Panama	I. *D. histrionicus*, Rio Palenque, Ecuador	XIV. *D. pumilio*, Isla Bastimentos, Panama
VI. *D. granuliferus*, Costa Rica	VIII. *D. lehmanni*, Valle, Colombia	XV. *D. silverstonei*, Loreto, Peru
VIIA. *D. histrionicus*, Santa Cecilia, Risaralda, Colombia	IX. *D. leucomelus*, Venezuela	XVI. *D. tinctorius*, Surinam
B. *D. histrionicus*, Playa de Oro, Choco, Colombia		XVII. *D. tricolor*, Ecuador
		XVIII. *D. trivittatus*, Peru
		XIX. *D. truncatus*, Tolima, Colombia
		XX. *D. viridis*, Valle, Colombia

[a] The skin samples analyzed for *D. auratus*, Isla Taboga; *D. histrionicus*, Guayacana, *D. pumilio*, Bastimentos, and *D. tricolor*, Ecuador, were very large samples and hence certain trace compounds present at very low levels were detected which might otherwise have been undetected. The data on *D. tricolor* are in part from (84) and are in part unpublished.

[b] The skin samples for *D. leucomelus*, *D. parvulus* and *D. viridis* were not analyzed by gas chromatography and hence trace compounds were much less readily detected.

[c] Another population of *D. minutus* (Playa de Oro, Choco, Colombia) contained **267A** as a major alkaloid, but was not analyzed by gas chromatography. A final population of *D. minutus* (Altos de Buey, Colombia, MYERS and DALY, unpublished) contained the following alkaloids: Major, **309A**, **267A**, **251D**, **217**, **325**. Minor, **307B**. Trace, **341A**, **307A**, **323B**, **181B**.

[d] Extracts from *D. bombetes* (Quebrada de la Chapa) contained isomers of **239A**, **239B** and **239C**. Extracts from *D. occultator* also contained isomers of **239A**, **239B**, **239C** and **239D**.

[e] Extracts from *D. trivittatus*, Surinam, had virtually identical alkaloid profiles to the Peruvian sample, but were lacking **181B**.

[f] Gephyrotoxin **167B** is known only from *D. species* (Isla Colon, Panama) as a trace compound (DALY and MYERS, unpublished; see

[g] Pumiliotoxins **307D** and **309C** are known only from *D.* species (Isla Colon, Panama) as a trace and minor alkaloid, respectively (DALY and MYERS, unpublished; see footnote h).

[h] Alkaloids **195C**, **209A**, **209B**, **223F** and **253B** (see text) are known only from *D.* species (Isla Colon, Panama) as trace compounds (DALY and MYERS, unpublished data). The alkaloids of this frog are as follows: Major, **281A**, **205**, **267A**. Minor, **181B**, **251D**, **297A**, **309C**. Trace, **167B**, **195C**, **209A**, **209B**, **223F**, **253B**, **265A**, **307A**, **307D**.

[i] Extracts from *P. terribilis* contained alkaloid **195B** (*191*). Other *Phyllobates* species contain certain alkaloids from this Table (unpublished results).

[j] It has been suggested that **257B** and **265A** might be degradation artefacts (*80*).

	I	IIA[a]	IIB	III	IVA[d]	IVB[b]	V	VI	VIIA	VIIB	VIIC	VIID	VIIE	VIIF	VIIG[a]	VIIH	VIII̲	VIII	IX[b]	XA[c]	XB	XC	XI[d]	XII[b]	XIII	XIV[a]	XV	XVI	XVII[a]	XVIII[c]	XIX	XX[b]
PTX C	+++	+++																+				+				+++				+		
HTX																																
235A	++	++						+																								
259	++	++						+++	++	++	++	++	++											+								
283A		+++		+++				+++	+++	+++	+++	+++	+++	+++	+++		++					+++	+++	+++	+++		+++	+++	+++	++	++	
285A		+++		+++				+++	+++	+++	+++	+++	+++	+++	+++		++					+++	+++	+++	+++		+++	+++	+++	+++	+++	
285B								++	++	++	++	++	++	++	++									+	+++							
285C								++	++	++	++	++	++	++	++		++							+	+++		+		++	++	++	
285E										+				+																		
287A	++										++	++	++	++	++										++							
287B								++	++	++	++	++	++	++	+								++				++	++		++	++	
287D								++	++																							
291A							++		++	++	++	++	++	++	++		++						++				++		++	++	++	
GTX[f]																																
223AB								+++	++	++	++	++	++	++	++		++					+++										
239AB																						+++										
239CD															+++																	
287C								++	++	++	++	+	+	++	+		+															
289B																																
PTX-A[g]																																
237A	+++														+																	
251D		++	+++															+									+++		+			++
281A		+																			++							+				
297B		+		+																+++												
307A		++						++										++	++	+	++	+	+			+++	+++	+	+++			+++
307B		++						++										++	++	+	+++		+			+	+					+++
309A																						++										
323A	++	+++		++	++	+++		+++										+++	+++			+++	+++	++		+++	+++	+++			+++	+++

Table 22 (continued)

	I	IIA	IIB	III	IVA	IVB	V	VI	VIIA	VIIB	VIIC	VIID	VIIE	VIIF	VIIG	VIIH	VIIĮ	VIII	IX	XA	XB	XC	XI	XII	XIII	XIV	XV	XVI	XVII	XVIII	XIX	XX
Allo-PTX-A																																
237B		+																														
253A	+++						+																									
267A	+++			++	+++		+++	+++			+						++	+++	++	+++	+++	+++					+++	+++	+			
297A	+			++																	+++											
307C				+		++												++							+	+	+					
323B	++			+++	‡		+++											++	++	+++	+++	+++				++	+++	‡	‡			
325																										+	+					
339	+																	‡														
341A	+++			++			‡											+									+++		‡			‡
341B	++			+																									+			
357.																																
Other[h]																																
167A	+																															
181A	+			+																												
181B	+			+																												
185	+						+																									
195B[i]							++			+																					+	
197				+																												
203	+													++							+					+	+++		++			
205											+			+																		
207											+				+																	
217	+++			‡				+			+++																					
219A	+++	+++					+++	++			‡			+++		‡			+++		+		+				+++	‡	‡	‡	++	
219B	+	+				‡		+						+		‡			+++	++	+		+							+	+++	
223A	+	+				+												+														
223B	++	+				‡		++						+		‡			+													
223C																‡				+												
223D																‡		+	+	+	+					+						
223E																‡		‡		‡												
225																																
231A				+										‡		‡					‡								+			

This page presents a data matrix in which rows are alkaloid compound numbers and columns are the Roman‑numeral structural classes (I, IIA, IIB, III, IVA, IVB, V, VI, VIIA, VIIB, VIIC, VIID, VIIE, VIIF, VIIG, VIIH, VIII_, VIII, IX, XA, XB, XC, XI, XII, XIII, XIV, XV, XVI, XVII, XVIII, XIX, XX). A "+" denotes a single occurrence; a "‡" denotes a double mark.

Left panel

Compound	I	IIA	IIB	III	IVA	IVB	V	VI	VIIA	VIIB	VIIC	VIID	VIIE	VIIF	VIIG	VIIH	VIII_	VIII	IX	XA	XB	XC	XI	XII	XIII	XIV	XV	XVI	XVII	XVIII	XIX	XX
231B		+										+				+			+				‡					‡	+			
233		+			+	‡																										
235B					+																					+						
237C						+									‡																	
237D				+		+																										
237E												+																				
239A					‡		‡																+									
239B					+		+																+									
239C												+																				
239D																																
239E																																
239F																																
239G																																
241														‡									+									
243		+		+		+	+	‡			‡	‡	‡						+				+		+	+		‡		‡	‡	
247					+		+																									
251A																			+													
251B																			+							+						
251C																					+	+										
251E																						‡										
251F						‡	+										+															
257A																		+					+							‡		
257B						‡	+																									
265A						+	+	+											+				+	+			+	+				
265B																																
267B																					+	+										
269A														+	+	+		‡	+				+							+	+	
269B																			+				+							+	+	
275							+	+																								
281B																																
283B													‡						‡				‡									
283C													‡										‡									
285D							+	+																								
289A																			+													

Right panel

Compound	I	IIA	IIB	III	IVA	IVB	V	VI	VIIA	VIIB	VIIC	VIID	VIIE	VIIF	VIIG	VIIH	VIII_	VIII	IX	XA	XB	XC	XI	XII	XIII	XIV	XV	XVI	XVII	XVIII	XIX	XX
291B																			+													
291C		+																	+													
295			+																													
301		+					+																									
309B		+						+																								
351																																

References

1. ADAMS, H. J., A. R. MASTRI, D. DOHERTY, JR., and D. CHARRON: Spinal anesthesia with batrachotoxin in sheep and microscopic examination of spinal cords and roots. Pharmacol. Res. Comm. **10**, 719—728 (1978).

2. ADLER, M., A. C. OLIVEIRA, E. X. ALBUQUERQUE, N. A. MANSOUR, and A. T. ELDEFRAWI: Reaction of tetraethylammonium with the open and closed conformations of the acetylcholine receptor ionic channel complex. J. Gen. Physiol. **74**, 129—152 (1979).

3. ADLER, M., A. C. OLIVEIRA, M. E. ELDEFRAWI, A. T. ELDEFRAWI, and E. X. ALBUQUERQUE: Tetraethylammonium: Voltage-dependent action on endplate conductance and inhibition of ligand binding to postsynaptic proteins. Proc. Nat. Acad. Sci. (USA) **76**, 531—535 (1979).

4. ALBUQUERQUE, E. X.: The mode of action of batrachotoxin. Federat. Proc. **31**, 1133—1138 (1972).

5. ALBUQUERQUE, E. X., and J. W. DALY: Batrachotoxin, a selective probe for channels modulating sodium conductances in electrogenic membranes. In: Receptors and Recognition **1**, series B (P.CUATRACASAS, ed.), pp. 297—338. London: Chapman and Hall. 1977.

6. ALBUQUERQUE, E.X., and P. W. GAGE: Differential effects of perhydrohistrionicotoxin on neurally and iontophoretically evoked endplate currents. Proc. Nat. Acad. Sci. (USA) **75**, 1596—1599 (1978).

7. ALBUQUERQUE, E. X., and A. C. OLIVEIRA: Physiological studies on the ionic channel of nicotinic neuromuscular synapses. Adv. Cytopharmacol. **3**, 197—211 (1979).

8. ALBUQUERQUE, E. X., and J. E. WARNICK: Pharmacology of batrachotoxin. IV. Interaction with tetrodotoxin on innervated and chronically denervated rat skeletal muscle. J. Pharmacol. Exp. Therapeut. **180**, 683—697 (1972).

9. ALBUQUERQUE, E. X., M. ADLER, C. E. SPIVAK, and L. AGUAYO: Mechanism of nicotinic channel activation and blockade. Ann. N. Y. Acad. Sci. **358**, 204—238 (1980).

10. ALBUQUERQUE, E. X., E. A. BARNARD, T. H. CHIU, A. J. LAPA, J. O. DOLLY, S.-E. JANSSON, J. DALY, and B. WITKOP: Acetylcholine receptor and ion conductance modulator sites at the murine neuromuscular junction: Evidence from specific toxin reactions. Proc. Nat. Acad. Sci. (USA) **70**, 949—953 (1973).

11. ALBUQUERQUE, E. X., N. BROOKES, R. ONUR, and J. E. WARNICK: Kinetics of interaction of batrachotoxin and tetrodotoxin on rat diaphragm muscle. Mol. Pharmacol. **12**, 82—91 (1976).

12. ALBUQUERQUE, E. X., J. W. DALY, and B. WITKOP: Batrachotoxin: Chemistry and pharmacology. Science **172**, 995—1002 (1971).

13. ALBUQUERQUE, E. X., A. T. ELDEFRAWI, M. E. ELDEFRAWI, N. A. MANSOUR, and M.-C. TSAI: Amantadine: Neuromuscular blockade by suppression of ionic conductance of the acetylcholine receptor. Science **199**, 788—790 (1978).

14. ALBUQUERQUE, E. X., P. W. GAGE, and A. C. OLIVEIRA: Differential effect of perhydrohistrionicotoxin on 'intrinsic' and 'extrinsic' end-plate responses. J. Physiol. **297**, 423—442 (1979).

15. ALBUQUERQUE, E. X., K. KUBA, and J. DALY: Effect of histrionicotoxin on the ionic conductance modulator of the cholinergic receptor: A quantitative analysis of the end-plate current. J. Pharmacol. Exp. Therapeut. **189**, 513—524 (1974).

16. ALBUQUERQUE, E. X., K. KUBA, A. J. LAPA, J. W. DALY, and B. WITKOP: Acetylcholine receptor and ionic conductance modulator of innervated and denervated muscle membranes. Effect of histrionicotoxins. In: Exploratory Concepts in Muscular Dystrophy, Vol. II (MOLHORAT, A. T., ed.), pp. 585—600. Amsterdam: Excerpta Medica. 1974.

17. ALBUQUERQUE, E. X., M. SASA, B. P. AVNER, and J. W. DALY: Possible site of action of batrachotoxin. Nature New Biology 234, 93—94 (1971).

18. ALBUQUERQUE, E. X., M. SASA, and J. M. SARVEY: Batrachotoxin has no effect on the electrogenic membranes of lobster and crayfish muscles. Life Sci. 11, 357—363 (1972).

19. ALBUQUERQUE, E. X., I. SEYAMA, and T. NARAHASHI: Characterization of batrachotoxin-induced depolarization of the squid giant axons. J. Pharmacol. Exp. Therapeut. 184, 308—314 (1973).

20. ALBUQUERQUE, E. X., M.-C. TSAI, R. S. ARONSTAM, A. T. ELDEFRAWI, and M. E. ELDEFRAWI: Sites of action of phencyclidine. II. Interaction with the ionic channel of the nicotinic receptor. Mol. Pharmacol. 18, 167—178 (1980).

21. ALBUQUERQUE, E. X., M.-C. TSAI, R. S. ARONSTRAM, B. WITKOP, A. T. ELDEFRAWI, and M. E. ELDEFRAWI: Phencyclidine interactions with the ionic channel of the acetylcholine receptor and electrogenic membrane. Proc. Nat. Acad. Sci. (USA) 77, 1224—1228 (1980).

22. ALBUQUERQUE, E. X., J. E. WARNICK, and L. GUTH: Spinal cord regeneration and paraplegia. Prog. Clin. Biol. Res. 39, 41—62 (1980).

23. ALBUQUERQUE, E. X., J. E. WARNICK, and F. M. SANSONE: The pharmacology of batrachotoxin. II. Effect on electrical properties of the mammalian nerve and skeletal muscle membranes. J. Pharmacol. Exp. Therapeut. 176, 511—528 (1971).

24. ALBUQUERQUE, E. X., J. E. WARNICK, M. A. MALEQUE, F. C. KAUFMANN, R. TAMBURINI, Y. NIMIT, and J. W. DALY: The pharmacology of pumiliotoxin-B. I. Interaction with calcium sites in the sarcoplasmic reticulum of skeletal muscle. Mol. Pharmacol. 19, 411—424 (1981).

25. ALBUQUERQUE, E. X., J. E. WARNICK, F. M. SANSONE, and J. DALY: The pharmacology of batrachotoxin. V. A comparative study of membrane properties and the effects of batrachotoxin on sartorius muscles of the frogs Phyllobates aurotaenia and Rana pipiens. J. Pharmacol. Exp. Therapeut. 184, 315—329 (1973).

26. ALBUQUERQUE, E. X., J. E. WARNICK, R. TAMBURINI, F. C. KAUFMANN, and J. W. DALY: Interaction of pumiliotoxin-B with calcium sites in the sarcoplasmic reticulum and nerve terminal of normal and dystrophic muscle. In: Exploratory Concepts in Muscular Dystrophy. Amsterdam: Excerpta Medica. In press (1982).

27. ANWYL, R., and T. NARAHASHI: Inhibition of the acetylcholine receptor by histrionico-toxin. Brit. J. Pharmacol. 68, 611—616 (1980).

28. ARATANI, M., L. V. DUNKERTON, T. FUKUYAMA, Y. KISHI, H. KAKOI, S. SUGIURA, and S. INOUE: Synthetic studies on histrionicotoxins. I. A stereocontrolled synthesis of ($\pm$)-perhydrohistrionicotoxin. J. Org. Chem. 40, 2009—2011 (1975).

29. ARONSTAM, R. S.: Interactions of tricyclic antidepressants with a synaptic ion channel. Life Sci. 28, 59—64 (1981).

30. ARONSTAM, R. S., A. T. ELDEFRAWI, and M. E. ELDEFRAWI: Similarities in the binding sites of the muscarinic receptor and the ionic channel of the nicotinic receptor. Biochem. Pharmacol. 29, 1311—1314 (1980).

31. ARONSTAM, R. S., A. T. ELDEFRAWI, I. N. PESSAH, J. W. DALY, E. X. ALBUQUERQUE, and M. E. ELDEFRAWI: Regulation of [^{3}H]-perhydrohistrionicotoxin binding to Torpedo ocellata electroplax by effectors of the acetylcholine receptor. J. Biol. Chem. 256, 2843—2851 (1981).

32. BARTELS-BERNAL, E., T. L. ROSENBERRY, and J. W. DALY: Effects of batrachotoxin on the electroplax of electric eel: Evidence for voltage-dependent interaction with sodium channels. Proc. Nat. Acad. Sci. (USA) 74, 951—955 (1977).

33. BARTELS DE BERNAL, E., M. I. LLANO, and E. DIAZ: Algunos efectos de la batrachotoxina sobre las electroplacas de la anguila electrica y su antagonismo con anesthesicos locales. Acta Med. Valle 6, 74—80 (1975).

34. BENNETT, G. B., and H. MINOR: 7,8-Dihydro-5-(6H)-quinolines: Potential intermediates for the synthesis of pumiliotoxin C. J. Heterocycl. Chem. **16**, 633—635
(1979).

35. BERNER, H., L. BERNER-FENZ, R. BINDER, W. GRAF, T. GRUTTER, C. PASCUAL, and H.
WEHRLI: Die Synthese von 5β0,19N-Ep-(oxyathanoimino)-Steroiden. Helv. Chim. Acta
53, 2252—2258 (1970).

36. BERNER-FENZ, L., H. BERNER, W. GRAF, and H. WEHRLI: Synthese von 14β0,18N-Ep-
(oxyathanoimino)-Steroiden. Helv. Chim. Acta **53**, 2258—2265 (1970).

37. BLANCHARD, S. G., and M. A. RAFTERY: Identification of the polypeptide chains in
Torpedo california electroplax membranes that interact with a local anesthetic analog.
Proc. Nat. Acad. Sci. (USA) **76**, 81—85 (1979).

38. BLANCHARD, S. G., J. ELLIOTT, and M. A. RAFERTY: Interaction of local anesthetics with
Torpedo californica membrane-bound acetylcholine receptor. Biochemistry **18**, 5880—
5885 (1979).

39. BOEGMAN, R. J., and E. X. ALBUQUERQUE: Axonal transport in rats rendered paraplegic
following a single subarachnoid injection of either batrachotoxin or 6-aminonicotin-
amide into the spinal cord. J. Neurobiol. **11**, 283—290 (1980).

40. BOEGMAN, R. J., and T. W. OLIVER: Neural influence on muscle hydrolase activity. Life
Sci. **27**, 1339—1344 (1980).

41. BOEGMAN, R. J., and R. J. RIOPELLE: The role of axonal transport and impulse
conduction on the uptake and retrograde transport of nerve growth factor and bovine
serum albumin in peripheral nerve. J. Neurobiol. **11**, 497—501 (1980).

42. — — Batrachotoxin blocks slow and retrograde axonal transport *in vivo*. Neurosci. Lett.
18, 143—147 (1980).

43. BOEGMAN, R. J., S. S. DESHPANDE, and E. X. ALBUQUERQUE: Consequences of axonal
transport blockade induced by batrachotoxin on mammalian neuromuscular junction I.
Early pre- and postsynaptic changes. Brain Res. **187**, 183—196 (1980).

44. BOND, F. T., J. E. STEMKE, and D. W. POWELL: Facile synthesis of l-azaspiro-(5.5)-
undecan-2,7-dione. Synthetic Commun. **5**, 427—433 (1975).

45. BROWN, G. B., and J. W. DALY: Interaction of batrachotoxinin-A benzoate with voltage
sensitive sodium channels. The effects of pH. Cell. Mol. Neurobiol., in press (1982).

46. BROWN, G. B., S. C. TIESZEN, J. W. DALY, J. E. WARNICK, and E. X. ALBUQUERQUE:
Batrachotoxin-A 20-α-benzoate: A new radioactive ligand for voltage sensitive sodium
channels. Cell. Mol. Neurobiol. **1**, 19—40 (1981).

47. BROWN, J. H.: Calcium-dependent blockade of cardiac cyclic AMP accumulation by
batrachotoxin and veratridine. Mol. Pharmacol. **20**, 113—117 (1981).

48. BURGERMEISTER, W., W. A. CATTERALL, and B. WITKOP: Histrionicotoxin enhances
agonist-induced desensitization of acetylcholine receptor. Proc. Nat. Acad. Sci. (USA)
74, 5754—5758 (1977).

49. BURGERMEISTER, W., W. L. KLEIN, M. NIRENBERG, and B. WITKOP: Comparative binding
studies with cholinergic ligands and histrionicotoxin at muscarinic receptors of neural
cell lines. Mol. Pharmacol. **14**, 751—767 (1978).

50. CATTERALL, W. A.: Inhibition of voltage-sensitive sodium channels in neuroblastoma
cells by antiarrhythmic drugs. Mol. Pharmacol.**20**, 356—362 (1981).

51. — Activation of the action potential Na$^+$ ionophore by neurotoxins: An allosteric
model. J. Biol. Chem. **252**, 8669—8676 (1977).

52. — Activation of the action potential Na$^+$ ionophore of cultured neuroblastoma cells by
veratridine and batrachotoxin. J. Biol. Chem. **250**, 4053—4059 (1975).

53. — Cooperative activation of action potential Na$^+$ ionophore by neurotoxins. Proc. Nat.
Acad. Sci. (USA) **72**, 1782—1786 (1975).

54. — Activation and inhibition of the action potential Na$^+$ ionophore of cultured rat
muscle cells by neurotoxin. Biochem. Biophys. Res. Comm. **68**, 136—142 (1976).

55. — Purification of a toxic protein from scorpion venom which activates the action potential Na⁺ ionophore. J. Biol. Chem. **251**, 5528—5536 (1976).

56. — Membrane potential-dependent binding of scorpion toxin to the action potential Na⁺ ionophore. Studies with a toxin derivative prepared by lactoperoxidase-catalyzed iodination. J. Biol. Chem. **252**, 8660—8668 (1977).

57. —Neurotoxins as allosteric modifiers of voltage-sensitive sodium channels. Adv. Cytopharmacol. **3**, 305—316 (1979).

58. — Pharmacologic properties of voltage-sensitive sodium channels in chick muscle fibers developing *in vitro*. Dev. Biol. **78**, 222—230 (1980).

59. — Neurotoxins that act on voltage-sensitive sodium channels in excitable membranes. Ann. Rev. Pharmacol. Toxicol. **20**, 15—44 (1980).

60. CATTERALL, W. A., and D. A. BENESKI: Interaction of polypeptide neurotoxins with a receptor site associated with voltage-sensitive sodium channels. J. Supramol. Struct. **14**, 295—304 (1980).

61. CATTERALL, W. A., and L. BERESS: Sea anemone toxin and scorpion toxin share a common receptor site associated with the action potential sodium ionophore. J. Biol. Chem. **253**, 7393—7396 (1978).

62. CATTERALL, W. A., and R. RAY: Interactions of neurotoxins with the action potential Na⁺ ionophore. J. Supramol. Struct. **5**, 397—407 (1976).

63. CATTERALL, W. A., and M. RISK: Toxin T₄₆ from *Ptychodiscus brevis* (formerly *Gymnodinium breve*) enhanced activation of voltage sensitive sodium channels by veratridine. Mol. Pharmacol. **19**, 345—348 (1981).

64. CATTERALL, W. A., C. S. MORROW, and R. P. HARTSHORNE: Neurotoxin binding to receptor sites associated with voltage-sensitive sodium channels in intact, lysed, and detergent-solubilized brain membranes. J. Biol. Chem. **254**, 11379—11388 (1979).

65. CATTERALL, W. A., R. RAY, and C. S. MORROW: Membrane potential dependent binding of scorpion toxin to action potential Na⁺ ionophore. Proc. Nat. Acad. Sci. (USA) **73**, 2682—2686 (1976).

66. CATTERALL, W. A., C. S. MORROW, G. B. BROWN, and J. W. DALY: Binding of batrachotoxinin A 20-α-benzoate to a receptor site associated with sodium channels in synaptic nerve ending particles. J. Biol. Chem. **256**, 8922—8927 (1981).

67. COCHRANE, C. S.: Journal of a residence and travels in Colombia during the years 1823 and 1824. London: Henry Colburn. 1825.

68. COLQUHOUN, D., R. HENDERSON, and J. M. RITCHIE: The binding of labelled tetrodotoxin to non-myelinated nerve fibres. J. Physiol. (London) **227**, 95—126 (1972).

69. CONN, P. J., and D. C. ROGERS: Gonadotropin release from pituitary cultures following activation of endogenous ion channels. Endocrinology **107**, 2133—2134 (1980).

70. COREY, E. J., and R. D. BALANSON: Studies directed toward the total synthesis of perhydrohistrionicotoxin. Heterocycles **5**, 445—470 (1976).

71. COREY, E. J., and R. A. RUDEN: Stereoselective methods for the synthesis of terminal *cis* and *trans* enyne units. Tetrahedron Letters **1973**, 1495—1499.

72. COREY, E. J., J. F. ARNETT, and G. N. WIDIGER: A simple total synthesis of (±)-perhydrohistrionicotoxin. J. Amer. Chem. Soc. **98**, 430—431 (1975).

73. COREY, E. J., M. PETRZILKA, and Y. UEDA: A new synthetic route to (±)-perhydro-histrionicotoxin. Tetrahedron Letters **1975**, 4343—4346.

74. COREY, E. J., M. PETRZILKA, and Y. UEDA: A new synthetic route to (±)-perhydrohistri-onicotoxin. Helv. Chim. Acta **60**, 2294—2302 (1977).

75. COREY, E. J., Y. UEDA, and R. A. RUDEN: Synthetic route to neurotoxins in the 2,7-*epi*-histrionicotoxin series. Tetrahedron Letters **1975**, 4347—4350.

76. CREVELING, C. R., E. T. MCNEAL, D. H. MCCULLOH, and J. W. DALY: Membrane potentials in cell-free preparations from guinea pig cerebral cortex: Effect of depolarizing agents and cyclic nucleotides. J. Neurochem. **35**, 922—932 (1980).

77. DALY, J. W., and C. W. MYERS: Toxicity of Panamanian poison frogs (*Dendrobates*): Some biological and chemical aspects. Science **156**, 970—973 (1967).

78. DALY, J., and B. WITKOP: Batrachotoxin, an extremely active cardio- and neurotoxin from the Colombian arrow poison frog. Clinical Toxiology **4**, 331—342 (1971).

79. DALY, J., E. X. ALBUQUERQUE, F. C. KAUFFMAN, and F. OESCH: Effects of batrachotoxin on electroplax Na^+-K^+-ATPase and levels of ATP in rat muscle. J. Neurochem. **19**, 2829—2833 (1972).

80. DALY, J. W., G. B. BROWN, M. MENSAH-DWUMAH, and C. W. MYERS: Classification of skin alkaloids from neotropical poison-dart frogs (Dendrobatidae). Toxicon **16**, 163—188 (1978).

81. DALY, J. W., I. KARLE, C. W. MYERS, T. TOKUYAMA, J. A. WATERS, and B. WITKOP: Histrionicotoxins: Roentgen-ray analysis of the novel allenic and acetylenic spiro-alkaloids isolated from a Colombian frog, *Dendrobates histrionicus*. Proc. Nat. Acad. Sci. (USA) **68**, 1870—1875 (1971).

82. DALY, J. W., E. MCNEAL, C. PARTINGTON, M. NEUWIRTH, and C. R. CREVELING: Accumulations of cyclic AMP in adenine-labeled cell-free preparations from guinea pig cerebral cortex: Role of α-adrenergic and H_1-histaminergic receptors. J. Neurochem. **35**, 326—337 (1980).

83. DALY, J. W., C. W. MYERS, J. E. WARNICK, and E. X. ALBUQUERQUE: Levels of batrachotoxin and lack of sensitivity to its action in poison-dart frogs (*Phyllobates*). Science **208**, 1383—1385 (1980).

84. DALY, J. W., T. TOKUYAMA, T. FUJIWARA, R. J. HIGHET, and I. L. KARLE: A new class of indolizidine alkaloids from the poison frog, *Dendrobates tricolor*. X-ray analysis of 8-hydroxy-8-methyl-6-(2'-methylhexylidene)-1-azabicyclo-[4.3.0]-nonane. J. Amer. Chem. Soc. **102**, 830—836 (1980).

85. DALY, J. W., B. WITKOP, P. BOMMER, and K. BIEMANN: Batrachotoxin. The active principle of the Colombian arrow poison frog, *Phyllobates bicolor*. J. Amer. Chem. Soc. **87**, 124—126 (1965).

86. DALY, J. W., B. WITKOP, T. TOKUYAMA, T. NISHIKAWA, and I. L. KARLE: Gephyrotoxins, histrionicotoxins and pumiliotoxins from the neotropical frog *Dendrobates histrionicus*. Helv. Chim. Acta **60**, 1128—1140 (1977).

87. DALY, J. W., T. TOKUYAMA, G. HABERMEHL, I. L. KARLE, and B. WITKOP: Froschgifte. Isolierung und Struktur von Pumiliotoxin C. Justus Liebigs Ann. Chem. **729**, 198—204 (1969).

88. DOLLY, J. O., E. X. ALBUQUERQUE, J. M. SARVEY, B. MALLICK, and E. A. BARNARD: Binding of perhydrohistrionicotoxin to the postsynaptic membrane of skeletal muscle in relation to its blockade of acetylcholine-induced depolarization. Mol. Pharmacol. **13**, 1—14 (1977).

89. ELDEFRAWI, M. E., and A. T. ELDEFRAWI: Biochemical studies on the ionic channel of *Torpedo* acetylcholine receptor. Adv. Cytopharmacol. **3**, 213—223 (1979).

90. — Coupling between the nicotinic acetylcholine receptor site and the ionic channel site. Ann. N. Y. Acad. Sci. **358**, 239—252 (1980).

91. ELDEFRAWI, A. T., N. M. BAKRY, M. E. ELDEFRAWI, M.-C. TSAI, and E. X. ALBUQUERQUE: Nereistoxin interaction with the acetylcholine receptor-ionic channel complex. Mol. Pharmacol. **17**, 172—179 (1980).

92. ELDEFRAWI, A. T., M. E. ELDEFRAWI, E. X. ALBUQUERQUE, A. C. OLIVEIRA, N. MANSOUR, M. ADLER, J. W. DALY, G. B. BROWN, W. BURGERMEISTER, and B. WITKOP: Perhydrohistrionicotoxin: A potential ligand for the ion conductance modulator of the acetylcholine receptor. Proc. Nat. Acad. Sci. (USA) **74**, 2172—2176 (1977).

93. ELDEFRAWI, M. E., R. S. ARONSTAM, N. M. BAKRY, A. T. ELDEFRAWI, and E. X. ALBUQUERQUE: Activation, inactivation, and desensitization of acetylcholine receptor

channel complex detected by binding of perhydrohistrionicotoxin. Proc. Nat. Acad. Sci. (USA) **77**, 2309—2313 (1980).

94. ELDEFRAWI, M. E., D. S. COPIO, C. S. HUDSON, J. RASH, N. A. MANSOUR, A. T. ELDEFRAWI, and E. X. ALBUQUERQUE: Effects of antibodies to *Torpedo* acetylcholine receptor on the acetylcholine receptor-ionic channel complex of *Torpedo* electroplax and rabbit intercostal muscle. Exp. Neurol. **64**, 428—444 (1979).

95. ELDEFRAWI, M. E., A. T. ELDEFRAWI, R. S. ARONSTAM, M. A. MALEQUE, J. E. WARNICK, and E. X. ALBUQUERQUE: [H-3]-phencyclidine — a probe of the ionic channel of the nicotinic receptor. Proc. Nat. Acad. Sci. USA **77**, 7458—7462 (1980).

96. ELDEFRAWI, M. E., A. T. ELDEFRAWI, N. A. MANSOUR, J. W. DALY, B. WITKOP, and E. X. ALBUQUERQUE: Acetylcholine receptor and ionic channel of *Torpedo* electroplax: Binding of perhydrohistrionicotoxin to membrane and solubilized preparations. Biochemistry **17**, 5474—5484 (1978).

97. ELDEFRAWI, M. E., N. SHAKER, N. A. MANSOUR, J. E. WARNICK, and E. X. ALBUQUERQUE: Detection of nicotinic cholinergic transmission in *Malapterurus electricus* electroplax. Life Science **29**, 1033—1037 (1981).

98. ELLIOTT, J., and M. A. RAFTERY: Interactions of perhydrohistrionicotoxin with postsynaptic membranes. Biochem. Biophys. Res. Comm. **77**, 1347—1353 (1977).

99. ELLIOTT, J., and M. A. RAFTERY: Binding of perhydrohistrionicotoxin to intact and detergent-solubilized membranes enriched in nicotinic acetylcholine receptor. Biochemistry **18**, 1868—1874 (1979).

100. ELLIOTT, J., S. M. J. DUNN, S. G. BLANCHARD, and M. A. RAFTERY: Specific binding of perhydrohistrionicotoxin to *Torpedo* acetylcholine receptor. Proc. Nat. Acad. Sci. (USA) **76**, 2576—2579 (1979).

101. EVANS, D. A., and E. W. THOMAS: A formal synthesis of (±)-perhydrohistrionicotoxin via α-acylimmonium ion-olefin cyclizations. Tetrahedron Letters **1979**, 411—414.

102. EZHOV, V. V., P. F. POTASHNIKOV, O. V. VERENIKIN, and G. A. SOKOLSKII: Bioactivity — a function of the structure. VII. Simulation of the bioactivity of batrachotoxin analogs. Khim.-Farm. Zh. **13**, 31—35 (1979).

103. FLIER, J., M. W. EDWARDS, J. W. DALY, and C. W. MYERS: Widespread occurrence in frogs and toads of skin compounds interacting with the ouabain site of Na^+, K^+-ATPase. Science **208**, 503—505 (1980).

104. FLIPPEN, J. L.: 2-n-Propyl-7-methyl-*trans*-decahydroquinoline hydrochloride, a synthetic isomer of pumiliotoxin C. Acta Crystallogr. **B30**, 2906—2907 (1974).

105. FORMAN, D. S., and W. G. SHAIN, JR.: Batrachotoxin blocks saltatory organelle movement in electrically excitable neuroblastoma cells. Brain Res. **211**, 242—247 (1981).

106. FRELIN, C., P. VIGNE, G. PONZIO, G. ROMEY, Y. TOURNEUR, H. P. HUSSON, and M. LAZDUNSKI: The interaction of ervatamine and epiervatamine with the action potential Na^+ ionophore. Mol. Pharmacol. **20**, 107—112 (1981).

107. FUJIMOTO, R., and Y. KISHI: On the absolute configuration of gephyrotoxin. Tetrahedron Letters **1981**, 4197—4198.

108. FUJIMOTO, R., Y. KISHI, and J. F. BLOUNT: Total synthesis of (±)-gephyrotoxin. J. Amer. Chem. Soc. **102**, 7154—7156 (1980).

109. FUKUYAMA, T., L. V. DUNKERTON, M. ARATANI, and Y. KISHI: Synthetic studies on histrionicotoxins. II. A practical synthetic route to (±)-perhydro- and (±)-octahydrohistrionicotoxin. J. Org. Chem. (USA) **40**, 2011—2012 (1975).

110. GARCIA, J. H., S. S. DESHPANDE, R. S. PENCE, and E. X. ALBUQUERQUE: Spinal myelopathy induced by subarachnoid batrachotoxin: Ultrastructure and electrophysiology. Brain Res. **140**, 75—87 (1978).

111. GARRISON, D. L., E. X. ALBUQUERQUE, J. E. WARNICK, J. W. DALY, and B. WITKOP: Antagonism of carbamylcholine-induced depolarization by batrachotoxin and veratridine. Mol. Pharmacol. **14**, 111—121 (1978).

112. GILARDI, R. D.: The absolute configuration of a steroidal substance, the O-p-bromobenzoate derivative of batrachotoxinin A. Acta Crystallogr. **B 26**, 440—441 (1970).

113. GILL, D. L., E. F. GROLLMAN, and L. D. KOHN: Calcium transport mechanism in membrane vesicles from guinea pig brain synaptosomes. J. Biol. Chem. **256**, 184—192 (1981).

114. GLAVINOVIC, M., J. L. HENRY, G. KATO, K. KRNJEVIC, and E. PUIL: Histrionicotoxin: Effects on some central and peripheral excitable cells. Canad. J. Physiol. Pharmacol. **52**, 1220—1226 (1974).

115. GOSSINGER, E., W. GRAF, R. IMHOF, and H. WEHRLI: Herstellung von 14β-Hydroxy-20-keto-Δ¹⁶-Steroiden: Ein neuer ergiebiger Zugang zu 3-O-Methyl-17α-20ζ-tetrahydro-batrachotoxinin A. Helv. Chim. Acta **54**, 2785—2788 (1971).

116. GOSSINGER, E., R. IMHOF, and H. WEHRLI: Modellversuche in der Histrionicotoxinreihe Synthese des (±)-Cis-1-azaspiro-[5.5]-undecan-8-ols. Helv. Chim. Acta **58**, 96—103 (1975).

117. GRAF, W., H. BERNER, L. BERNER-FENZ, F. E. GOSSINGER, R. IMHOF, and H. WEHRLI: Die Synthese von 3 β - Methoxy - 3 α, 9 α - oxido - 11 α, 20 ζ - dihydroxy - 14 β O, 18 N - [ep (oxy - athano - N - methylimino)] - 5 β, 17 α - pregnan (3 - O - methyl - 17 α, 20 ζ - tetrahydrobatrachotoxinin A). Helv. Chim. Acta **53**, 2267—2275 (1970).

118. GRAF, W., E. GOSSINGER, R. IMHOF, and H. WEHRLI: Die Partialsynthese von 3-O-Methyl-20ζ-7,8-dihydrobatrachotoxinin A. Helv. Chim. Acta **54**, 2789—2793 (1971).

119. GRAF, W., F. E. GOSSINGER, R. IMHOF, and H. WEHRLI: Synthese der C-20 epimeren 7,8-Dihydrobatrachotoxinine. Helv. Chim. Acta **55**, 1545—1560 (1972).

120. HABERMEHL, G., H. ANDRES, and B. WITKOP: Synthese von rac.-pumiliotoxin C. Naturwiss. **62**, 345—346 (1975).

121. HABERMEHL, G., and W. KISSING: Pumiliotoxins and related compounds. 2. 2-Propyl-cis-perhydroquinolin-5.alpha.-ol. Chem. Ber. **107**, 2326—2328 (1974).

122. HABERMEHL, G., H. ANDRES, K. MIYAHARA, B. WITKOP, and J. W. DALY: Synthese von pumiliotoxin C. Justus Liebigs Ann. Chem. **1976**, 1577—1583.

123. HART, D. J.: The effect of A^(1,3) strain on the stereochemical course of N-acyliminium ion cyclizations. J. Amer. Chem. Soc. **102**, 397—398 (1980).

124. — A synthesis of (±)-gephyrotoxin. J. Organ. Chem. (USA) **46**, 3576—3578 (1981).

125. HEIDMANN, T., and J.-P. CHANGEUX: Structural and functional properties of the acetylcholine receptor protein in its purified and membrane-bound states. Ann. Rev. Biochem. **47**, 317—357 (1978).

126. HENDERSON, R., and G. STRICHARTZ: Ion fluxes through the sodium channels of garfish olfactory nerve membranes. J. Physiol. (London) **238**, 329—342 (1974).

127. HERY, F., G. SIMONNET, S. BOURGOIN, P. SOUBRIE, F. ARTAUD, M. HAMON, and J. GLOWINSKI: Effect of nerve activity on the in vivo release of [³H]-serotonin continuously formed from L-[³H]-tryptophan in the caudate nucleus of the cat. Brain Res. **169**, 317—334 (1979).

128. HIGHET, R. J., J. W. DALY, T. FUJIWARA, and T. TOKUYAMA: Indolizidine alkaloids from poison frogs of *Dendrobates* spp. Planta Medica **39**, 260—261 (1980).

129. HOGAN, P. M., and E. X. ALBUQUERQUE: The pharmacology of batrachotoxin. III. Effect on the heart Purkinje fibers. J. Pharmacol. Exp. Therapeut. **176**, 529—537 (1971).

130. HOLMES, A. B., R. A. RAPHAEL, and N. K. WELLARD: Model studies in the histrionicotoxin series: A highly stereoselective synthesis of terminal *cis* enyne units. Tetrahedron Letters **1976**, 1539—1542.

131. HOLZ, R. W., and J. T. COYLE: The effects of various salts, temperature, and the alkaloids veratridine and batrachotoxin on the uptake of [³H]-dopamine into synaptosomes from rat striatum. Mol. Pharmacol. **10**, 746—758 (1974).

132. HONERJAEGER, P., and M. REITER: Batrachotoxin: Activity-dependent prolongation of

the cardiac action potential and positive inotropic effect. Brit. J. Pharmacol. **58**, 415P (1976).

133. — — The cardiotoxic effect of batrachotoxin. Naunyn-Schmiedeberg's Arch. Pharmacol. **299**, 239—252 (1977).

134. HUANG, L.-Y. M., W. A. CATTERALL, and G. EHRENSTEIN: Comparison of ionic selectivity of batrachotoxin-activated channels with different tetrodotoxin dissociation constants. J. Gen. Physiol. **73**, 839—854 (1979).

135. HUANG, L.-Y. M., and G. EHRENSTEIN: Local anesthetics QX 572 and benzocaine act at separate sites on the batrachotoxin-activated sodium channel. J. Physiol. **77**, 137—154 (1981).

136. HUANG, L.-Y. M., G. EHRENSTEIN, and W. A. CATTERALL: Interaction between batrachotoxin and yohimbine. Biophys. J. **23**, 219—231 (1978).

137. HUANG, M., and J. W. DALY: Interrelationship among levels of ATP, adenosine and cyclic AMP in incubated slices of guinea pig cerebral cortex: Effect of depolarizing agents, psychotropic drugs and metabolic inhibitors. J. Neurochem. **23**, 393—404 (1974).

138. HUANG, M., H. SHIMIZU, and J. W. DALY: Accumulation of cyclic adenosine monophosphate in incubated slices of brain tissue. 2. Effect of depolarizing agents, membrane stabilizers, phosphodiesterase inhibitors and adenosine analogs. J. Med. Chem. **15**, 462—466 (1972).

139. IBUKA, T., Y. INUBUSHI, I. SAJI, K. TANAKA, and N. MASAKI: Total synthesis of dl-pumiliotoxin C hydrochloride and its crystal structure. Tetrahedron Letters **1975**, 323—326.

140. IBUKA, T., N. MASAKI, I. SAJI, K. TANAKA, and Y. INUBUSHI: Synthesis of dl-pumiliotoxin C hydrochloride and its crystal structure. Chem. Pharm. Bull. (Japan) **23**, 2779—2790 (1975).

141. IBUKA, T., H. MINAKATA, Y. MITSUI, E. TABUSHI, T. TAGA, and Y. INUBUSHI: Efficient stereoselective synthesis of rel-(6S,7S,8S)-7-butyl-8-hydroxyl-1-azaspiro[5,5]-undecan-2-one, a key intermediate for perhydrohistrionicotoxin and its rel-(6R) isomer. Chemistry Letters **1981**, 1409—1412.

142. IBUKA, T., Y. MORI, and Y. INUBUSHI: A new stereoselective synthesis of dl-pumiliotoxin C using novel 1,2-bis-(trimethylsilyloxy)-1,2-dienes. Tetrahedron Letters **1976**, 3169—3172.

143. — — — A stereoselective synthesis of dl-pumiliotoxin C. Chem. Pharm. Bull. (Japan) **26**, 2442—2448 (1978).

144. IMHOF, R., E. GOSSINGER, W. GRAF, H. BERNER, L. BERNER-FENZ, and H. WEHRLI: Partial synthesis of batrachotoxinin A. Helv. Chim. Acta **55**, 1151—1153 (1972).

145. IMHOF, R., E. GOSSINGER, W. GRAF, L. BERNER-FENZ, H. BERNER, R. S. CHANFELBERGER, and H. WEHRLI: Die Partialsynthese von Batrachotoxinin A. Helv. Chim. Acta **56**, 139—162 (1973).

146. IMHOF, R., E. GOSSINGER, W. GRAF, W. SCHNURIGER, and H. WEHRLI: Die Partialsynthese von 3β-Methoxy-3α,9α-oxido-7α-hydroxy-11α-acetoxy-5β-steroiden. Helv. Chim. Acta **54**, 2775—2785 (1971).

147. INUBUSHI, Y., and T. IBUKA: Synthesis of pumiliotoxin C. A toxic alkaloid from Central American arrow poison frogs, *Dendrobates pumilio* and *D. auratus*. Heterocycles **8**, 633—660 (1977).

148. JACQUES, Y., G. ROMEY, and M. LAZDUNSKI: Toxin-induced K⁺ efflux through the Na⁺ channel of neuroblastoma cells. Eur. J. Biochem. **111**, 265—273 (1980).

149. JANSSON, S.-E., E. X. ALBUQUERQUE, and J. DALY: The pharmacology of batrachotoxin. VI. Effects on the mammalian motor nerve terminal. J. Phamacol. Exp. Therapeut. **189**, 525—537 (1974).

150. JOHNSON, D. F., and J. W. DALY: Biosynthesis of cholesterol and cholesterol acetate in dendrobatid arrow poison frogs. Biochem. Pharmacol. **20**, 2555—2559 (1971).

151. KARLE, I. L.: The structure of dihydroisohistrionicotoxin, a unique unsaturated alkaloid and anticholinergic agent. J. Amer. Chem. Soc. **95**, 4036—4040 (1973).
152. — Configuration of the C(20) epimer of 7,8-dihydrobatrachotoxinin A. Proc. Nat. Acad. Sci. (USA) **69**, 2932—2936 (1972).
153. KARLE, I. L., and J. KARLE: The structural formula and crystal structure of the O-p-bromobenzoate derivative of batrachotoxinin A, $C_{31}H_{38}NO_6Br$, a frog venom and steroidal alkaloid. Acta Crystallogr. **B25**, 428—434 (1969).
154. KATO, G., and J.-P. CHANGEUX: Studies on the effect of histrionicotoxin on the monocellular electroplax from *Electrophorus electricus* and on the binding of [^{3}H]-acetylcholine to membrane fragments from *Torpedo marmorata*. Mol. Pharmacol. **12**, 92—100 (1976).
155. KATO, G., M. GLAVINOVIC, J. HENRY, K. KRNJEVIC, E. PUIL, and B. TATTRIE: Actions of histrionicotoxin on acetylcholine receptors. Croatica Chemica Acta **47**, 439—447 (1975).
156. KAYAALP, S. O., E. X. ALBUQUERQUE, and J. E. WARNICK: Ganglionic and cardiac actions of batrachotoxin. Eur. J. Pharmacol. **12**, 10—18 (1970).
157. KEANA, J. F. W., and R. R. SCHUMAKER: Synthesis of the ABC ring system of batrachotoxin and several related highly functionalized cholane derivatives. J. Organ. Chem. (USA) **41**, 3840—3846 (1976).
158. KERB, U., H.-D. BERNDT, U. EDER, R. WIECHERT, P. BUCHSCHACHER, A. FURLENMEIER, A. FÜRST, and M. MULLER: Zur Synthese des Batrachotoxinins: Synthese von 3β-Acetoxy-16β,20β-dihydroxy-4′-methyl-18-nor-5β,14β-pregnano-[13,14-f]-hexahydro-1′,4′-oxazepin-3′-on-(20a). Experientia **15**, 759—761 (1971).
159. KHODOROV, B., I.: Some aspects of the pharmacology of sodium channels in nerve membrane. Process of inactivation. Biochem. Pharmacol. **28**. 1451—1459 (1979).
160. — Chemicals as tools to study nerve fiber sodium channels: Effects of batrachotoxin and some local anesthetics. In: Membrane Transport Processes, Vol. II (D.C. TOSTESON, Y. A. DUCHINNIKOV, and R. LATORRE, eds.), pp. 153—174. New York: Raven Press. 1978.
161. — Pharmacologic analysis of sodium channel inactivation in a nerve fiber membrane. Neirofiziologiya **12**, 317—331 (1980).
162. KHODOROV, B. I., and S. V. REVENKO: Further analysis of the mechanisms of action of batrachotoxin on the membrane of myelinated nerve. Neuroscience **4**, 1315—1330 (1979).
163. KHODOROV, B. I., E. M. PEGANOV, S. V. REVENKO, and L. D. SHISHKOVA: Sodium currents in voltage clamped nerve fiber of frog under the combined action of batrachotoxin and procaine. Brain Res. **84**, 541—546 (1975).
164. KILPATRICK, D. L., R. SLEPETIS, and N. KIRSHNER: Inhibition of catecholamine secretion from adrenal medulla cells by neurotoxins and cholinergic antagonists. J. Neurochem. **37**, 125—131 (1981).
165. — — — Ion channels and membrane potential in stimulus-secretion coupling in adrenal medulla cells. J. Neurochem. **36**, 1245—1255 (1981).
166. KISSING, W., and B. WITKOP: Ein einfacher Zugang zu 1-Azaspiro-[5.5]-undecanen. Chem. Ber. **108**, 1623—1629 (1975).
167. KRODEL, E. K., R. A. BECKMAN, and J. B. COHEN: Identification of a local anesthetic binding site in nicotinic post-synaptic membranes isolated from *Torpedo marmorata* electric tissue. Mol. Pharmacol. **15**, 294—312 (1979).
168. KRUEGER, B. K., and M. P. BLAUSTEIN: Sodium channels in presynaptic nerve terminals. J. Gen. Physiol. **76**, 287—313 (1980).
169. KUMARA-SIRI, M. H.: Batrachotoxin inhibits axonal transport without affecting membrane potential in single neurons of *Aplysia californica*. J. Neurobiol. **10**, 509—512 (1979).

170. Lapa, A. J., E. X., Albuquerque, J. M. Sarvey, J. Daly, and B. Witkop: Effects of histrionicotoxins on the chemosensitive and electrical properties of skeletal muscle. Exp. Neurol. **47**, 558—578 (1975).

171. Lawrence, J. C., and W. A. Catterall: Textrodotoxin-insensitive sodium channels. Binding of polypeptide neurotoxins in primary cultures of rat muscle cells. J. Biol. Chem. **256**, 6223—6229 (1981).

172. Tetrodotoxin-insensitive sodium channels. Ion flux studies of neurotoxin action in a clonal rat muscle cell line. J. Biol. Chem. **256**, 6213—6222 (1981 b).

173. Lazdunski, M., M. Balerna, J. Barhanin, R. Chicheportiche, M. Fosset, C. Frelin, Y. Jacques, A. Lombet, J. Pouyssegur, J. F. Renaud, G. Romey, H. Schweitz, and J. P. Vincent: Molecular aspects of the structure and mechanism of the voltage-dependent sodium channel. Ann. N. Y. Acad. Sci. **358**, 169—182 (1980).

174. Lester, H. A.: Analysis of sodium and potassium redistribution during sustained permeability increases at the innervated face of *Electrophorus* electroplaques. J. Gen. Physiol. **72**, 847—862 (1978).

175. MacDonald, T. L.: Indolizidine alkaloid synthesis. Preparation of the pharaoh ant trail pheromone and gephyrotoxin 223 stereoisomers. J. Organ. Chem. (USA) **45**, 193—195 (1980).

176. Märki, F., and B. Witkop: The venom of the Colombian arrow poison frog *Phyllobates bicolor*. Experientia **19**, 329—338 (1963).

177. Masukawa, L. M., and E. X. Albuquerque: Voltage- and time-dependent action of histrionicotoxin on the endplate current of the frog muscle. J. Gen. Physiol. **72**, 351—367 (1978).

178. Matthews, J. C., E. X. Albuquerque, and M. E. Eldefrawi: Influence of batrachotoxin, veratridine, grayanotoxin I and tetrodotoxin on uptake of Na-22 by rat brain membrane preparations. Life Sci. **25**, 1651—1658 (1979).

179. Matthews, J. C., J. E. Warnick, E. X. Albuquerque, and M. E. Eldefrawi: Characterization of the electrogenic sodium channel from rat brain membranes using neurotoxin-dependent ^{22}Na$^+$ uptake. Membrane Biochem. **4**, 71—104 (1981).

180. Max, S. R., S. S. Deshpande, and E. X. Albuquerque: Neural regulation of muscle acetylcholinesterase: Effects of batrachotoxin and 6-aminonicotinamide. Brain Res. **130**, 101—107 (1977).

181. McCarthy, K. D., and T. K. Harden: Identification of two benzodiazepine binding sites on cells cultured from rat cerebral cortex. J. Pharmacol. Exp. Therapeut. **216**, 183—191 (1981).

182. McNeal, E. T., C. R. Creveling, and J. W. Daly: Cyclic AMP-generating systems in cell-free preparations from guinea pig cerebral cortex: Loss of adenosine and amine responsiveness due to low levels of endogenous adenosine. J. Neurochem. **35**, 338—342 (1980).

183. Mensah-Dwumah, M., and J. W. Daly: Pharmacological activity of alkaloids from poison-dart frogs (Dendrobatidae). Toxicon **16**, 189—194 (1978).

184. Mezey, K.: Venenos de flecha de Colombia. Acad. Colombiana de Ciencias Exactas físicas y naturales (Bogota) Revista **7**, 319—323 (1947).

185. — "Fiù-Fiù". Estudio toxicològica y farmacodinamico de un veneno de flechas y dardos, obtenido de la secreciòn cutànea de una "Rana del Choco" (*Dendrobates* sp.) [Cèsar Uribe-Piedrahita]. Bogota, Colombia (1947).

186. Moore, G. R. W., R. J. Boegman, D. M. Robertson, and R. J. Riopelle: Batrachotoxin induced axonal necrosis in peripheral nerves. Brain Res. **207**, 481—485 (1981).

187. Myers, C. W., and J. W. Daly: A new species of poison frog (*Dendrobates*) from Andean Ecuador, including an analysis of its skin toxins. Occasional Papers of the Museum of Natural History Univ. Kansas **1976**, 1—12.

188. — — Preliminary evaluation of skin toxins and vocalizations in taxonomic and evolutionary studies of poison-dart frogs (Dendrobatidae). Bull. Amer. Museum Natural History **157**, 173—262 (1976).

189. MYERS, C. W., and J. W. DALY: A name for the poison frog of Cordillera Azul, eastern Peru, with notes on its biology and skin toxins (Dendrobatidae). American Museum Novitates No. 2674, **1979**, 1—24.

190. —— —— Taxonomy and ecology of *Dendrobates bombetes,* a new Andean poison frog with new skin toxins. American Museum Novitates No. 2692 **1980**, 1—23.

191. MYERS, C. W., J. W. DALY, and B. MALKIN: A dangerously toxic new frog (*Phyllobates*) used by Embera·Indians of western Colombia, with discussion of blowgun fabrication and dart poisoning. Bull. Amer. Museum Natural History **161**, 307—366 (1978).

192. NARAHASHI, T.: Toxic chemicals as probes of nerve membrane function. Adv. Exp. Med. Biol. **84**, 407—445 (1977).

193. NARAHASHI, T., E. X. ALBUQUERQUE, and T. DEGUCHI: Effects of batrachotoxin on membrane potential and conductance of squid giant axons. J. Gen. Physiol. **58**, 54—70 (1971).

194. NARAHASHI, T., T. DEGUCHI, and E. X. ALBUQUERQUE: Effects of batrachotoxin on nerve membrane potential and conductances. Nature New Biology **229**, 221—222 (1971).

195. NEUBIG, R. R., E. K. KRODEL, N. D. BOYD, and J. B. COHEN: Acetylcholine and local anesthetic binding to *Torpedo* nicotinic postsynaptic membranes after removal of nonreceptor peptides. Proc. Nat. Acad. Sci. (USA) **76**, 690—694 (1979).

196. NEUWIRTH, M., J. W. DALY, C. W. MYERS, and L. W. TICE: Morphology of the granular secretory glands in skin of poison-dart frogs (Dendrobatidae). Tissue & Cell **11**, 755—771 (1979).

197. NIMITKITPAISAN, Y., J. W. DALY, P. J. JESSUP, L. E. OVERMAN, J. W. WARNICK, and E. X. ALBUQUERQUE: Pumiliotoxin-C and synthetic analogs: A new class of nicotinic antagonists. Trans. Am. Soc. Neurochem. **11**, 233 (1980).

198. OCHS, S., and R. WORTH: Batrachotoxin block of fast axoplasmic transport in mammalian nerve fibers. Science **187**, 1087—1089 (1975).

199. OTTEN, U., and H. THOENEN: Role of membrane depolarization in transsynaptic induction of tyrosine hydroxylase in organ cultures of sympathetic ganglia. Neurosci. Lett. **2**, 93—96 (1976).

200. OPPOLZER, W., and E. FLASKAMP: An enantioselective synthesis and the absolute configuration of natural pumiliotoxin-C. Helv. Chim. Acta **60**, 204—207 (1977).

201. OPPOLZER, W., and W. FROSTL: A stereoselective approach to *cis-* and *trans-*1, 2, 3, 4, 4a, 5, 6, 8a-octahydroquinolines by intramolecular *Diels-Adler* reactions. Helv. Chim. Acta **58**, 590—593 (1975).

202. OPPOLZER, W., C. FEHR, and J. WARNEKE: A new total synthesis of dl-pumiliotoxin-C via an indanone. Helv. Chim. Acta **60**, 48—58 (1977).

203. OPPOLZER, W., W. FROSTL, and H. P. WEBER: The total synthesis of (±)-pumiliotoxin-C. Helv. Chim. Acta **58**, 593—595 (1975).

204. OSWALD, R., and J.-P. CHANGEUX: Ultraviolet light-induced labeling by noncompetitive blockers of the acetylcholine receptor from *Torpedo mamorata.* Proc. Nat. Acad. Sci. (USA) **78**, 3925—3929 (1981).

205. OVERMAN, L. E.: Synthesis of 1-azaspiro-[5.5]-undec-7-en-2-one. Tetrahedron Letters **1975**, 1149—1152.

206. OVERMAN, L. E., and K. L. BELL: Enantiospecific total synthesis of dendrobatid toxin 251 D. A short chiral entry to the cardiac-active pumiliotoxin A alkaloids *via* stereospecific iminium ion — vinylsilane cyclizations. J. Amer. Chem. Soc. **103**, 1851—1852 (1981).

207. OVERMAN, L. E., and C. FUKAYA: Stereoselective total synthesis of (±)-perhydro-gephyrotoxin. Synthetic applications of directed 2-azonia-[3,3]-sigmatropic rearrangements. J. Amer. Chem. Soc. **102**, 1454—1456 (1980).

208. OVERMAN, L. E., and P. J. JESSUP: A short stereospecific total synthesis of dl-pumiliotoxin C. Tetrahedron Letters **1977**, 1253—1256.

209. — — Synthetic applications of N-acylamino-1,3-dienes. An efficient stereospecific total synthesis of dl-pumiliotoxin C, and a general entry to *cis*-decahydroquinoline alkaloids. J. Amer. Chem. Soc. **100**, 5179—5185 (1978).

210. OVERMAN, L. E., and T. YOKOMATSU: A new method for stereoselective piperidine annulation. Directing the 2-azonia-[3,3]-sigmatropic rearrangement by irreversible hydrolysis. J. Organ. Chem. (USA) **45**, 5229—5230 (1980).

211. OVERMAN, L. E., and R. L. FREERKS: Short total synthesis of (±) perhydrogephyrotoxin. J. Org. Chem. **46**, 2833—2835 (1981).

212. PEGANOV, E. M., S. V. REVENKO, B. I. KHODOROV, and L. D. SHISHKOVA: Batrachotoxin and aconitine, modifiers of rapid sodium channels in the nerve fiber membrane. Mol. Biol. (Kiev) **15**, 42—56 (1976).

213. PONZIO, G., Y. JACQUES, C. FRELIN, R. CHICHEPORTICHE, and M. LAZDUNSKI: An *in vitro* system to study the action potential sodium channel. FEBS Lett. **121**, 265—269 (1980).

214. POSADA ARANGO, A.: El veneno de rana de los indios del Chocò. In: Estudios cientificos del doctor Andrès Posado con algunos otros escritos suyos sobre diversos temas (MOLINA, C. A., ed.), pp. 78—88. Medellìn, Colombia: Imprenta Oficial. 1909.

215. RAMOS, S., E. F. GROLLMAN, P. S. LAZO, S. A. DYER, W. H. HABIG, M. C. HARDEGREE, H. R. KABACK, and L. D.. KOHN: Effect of tetanus toxin on the accumulation of the permeant lipophilic cation tetraphenylphosphonium by guinea pig brain synaptosomes. Proc. Nat. Acad. Sci. (USA) **76**, 4783—4787 (1979).

216. RAY, R., and W. A. CATTERALL: Membrane potential dependent binding of scorpion toxin to the action of potential sodium ionophore. Studies with a 3-(4-hydroxy-3-[125I]-iodophenyl)-propionyl derivative. J. Neurochem. **31**, 397—407 (1978).

217. RAY, R., C. S. MORROW, and W. A. CATTERALL: Binding of scorpion toxin to receptor sites associated with voltage-sensitive sodium channels in synaptic nerve ending particles. J. Biol. Chem. **253**, 7307—7313 (1978).

218. REVENKO, S. V.: Effect of electrical stimulation of nodes of Ranvier on the rate of sodium channel modification by batrachotoxin under conditions of potential fixation. Neirofiziologiya **9**, 546—549 (1977).

219. REVENKO, S. V., and B. I. KHODOROV: Effect of batrachotoxin on the selectivity of sodium channels in myelinated nerve fibre membrane. Neirofiziologiya **9**, 313—316 (1977).

220. ROSEEN, J. S., and F. A. FUHRMAN: Comparison of the effects of atelopidtoxin with those of tetrodotoxin, saxitoxin and batrachotoxin on beating of cultured chick heart cells. Toxicon **9**, 411—415 (1971).

221. SAFFRAY, LE DOCTEUR: Voyage à la Novelle-Grenade, X., Paris. In: Le Tour du Monde, Noveau Journal des Voyages **26**, 97—112 (1873).

222. SANTESSON, C. G.: An arrow poison with cardiac effect from the New World. In: Comparative Ethnographical Studies, 9 (E. Nordenskiold, ed.), pp. 157—187. Goteborg: Elanders Boktryckeri. 1931.

223. — Froschgift aus Columbia. Naunyn-Schmiedeberg's Arch. Pharmacol. **181**, 180 (1936).

224. SCHIMERLIK, M. I., U. QUAST, and M. A. RAFTERY: Ligand-induced changes in membrane-bound acetylcholine receptor observed by ethidium fluorescence. 3. Stopped-flow studies with histrionicotoxin. Biochemistry **18**, 1902—1906 (1979).

225. SCHOEMAKER, H. E., and W. N. SPECKAMP: A novel synthetic approach to perhydro-histrionicotoxin. Stereoselective synthesis of 1-aza-spiranes. Tetrahedron Letters **1978**, 1515—1518.

226. — — A short and stereoselective synthesis of perhydrohistrionicotoxin. Tetrahedron Letters **1978**, 4841—4844.

227. SCHOEMAKER, H. E., and W. N. SPECKAMP: Stereocontrolled synthesis of functionalized 1-azaspirans. Efficient synthesis of perhydrohistrionicotoxin. Tetrahedron 36, 951—958 (1980).

228. SCHUMAKER, R. R., and J. F. W. KEANA: Synthesis of the ABC ring system of the steroid batrachotoxin. J. Chem. Soc. Chem. Commun. 1972, 622.

229. SHIMIZU, H., and J. W. DALY: Effect of depolarizing agents on accumulation of cyclic adenosine 3',5'-monophosphate in cerebral cortical slices. Eur. J. Pharmacol. 17, 240—252 (1972).

230. SHIMIZU, H., C. R. CREVELING, and J. W. DALY: Stimulated formation of adenosine 3'-5'-cyclic phosphate in cerebral cortex: Synergism between electrical activity and biogenic amines. Proc. Nat. Acad. Sci. (USA) 65, 1033—1040 (1970).

231. — — — Cyclic adenosine 3',5'-monophosphate formation in brain slices: Stimulation by batrachotoxin, ouabain, veratridine, and potassium ions. Mol. Pharmacol. 6, 184—188 (1970).

232. — — — Effect of membrane depolarization and biogenic amines on the formation of cyclic AMP in incubated brain slices. Adv. Biochem. Psychopharmacol. 3, 135—154 (1970).

233. SHOTZBERGER, G. S., E. X. ALBUQUERQUE, and J. W. DALY: The effects of batrachotoxin on cat papillary muscle. J. Pharmacol. Exp. Therapeut. 196, 433—444 (1976).

234. SILVERSTONE, P. A.: A revision of the poison-arrow frogs of the genus *Dendrobates Wagler*. Natural History Museum of Los Angeles County, Science Bulletin 21, 1—55 (1975).

235. — A revision of the poison-arrow frogs of the genus *Phyllobates* Bibron *in* Sagra (family Dendrobatidae). Natural History Museum of Los Angeles, Science Bulletin 27, 1—53 (1976).

236. SIMPSON, L. L.: Pharmacological studies on the subcellular site of action of botulinum toxin type A. J. Pharmacol. Exp. Therapeut. 206, 661—669 (1978).

237. SMYTHIES, J. R.: A simplified model of the molecular structure of the sodium channel. Ala. J. Med. Sci. 15, 372—382 (1978).

238. — A model of the molecular structure of part of the sodium channel. Adv. Cytopharmacol. 3, 317—324 (1979).

239. SMYTHIES, J. R., F. BENINGTON, R. J. BRADLEY, W. F. BRIDGERS, and R. D. MORIN: The molecular structure of the sodium channel. J. Theor. Biol. 43, 29—42 (1974).

240. SMYTHIES, J. R., F. BENINGTON, and R. D. MORIN: Model for the action of tetrodotoxin and batrachotoxin. Nature 231, 188—190 (1971).

241. SOBEL, A., T. HEIDMANN, and J.-P. CHANGEUX: Purification of a protein binding quinacrine and histrionicotoxin from membrane fragments rich in acetylcholine receptor from *Torpedo-marmorata*. Comp. Rendus Acad. Sci. Paris 285D, 1255—1258 (1977).

242. SOBEL, A., T. HEIDMANN, J. HOFLER, and J.-P. CHANGEUX: Distinct protein components from *Torpedo marmorata* membranes carry the acetylcholine receptor site and the binding site for local anesthetics and histrionicotoxin. Proc. Nat. Acad. Sci. (USA) 75, 510—514 (1978).

243. SONNETT, P. E., D. A. NETZEL, and R. MENDOZA: ^{13}C Nmr assignments of selected octahydroindolizines. J. Heterocyclic Chem. 16, 1041—1047 (1979).

244. SPANDE, T. F., J. W. DALY, D. J. HART, Y.-M. TSAI, and T. L. MACDONALD: The structure of gephyrotoxin 223AB. Experientia 37, 1242—1245 (1981).

245. SPIVAK, C. E., M. A. MALEQUE, A. C. OLIVEIRA, L. MASUKAWA, T. TOKUYAMA, J. W. DALY, and E. X. ALBUQUERQUE: Actions of histrionicotoxins at the ion channel of the nicotinic acetylcholine receptor and the voltage sensitive ion channels of muscle membranes. Mol. Pharmacol., in press (1982).

246. TAKAHASHI, K., B. WITKOP, A. BROSSI, A. C. MALEQUE, and E. X. ALBUQUERQUE: Total synthesis and electrophysiological properties of natural (−)-perhydrohistrionicotoxin,

its unnatural (+)-antipode and their desamyl analogs. Helv. Chim. Acta, submitted (1982).

247. TAMBURINI, R., E. X. ALBUQUERQUE, J. W. DALY, and F. C. KAUFFMAN: Inhibition of calcium-dependent ATPase from sarcoplasmic reticulum by a new class of indolizidine alkaloids pumiliotoxins A, B and 251D. J. Neurochem. **37**, 775—780 (1981).

248. TAMKUN, M. M., and W. A. CATTERALL: Ion flux studies of voltage-sensitive sodium channels in synaptic nerve-ending particles. Mol. Pharmacol. **19**, 78—86 (1981).

249. TANG, C. M., G. R. STRICHARTZ, and R. K. ORKAND: Sodium channels in axons and glial cells of the optic nerve of *Necturus maculosa*. J. Gen. Physiol. **74**, 629—642 (1979).

250. TIEDT, T. N., E. X. ALBUQUERQUE, N. M. BAKRY, M. E. ELDEFRAWI, and A. T. ELDEFRAWI: Voltage- and time-dependent actions of piperocaine on the ion channel of the acetylcholine receptor. Mol. Pharmacol. **16**, 909—921 (1979).

251. TOKUYAMA, T., J. DALY, and B. WITKOP: The structure of batrachotoxin, a steroidal alkaloid from the Colombian arrow poison frog, *Phyllobates aurotaenia*, and partial synthesis of batrachotoxin and its analogs and homologs. J. Amer. Chem. Soc. **91**, 3931—3938 (1969).

252. TOKUYAMA, T., J. DALY, B. WITKOP, I. L. KARLE, and J. KARLE: The structure of batrachotoxinin A, a novel steroidal alkaloid from the Colombian arrow poison frog. J. Amer. Chem. Soc. **90**, 1917—1918 (1968).

253. TOKUYAMA, T., K. UENOYAMA, G. BROWN, J. W. DALY, and B. WITKOP: Allenic and acetylenic spiropiperidine alkaloids from the neotropical frog, *Dendrobates histrionicus*. Helv. Chim. Acta **57**, 2597—2604 (1974).

254. TOKUYAMA, T., and J. W. DALY: Steroidal alkaloids (batrachotoxins and 4-hydoxy-batrachotoxins), indole alkaloids (calycanthine and chimonanthine) and a piperidinyl-dipyridine alkaloid (noranabasamine) in skin extracts from the Colombian poison-dart frog *Phyllobates terribilis* (Dendrobatidae). Tetrahedron Letters, in preparation (1982).

255. TOKUYAMA, T., J. W. DALY, and R. J. HIGHET: Histrionicotoxins: Carbon-13 magnetic resonance spectral assignments and structural definition of further alkaloids from poison frogs (Dendrobatidae). Tetrahedron Letters, in preparation (1982).

256. TOKUYAMA, T., R. J. HIGHET, and J. W. DALY: Pumiliotoxins: Magnetic resonance spectral assignments and further structural definition of pumiliotoxin A and B and the isomeric allopumiliotoxins (7-hydroxypumiliotoxin A). Tetrahedron Letters, in preparation (1982).

257. TSAI, M.-C., N. A. MANSOUR, A. T. ELDEFRAWI, M. E. ELDEFRAWI, and E. X. ALBUQUERQUE: Mechanism of action of amantadine on neuromuscular transmission. Mol. Pharmacol. **14**, 787—803 (1978).

258. TSAI, M.-C., A. C. OLIVEIRA, E. X. ALBUQUERQUE, M. E. ELDEFRAWI, and A. T. ELDEFRAWI: Mode of action of quinacrine on the acetylcholine receptor ionic channel complex. Mol. Pharmacol. **16**, 382—392 (1979).

259. TUFARIELLO, J. J., and E. J. TRYBULSKI: A synthetic approach to the skeleton of histrionicotoxin. J. Org. Chem. (USA) **39**, 3378—3384 (1974).

260. VENIT, J. J., and P. MAGNUS: Studies on histrionicotoxin: Rearrangement of the spirocyclic histrionicotoxin carbon skeleton into the fused pumiliotoxin skeleton. Tetrahedron Letters **1980**, 4815—4818.

261. VINCENT, J. P., M. BALERNA, J. BARHANIN, M. FOSSET, and M. LAZDUNSKI: Binding of sea anemone toxin to receptor sites associated with gating system of sodium channel in synaptic nerve endings *in vitro*. Proc. Nat. Acad. Sci. (USA) **77**, 1646—1650 (1980).

262. VON BRAUN, J., W. GMELIN, and A. PETZOLD: Über Bz-Tetrahydro-chinoline und ihre Derivate (IV). Ber. dtsch. chem. Ges. **57**, 382—391 (1924).

263. WAGNER, H. R., and J. N. DAVIS: β-Adrenergic receptor regulation by agonists and membrane depolarization in rat brain slices. Proc. Nat. Acad. Sci. (USA) **76**, 2057—2061 (1979).

264. WAN, K. K., and R. J. BOEGMAN: Changes in rat muscle sarcoplasmic reticulum following neural application of batrachotoxin or tetrodotoxin. Exp. Neurol. **70**, 475—486 (1980).
265. — — Calcium uptake by muscle sarcoplasmic reticulum following neural application of batrachotoxin or tetrodotoxin. FEBS Lett. **112**, 163—167 (1980).
266. WARNICK, J. E., and E. X. ALBUQUERQUE: Models of paraplegia in animals: Trophic relationships. Federat. Proc. **37**, 2811—2817 (1978).
267. WARNICK, J. E., E. X. ALBUQUERQUE, A. J. LAPA, J. DALY, and B. WITKOP: Actions of neurotoxins on the acetylcholine receptor-ionic conductance modulator unit and on sodium channels. Proc. Sixth Int. Congr. Pharmacol. **1**, 67—76 (1976).
268. WARNICK, J. E., E. X. ALBUQUERQUE, R. ONUR, S.-E. JANSSON, J. DALY, T. TOKUYAMA, and B. WITKOP: The pharmacology of batrachotoxin. VII. Structure-activity relationships and the effects of pH. J. Pharmacol. Exp. Therapeut. **193**, 232—245 (1975).
269. WARNICK, J. E., E. X. ALBUQUERQUE, and F. M. SANSONE: The pharmacology of batrachotoxin. I. Effects on the contractile mechanism and on neuromuscular transmission of mammalian skeletal muscle. J. Pharmacol. Exp. Therapeut. **176**, 497—510 (1971).
270. WASSEN, S. H.: Notes on southern groups of Chocò Indians in Colombia. Etnograsfiska Museum, Göteborg Etnologiska Studier, pp. 35—182 (1935).
271. — On Dendrobates-frog-poison material among Emperà (Chocò)-speaking Indians in western Caldas, Colombia. Ethnografiska Museum, Göteborg, Arstryck 1955—1956, pp. 73—94 (1957).
272. WATERS, J. A., C. R. CREVELING, and B. WITKOP: 2,4,5-Trimethylpyrrole-3-carboxylic acid esters of various alkaloids. J. Med. Chem. **17**, 488—491 (1974).
273. WENNOGLE, L. P., R. OSWALD, T. SAITOH, and J.-P. CHANGEUX: Dissection of the 66000-Dalton subunit of the acetylcholine receptor. Biochemistry **20**, 2492—2496 (1981).
274. WEST, G. J., and W. A. CATTERALL: Selection of variant neuroblastoma clones with missing or altered sodium channels. Proc. Nat. Acad. Sci. (USA) **76**, 4136—4140 (1979).
275. WINTERFELDT, E.: Recent progress in alkaloid synthesis. Heterocycles **12**, 1631—1650 (1979).
276. WITZEMANN, V., and M. RAFTERY: Ligand binding sites and subunit interactions of Torpedo californica acetylcholine receptor. Biochemistry **17**, 3598—3604 (1978).
277. WU, W. C.-S., and M. A. RAFTERY: Functional properties of acetylcholine receptor monomeric and dimeric forms in reconstituted membranes. Biochem. Biophys. Res. Commun. **99**, 436—444 (1981).

(Received September 8, 1981)

Author Index

Page numbers printed in *italics* refer to References

Subject Index

By

A. Siegel, Wien

Fortschritte der Chemie organischer Naturstoffe

Progress in the Chemistry of Organic Natural Products

All Volumes and Cumulative Index 1—20 available / Alle Bände und Generalregister 1—20 lieferbar.

Price reduction for subscribers / Preisermäßigung für Subskribenten: 10%.

Special reduced price (20% reduction) for the complete Series Vols. 1—40 incl. the Cumulative Index to Vols. 1—20 / Vorzugspreis (20% Nachlaß) bei Bezug der Bände 1—40 inklusive Generalregister (Band 1—20).

Springer-Verlag Wien · New York